INTRODUCTION TO FOCUSED ION BEAMS

Instrumentation, Theory, Techniques and Practice

INTRODUCTION TO FOCUSED ION BEAMS

Instrumentation, Theory, Techniques and Practice

Edited by

Lucille A. Giannuzzi
FEI Company

Fred A. Stevie
North Carolina State University

Library of Congress Cataloging-in-Publication Data

A C.I.P. Catalogue record for this book is available
from the Library of Congress.

ISBN 0-387-23116-1 e-ISBN 0-387-23313-X Printed on acid-free paper.

Printed in the United States of America.

9 8 7 6 5 4 3 2 1 SPIN 11312024

springeronline.com

Dedication

This book is dedicated to Jeff Bindell, whose insight made it possible to have leading edge instrumentation available and whose inclusiveness fostered the interactions that provided much of the material for this work.

Contents

Contributing Authors

Ron Anderson, Microscopy Today
Derren N. Dunn, IBM
Lucille A. Giannuzzi, FEI Company
Peter Gnauck, Carl Zeiss SMT, Inc.
Dieter P. Griffis, North Carolina State University
T.Hashimoto, Hitachi High-Technologies
Peter Hoffrogge, LEO Elektronenmikroskopie GmbH
Becky Holdford, Texas Instruments, Inc.
Kultaransingh (Bobby) N. Hooghan, Agere Systems
Robert Hull, University of Virginia
Takashi Kaito, Seiko Instruments, Inc.
Takeo Kamino, Hitachi Science Systems
Brian W. Kempshall, NanoSpective Inc.
Stanley J. Klepeis, IBM Microelectronics Division
A.J. Kubis, University of Virginia
Janice K. Lomness, University of Central Florida
Mary V. Moore, FEI Company
T.Ohnishi, Hitachi High-Technologies
Mike W. Phaneuf, Fibics Inc.
Brenda I. Prenitzer, NanoSpective Inc.
Edward Principe, Carl Zeiss SMT, Inc.
Phil E. Russell, North Carolina State University
Fred A. Stevie, North Carolina State University
M.Schumann, Carl Zeiss SMT, Inc.
Stephen M. Schwarz, University of Central Florida, NanoSpective Inc.
K.Umemura, Hitachi Central Laboratory

T.Yaguchi, Hitachi Science Systems
Richard Young, FEI Company

Preface

The focused ion beam (FIB) instrument has experienced an intensive period of maturation since its inception. Numerous new techniques and applications have been brought to fruition by the tireless efforts of some very innovative scientists with the foresight to recognize the potential of this upstart apparatus. Over the past few years, the FIB has gained acceptance as more than just an expensive sample preparation tool, and has taken its place among the suite of other instruments commonly available in analytical and forensic laboratories, universities, geological, medical and biological research institutions, manufacturing plants, and more. The applications for FIB that have yet to be realized are endless. The future for this instrument is certain to be filled with innovation and excitement.

Although the utility of the FIB is not limited to the preparation of specimens for subsequent analysis by other analytical techniques, it has revolutionized the area of TEM specimen preparation. One anecdotal example is relayed by Lucille Giannuzzi, one of the editors of this book. Approximately 18 months of Lucille's graduate research effort was devoted to the development of a TEM specimen preparation technique for the cross-section analysis of galvanized steel. Upon her introduction to an FEI 611 FIB in 1995, the value of the FIB instrument, which was then capable of preparing TEM specimens of semiconductor materials in about five hours, was overwhelmingly and immediately apparent.

Today's FIB instruments can prepare TEM specimens in less than an hour. The FIB has also been used to prepare samples for numerous other analytical techniques, and offers a wide range of other capabilities. While the mainstream of FIB usage remains within the semiconductor industry, FIB usage has expanded to applications in metallurgy, ceramics, composites, polymers, geology, art, biology, pharmaceuticals, forensics, and other disciplines. In addition, the FIB has been used to prepare samples for numerous other analytical techniques. Computer automated procedures have been configured for unattended use of FIB and dual platform instruments. New applications of FIB and dual platform instrumentation are constantly being developed for materials characterization and nanotechnology. The site specific nature of the FIB milling and deposition capabilities allows preparation and processing of materials in ways that are limited only by one's imagination. Additional uses and applications will likely have been discovered by the time that this volume hits the shelves. The hardest task in editing this compilation was to decide when to stop and send it to press.

In this book we have attempted to produce a reference on FIB geared towards techniques and applications. The first portion of this book introduces the basics of FIB instrumentation, milling, and deposition capabilities. The chapter dedicated to ion-solid interactions is presented so that the FIB user can understand which parameters will influence FIB milling behavior. The remainder of the book focuses on how to prepare and analyze samples using FIB and related tools, and presents specific applications and techniques of the uses of FIB milling, deposition, and dual platform techniques. May you have as much fun working with FIB instruments as we continue to have!!

The Editors

Lucille A. Giannuzzi recently changed positions from Professor, Mechanical Materials and Aerospace Engineering at the University of Central Florida, to Field Product Marketing Engineer for FEI Company. Her research endeavors have broadly focused on structure/property relationships of materials using FIB/TEM methods, ion-solid interactions, and FIB and DualBeam applications and development. She has co-taught short courses on FIB and FIB/TEM specimen preparation at UCF and Lehigh University and has been a local affiliates and traveling speaker for both the Microscopy Society of America and the Microbeam Analysis Society. Dr. Giannuzzi is on the editorial board of Microscopy and Microanalysis, on MAS Council, on the MSA education committee, and is also a member of ACerS, ASM International, AVS, MRS, and TMS.

Fred A. Stevie is a Senior Researcher at North Carolina State University. His career in materials characterization spans more than 30 years with a range of techniques, principally with SIMS and FIB. His FIB work has concentrated on the interaction of FIB with other analytical methods, particularly in sample preparation for TEM analysis and EDS quantification. He is the AVS instructor for SIMS and co-instructor for FIB, and has co-taught short courses on FIB/TEM specimen preparation. He is on the advisory board of Surface and Interface Analysis, an Associate Editor of Surface Science Spectra, a Fellow of AVS, and a member of ASM International and MAS.

Acknowledgments

First and foremost, we are indebted to the hard work, dedication, and patience from the numerous authors who made contributions to this book. Funding agencies that have contributed to this work over the years are gratefully acknowledged. In addition, we thank our employers, both past and present, for allowing us the opportunity to dedicate time and resources in this area. Last but not least, we also recognize the encouragement and support of our families.

Chapter 1

THE FOCUSED ION BEAM INSTRUMENT

F. A. Stevie[1], L. A. Giannuzzi[2], and B. I. Prenitzer[3]
[1]North Carolina State University, Analytical Instrumentation Facility, Raleigh, NC 27695. [2]FEI Company, Hillsboro, OR 97124; [3]NanoSpective Inc., Orlando, FL 32826.

Abstract: The typical focused ion beam (FIB) instrument consists of a vacuum system, liquid metal ion source, ion column, stage, detectors, gas inlets, and computer. The liquid metal ion source provides the finely focused ion beam that makes possible high lateral resolution removal of material. Five axis motorized eucentric stage motion allows rapid sputtering at various angles to the specimen. The ion beam interaction with organo-metallic species facilitates site specific deposition of metallic or insulating species. Other gases may be used for enhanced etching of materials. The combination of a scanning electron microscope column and a FIB column forms a dual platform system that provides enhanced capabilities.

Key words: Focused ion beam, FIB, liquid metal ion source, LMIS, Ga

1. THE BASIC FIB INSTRUMENT

The basic FIB instrument consists of a vacuum system and chamber, a liquid metal ion source, an ion column, a sample stage, detectors, gas delivery system, and a computer to run the complete instrument as shown schematically in Figure 1. The instrument is very similar to a scanning electron microscope (SEM). FIB instruments may be stand-alone single beam instruments. Alternatively, FIB columns have been incorporated into other analytical instruments (either commercially or in research labs) such as an SEM, Auger electron spectroscopy, transmission electron microscopy, or secondary ion mass spectrometry, the most common of which is a FIB/SEM dual platform instrument. The ion column in a single beam FIB instrument is typically mounted vertically. In contrast, dual platform instruments usually have the FIB mounted at some angle with respect to vertical (i.e., the SEM column). Details on combined FIB/SEM applications are discussed elsewhere in this volume.

What follows below is a basic description of how a FIB instrument works. The interested reader may visit the references listed for additional details on ion optics and on the physics of liquid metal ion sources.

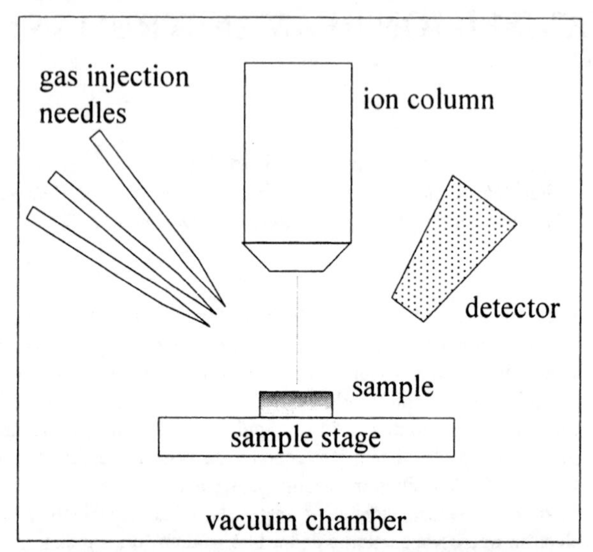

Figure 1. (a) A schematic diagram of a basic FIB system. (b) A single beam FEI 200TEM FIB instrument located at the University of Central Florida.

2. THE VACUUM SYSTEM

A vacuum system is required to make use of the ion beam for analysis. The typical FIB system may have three vacuum pumping regions, one for the source and ion column, one for the sample and detectors, and a third for sample exchange. The source and column require a vacuum similar to that used for field emission SEM sources (i.e., on the order of 1×10^{-8} torr) to avoid contamination of the source and to prevent electrical discharges in the high voltage ion column. The sample chamber vacuum can be at higher pressure and the system can be used with this chamber in the 1×10^{-6} torr range. Pressures in the 1×10^{-4} torr range will show evidence of interaction of the ion beam with gas molecules because the mean free path of the ions decreases as the chamber pressure is increased. The mean free path at high pressure is reduced to the point where the ions can no longer traverse the distance to the sample without undergoing collisions with the gas atoms or molecules. Ion pumps are normally used for the primary column and turbomolecular pumps backed by oil or dry forepumps are typically used for the sample and sample exchange chambers.

3. THE LIQUID METAL ION SOURCE

The capabilities of the FIB for small probe sputtering are made possible by the liquid metal ion source (LMIS). The LMIS has the ability to provide a source of ions of ~ 5 nm in diameter. Figure 2a shows a schematic diagram of a typical LMIS which contains a tungsten needle attached to a reservoir that holds the metal source material. There are several metallic elements or alloy sources that can be used in a LMIS. Gallium (Ga) is currently the most commonly used LMIS for commercial FIB instruments for a number of reasons: (i) its low melting point (T_{mp} = 29.8 °C) minimizes any reaction or interdiffusion between the liquid and the tungsten needle substrate, (ii) its low volatility at the melting point conserves the supply of metal and yields a long source life, (iii) its low surface free energy promotes viscous behaviour on the (usually W) substrate, (iv) its low vapor pressure allows Ga to be used in its pure form instead of in the form of an alloy source and yields a long lifetime since the liquid will not evaporate, (v) it has excellent mechanical, electrical, and vacuum properties, and (vi) its emission characteristics enable high angular intensity with a small energy spread.

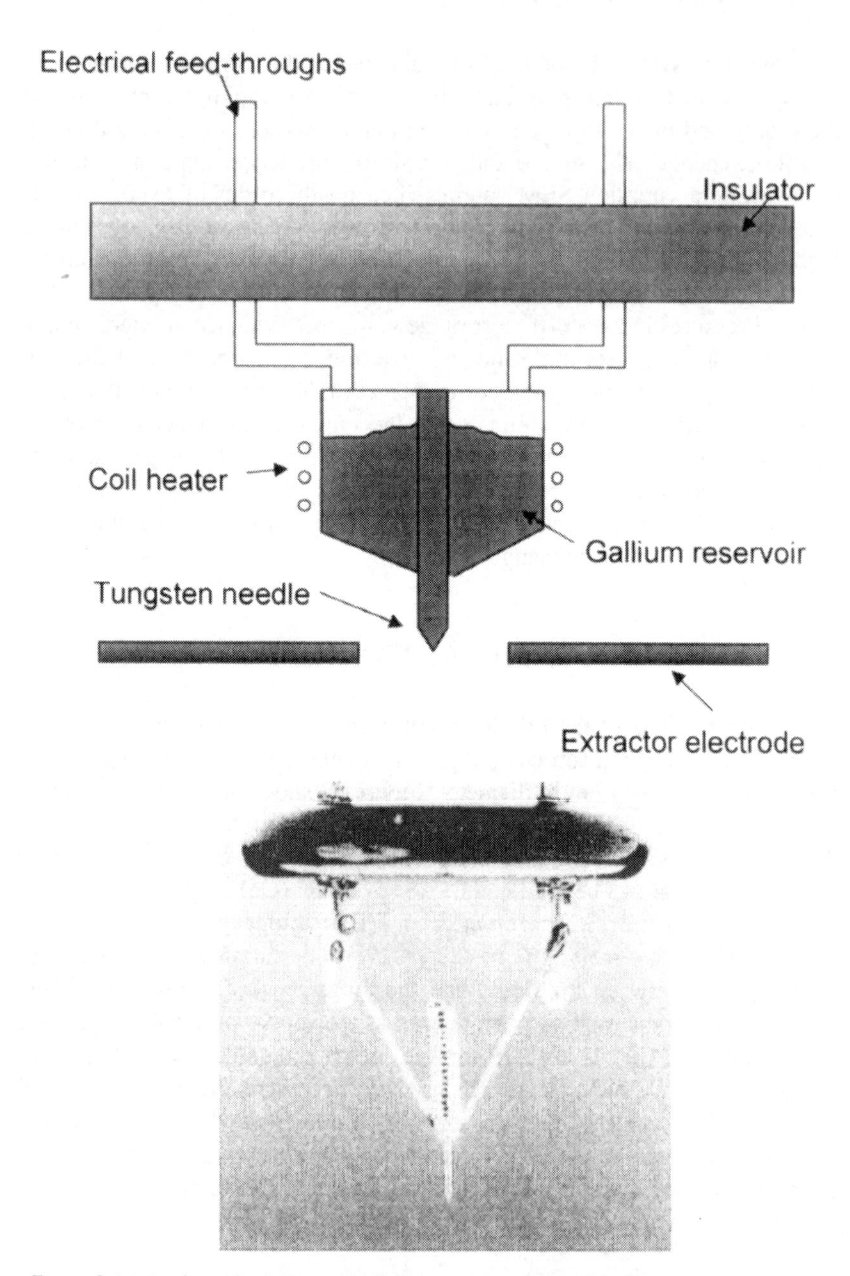

Figure 2. (a) A schematic diagram of a liquid metal ion source (LMIS) and (b) an actual commercial Ga LMIS (courtesy of FEI Company).

Ga^+ ion emission occurs via a two step process as described below: (i) The heated Ga flows and wets a W needle having a tip radius of ~ 2-5 μm. Once heated, the Ga may remain molten at ambient conditions for weeks due to its super-cooling properties. An electric field (10^8 V/cm) applied to the end of the wetted tip causes the liquid Ga to form a point source on the order of 2-5 nm in diameter in the shape of a "Taylor cone." The conical shape forms as a result of the electrostatic and surface tension force balance that is set up due to the applied electric field. (ii) Once force balance is achieved, the cone tip is small enough such that the extraction voltage can pull Ga from the W tip and efficiently ionize it by field evaporation of the metal at the end of the Taylor cone. The current density of ions that may be extracted is on the order of ~ 1×10^8 A/cm^2. A flow of Ga to the cone continuously replaces the evaporated ions.

The applied voltage/emission current output characteristics of a LMIS are non-linear. In FEI instruments, the extraction voltage is typically set to a constant value and the suppressor voltage is used to generate emission current from the LMIS. A finite voltage is needed to create the Taylor cone shape and result in emission current. The emission current will then rise with applied voltage on the order of 20 μA/kV. The source is generally operated at low emission currents (~ 1-3 μA) to reduce the energy spread of the beam and to yield a stable beam. At low emission current, the beam may consist of singly or doubly charged monomer ions, and neutral atoms which are not ionized. As the current increase, the propensity for the formation of dimmers, trimers, charged clusters, and charged droplets increases.

As the source ages, the suppressor voltage is gradually increased to maintain the beam current necessary for the constant extraction voltage. When an increase in suppressor voltage will no longer yield a working beam current (i.e., will no longer start the source), the source must be re-heated. Sometimes a larger extraction voltage (i.e., an "over voltage") may be used to start the source which is then reduced as soon as emission begins. The reservoir size is quite small and thus, the source can only be heated a limited number of times. To extend the source lifetime, heating should be performed only when necessary. The lifetime of a LMIS depends on its inherent material properties and the amount of material in the reservoir and is measured in terms of μA-hours per mg of source material. Typical source lifetimes for Ga are ~ 400 μA-hours/mg.

4. THE ION COLUMN

Once the Ga^+ ions are extracted from the LMIS, they are accelerated through a potential down the ion column. Typical FIB accelerating voltages range from 5-50 keV. A schematic diagram of the FIB column is shown in figure 3. The ion column typically has two lenses, i.e., a condenser lens and an objective lens. The condenser lens (lens 1) is the probe forming lens and the objective lens (lens 2) is used to focus the beam of ions at the sample surface. A set of apertures of various diameters also help in defining the probe size and provides a range of ion currents that may be used for different applications. Beam currents from a few pA to as high as 20 or 30 nA can be obtained. Methods for manual or automatic aperture selection have been developed. Optimizing the beam shape is obtained by centering each aperture, tuning the column lenses, and fine tuning the beam with the use of stigmators. Cylindrical octopole lenses may be used to perform multiple functions such as beam deflection, alignment, and stigmation correction. In addition, the scan field can be rotated using octopole lenses. Beam blankers are used to prevent unwanted erosion of the sample by deflecting the beam away from the center of the column.

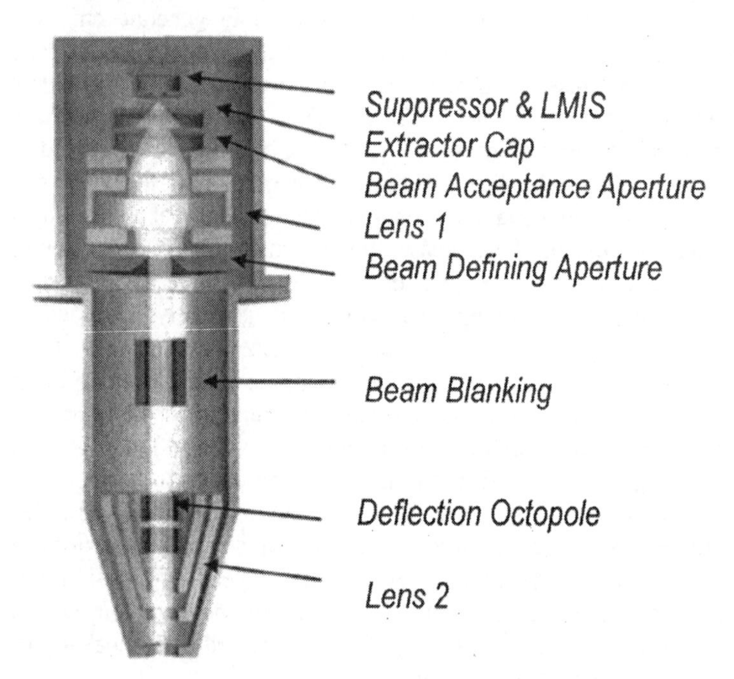

Figure 3. The basic FIB column. (courtesy FEI Company)

The FIB has a relatively large working distance with a typical value of about 2 cm or less. This large working distance permits the introduction of samples with varied topography without concern for field variations. When the Ga^+ beam strikes the sample surface, many species are generated including sputtered atoms and molecules, secondary electrons, and secondary ions. A more detailed description of the sputtering process is presented elsewhere in this volume.

The energy spread of a finely confined ion beam is generally larger than the energy spread of an electron beam and is ~ 5 eV. Since ions are much more massive than electrons, space charge effects limit the apparent source size and increase the width of the energy distribution of the emitting ions. Thus, chromatic aberration is often the limiting factor in the resolution of a FIB system.

5. THE STAGE

The sample stage typically has the ability to provide 5-axis movement (X, Y, Z, rotation and tilt). All five axis stage motions may be motorized for automatic positioning. The stage can be of sufficient size to handle 300 mm wafers. New FIB stages often have the capability for eucentric motion to avoid having to realign the sample every time the stage is moved. The large stage must be very stable and not be subject to significant heating as a result of the mechanical action required for stage movement. Thermal stability prevents specimen drift during FIB milling or deposition. Automated stage navigation can be used for precise location of sites on large wafers or for device repairs that involve multiple layers of material.

6. IMAGING DETECTORS

Two different types of detectors are typically used to collect secondary electrons for image formation, a multi-channel plate or an electron multiplier. A multi-channel plate is generally mounted directly above the sample. The electron multiplier is usually oriented to one side of the ion column. A typical alignment of the electron multiplier would put the detector at an angle of 45° to the incident beam. The electron multiplier detector can be biased to detect either secondary electrons or secondary positive ions emitted from the sample. It is important to note that the sample is being sputtering during the FIB imaging process. Therefore, small beam currents (< 100 pA) are typically used for FIB observation and image capture to minimize material removal during imaging.

Several contrast mechanisms can be used to provide various imaging capabilities. The secondary electron images provide images with good depth of field. The penetration of the ion beam into the specimen varies with different materials and for different grain orientations. Channelling contrast is observed for different grain orientations such that polycrystalline microstructures can be easily delineated. Secondary ion imaging provides a different type of contrast compared to secondary electron imaging. Note that the penetration of a 30 to 50 keV Ga^+ beam is limited to only a few tens of nanometers and therefore the secondary electron imaging capabilities are directly related to the surface effects of the collision cascade. Thus, observation of a region of interest under an oxide in an FIB system will require use of an optical microscope, or extensive pre-marking of the specimen before entry into the system.

The bombardment of charged species to the surface of an insulator can cause sample charging. The bombardment of an insulator with Ga^+ will cause the specimen to accumulate excess positive charge. Thus, any emitted secondary electrons will be attracted back to the surface, and will not be detected. Hence, charging effects in secondary electron FIB images are defined by dark regions in the image because the secondary electrons in these regions do not reach the detector. Analyzing a specimen that contains both insulating and charging regions is possible using because the insulating and conductive regions will appear dark and light, respectively. "Passive voltage contrast" is often used in semiconductor applications to test for circuit failure sites (see chapter by Holdford).

If the sample charges significantly, charge reduction methods may be required. Sometimes, good grounding of the specimen may be sufficient and it may be necessary to ground the inputs or bond pads of a semiconductor device. Charging can also be reduced or eliminated by coating the sample, or with the use of an electron flood gun (i.e., charge neutralizer), or by also imaging or milling with the SEM column turned on in the case of a dual platform instrument. Carbon coating can be removed in a plasma etcher after the sample is modified in the FIB, whereas metal coatings such as Au, Au/Pd, or Cr are not easily removed. It is interesting to note that for a wide range of materials and applications, coating is normally not required for analysis.

In addition, secondary ion (SI) imaging may be used to image insulating materials. The implementation of a flood gun (a charge neutralizer) in addition to SI imaging may also aid in observing or milling insulating materials. Figure 4 shows FIB images of a semiconductor device that is charging. Figure 4a shows a SE FIB image. Note that nearly the entire image is dark in contrast indicating that charging of the sample has tken place. Figure 4b shows a SI FIB image. Note that the details of the circuitry are now very will defined. In Figure 4c, a charge neutralizer is used in conjunction with SI FIB imaging. A brighter contrast image is observed with slightly better image details in figure 4c.

Charging can also affect the quality of FIB milling in insulating materials. Figure 5 shows an SEM image of two FIB milled square trenches in an insulator. The trench on the left was performed without charge neutralization and the trench on the right was performed with charge neutralization. Note that the use of FIB milling of an insulator without charge neutralization shows an irregular FIB milled box plus charging artifacts that are observed on the surface of the specimen. Conversely, the FIB milled trench performed with charge neutralization shows a uniformly milled box with no other apparent charging artifacts.

7. THE USE OF GAS SOURCES

Gas delivery systems can be used in conjunction with the ion beam to produce site specific deposition of metals or insulators or to provide enhanced etching capabilities. Metals, such as W or Pt, are deposited by ion beam assisted chemical vapor deposition of a precursor organometallic gas. A controlled amount of gas is introduced into the chamber by opening a valve that separates the reservoir and an inlet capillary that is positioned ~100 μm above the sample surface. The gas molecules are adsorbed on the surface in the vicinity of the gas inlet, but decompose only where the ion beam strikes. Repeated adsorption and decomposition result in the buildup of material in the ion scanned region. The ion beam assisted chemical vapor deposition process consists of a fine balance between sputtering and deposition. If the primary beam current density is too high for the deposition region, milling will occur. In addition to the CVD deposition of material, chemically enhanced sputtering is facilitated by the introduction of select species into the FIB chamber. For example, halogen-based species may enhance sputtering rates for specific substrate materials in the presence of the Ga ion beam. In addition, water has been shown to provide enhanced etching for carbonaceous materials. More details on deposition and etch are provided in a chapter dedicated to those processes.

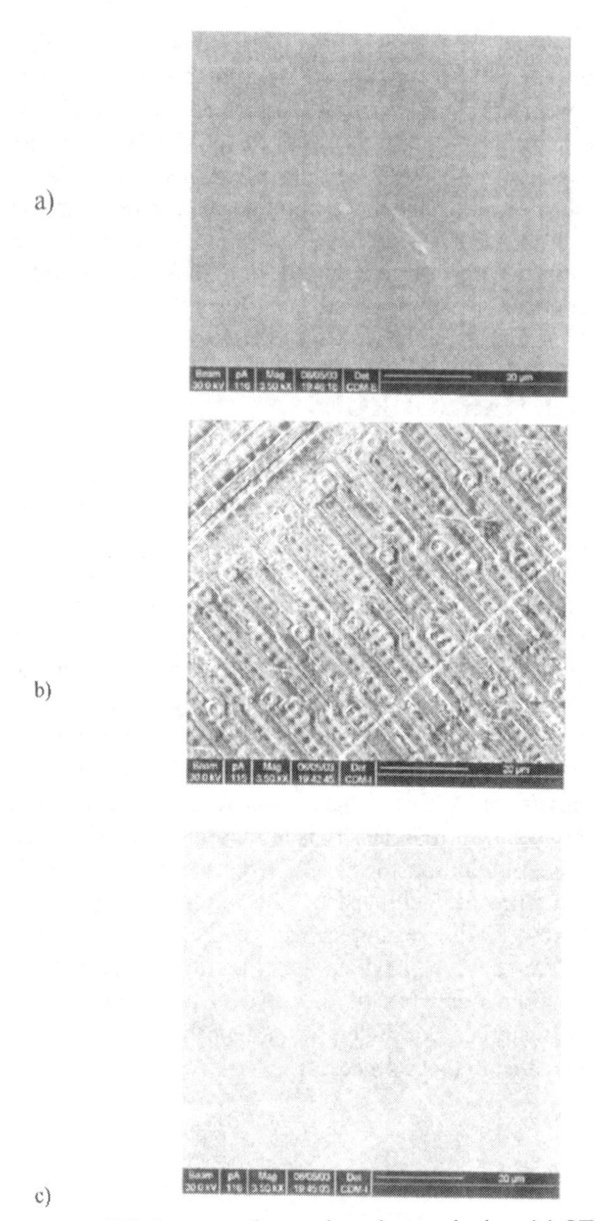

Figure 4. FIB images of a semiconductor device. (a) SE FIB image (b) SI FIB image (c) SI FIB image + charge neutralization.

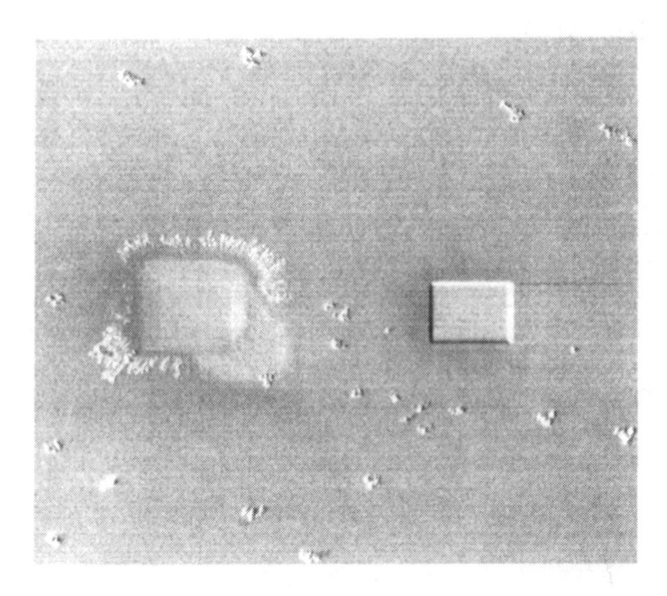

Figure 5. SEM image comparing two FIB milled 20 μm x 20 μm box trenches in an insulator. (left) A FIB milled trench without the use of charge neutralization and (right) a FIB milled trench with the use of charge neutralization. (courtesy of FEI Company)

8. DUAL PLATFORM INSTRUMENTS

The most common dual platform system incorporates an ion column and an electron column (i.e., an SEM) and has advanced capabilities. The electron beam can be used for imaging without concern of sputtering the sample surface. As a result, very creative ion beam milling and characterization can be obtained. In addition, electron beam deposition of materials can be used to produce very low energy deposition that will not affect the underlying surface of interest as dramatically as ion beam assisted deposition. It is possible to integrate the electron and ion beam operation to provide three dimensional information by sputtering the sample in increments and obtaining an SEM image of the specimen after each sputtering cycle. An energy dispersive spectrometry (EDS) detector can be added to provide elemental analysis. An electron backscatter diffraction detector (EBSD) can also be added to provide crystallographic analysis. In other cases, a combination FIB/SIMS instrument has been available for site specific specimen preparation plus elemental analysis at the trace impurity level. The benefits of FIB/SEM instruments are discussed in later chapters.

9. SUMMARY

The fundamental principles of a generalized FIB source, column, and detection systems have been summarized. The basic FIB system is comprised of a liquid metal ion source, stage, computer system, and detectors. Additional options can include a gas injector system used for either CVD deposition or enhanced etching. FIB platforms can be single ion beam or multi-column systems. There are certain advantages unique to combined FIB/SEM systems. Non-destructive imaging of the sample may be accomplished with an integrated SEM and additional peripherals such as EDS and EBSD may be used for elemental or crystallographic information. Specific details regarding the operation of the LMIS and ion optics may be found in the references listed below.

REFERENCES

Orloff J, Handbook of Charged Particle Optics, CRC Press, Boca Raton (1997)

Orloff J, Utlaut M, and Swanson L, High Resolution Focused Ion Beams: FIB and Its Applications, Kluwer Academic/Plenum Publishers, NY (2003)

Chapter 2

ION - SOLID INTERACTIONS

Lucille A. Giannuzzi[1], Brenda I. Prenitzer,[2] Brian W. Kempshall[2]
[1]FEI Company, Hillsboro, OR 97124, [2]Nanospective Inc., Orlando FL 32826.

Abstract: In this chapter we summarize reactions that take place when an energetic ion impinges on a target surface. The results based on equations that are usually used to estimate ion range and ion sputtering in amorphous materials are presented. A discussion on ion channeling and ion damage in crystalline materials is presented. The problems of redeposition associated with an increase in sputtering yield within a confined trench are presented. Knowledge of ion - solid interactions may be used to prepare excellent quality FIB milled surfaces.

Key words: secondary electron imaging, secondary ion imaging, ion range, incident angle, ion energy, sputtering, backsputtering, channeling, ion damage, amorphization, redeposition

1. INTRODUCTION

The ability to mill, image, and deposit material using a focused ion beam (FIB) instrument depends critically on the nature of the ion beam - solid interactions. Figure 1 shows a schematic diagram illustrating some of the possible ion beam/material interactions that can result from ion bombardment of a solid. Milling takes place as a result of physical sputtering of the target. An understanding of sputtering requires consideration of the interaction between an ion beam and the target. Sputtering occurs as the result of a series of elastic collisions where momentum is transferred from the incident ions to the target atoms within a collision cascade region. A surface atom may be ejected as a sputtered particle if it receives a component of kinetic energy that is sufficient to overcome the surface binding energy (SBE) of the target material. A portion of the ejected atoms may be ionized and collected to either form an image or be mass analyzed (see chapters on FIB/SIMS). Inelastic interactions also occur as the result of ion bombardment. Inelastic scattering events can result in the production of phonons, plasmons (in metals), and the emission of secondary electrons (SE). Detection of the emitted SE is the standard mode for imaging in the FIB; however, as previously mentioned secondary ions (SI) can also be detected and used to form images.

In general, the number of secondary electrons generated per incident ion is ~ 1 and is 10-1000x greater than the number of secondary ions generated per incident ion (Orloff et al., 2003). A comparison of an ion beam induced SE image and an ion beam induced SI image from the same region of the eye of a typical Florida bug is shown in figure 2. Note the complementary information that may be obtained using both of these imaging conditions. Non-conducting regions of a sample will accumulate a net positive charge as a result of the impinging Ga^+ ions. The net positive charge will inhibit the escape of SEs emitted from the surface. This type of charging artifact is observed as dark contrast in the image. For example, the regions around the bug eye in the lower left shows charging artifacts in the SE image, but are clearly delineated in the SI image. In addition, the dark feature on top of the eye shown in the SE image also shows evidence of charging, while the SI image clearly shows the details of the feature. Thus, secondary ion imaging is a useful alternative to circumvent charging artifacts during FIB imaging and milling of non-conducting samples.

Interactions between the incident ion and the solid occur at the expense of the initial kinetic energy of the ion. Consequently, if the ion is not backscattered out of the target surface, the ion will eventually come to rest, implanted within the target at some depth (i.e., R_p as shown in Figure 1) below the specimen surface.

The quality of the milled cuts or CVD deposited regions depends critically on the interactions between the impinging ion beam and the target. Thus, understanding the basics of ion beam-solid interactions may greatly enhance the ability to achieve optimum results using an FIB system. In this chapter, we present a brief introduction to the interactions that occur when an energetic ion impinges on a solid target surface. The interactions summarized below are those that are important within the energy regime that is characteristically used in the FIB (~ 5-50 keV). More extensive details on ion-solid interactions are available elsewhere (see e.g., Orloff et al., 2003; Nastasi et al., 1996).

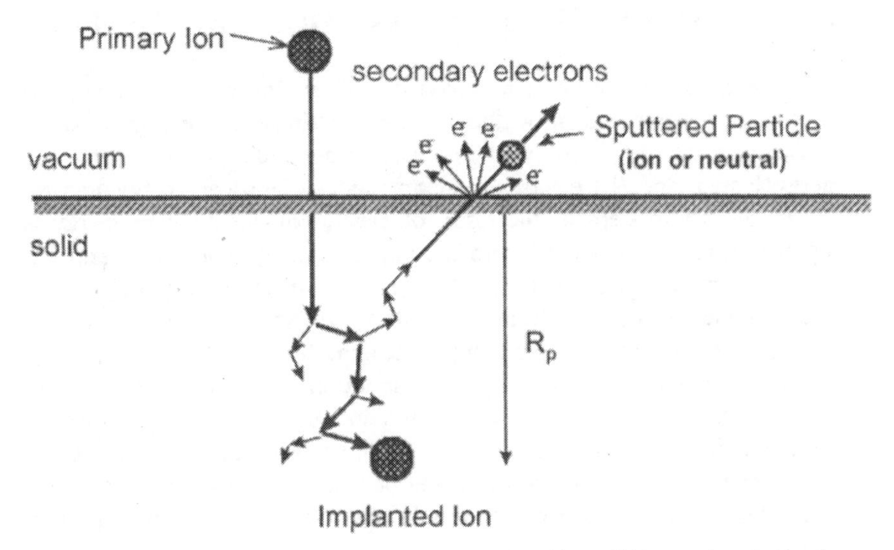

Figure 2-1. Schematic diagram of the sputtering process and ion-solid interactions (adapted from Nastasi et al., 1996).

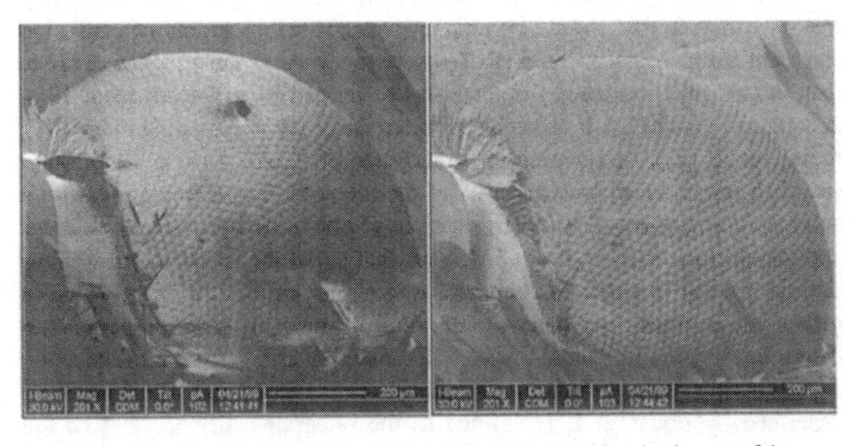

Figure 2-2. Ion induced (left) secondary electron and (right) secondary ion images of the eye of a typical Florida bug.

2. THE RANGE OF IONS IN AMORPHOUS SOLIDS

When a solid material is bombarded with an ion beam, a number of mechanisms operate to slow the ion and dissipate the energy. These mechanisms can be subdivided into two general categories: (i) nuclear energy losses, and (ii) electronic energy losses. Nuclear energy transfer occurs in discrete steps as the result of elastic collisions where energy is imparted from the incident ion to the target atom by momentum transfer. Electronic energy losses occur as the result of inelastic scattering events where the electrons of the ion interact with the electrons of the target atoms. The rate of ion energy loss per unit path length, dE/dx, and has both nuclear and electronic contributions. However, sputtering in typical FIB processes occur in energy ranges that are dominated by nuclear energy losses. Therefore, it is sufficient to present ion-solid interactions due only to the nuclear energy loss of ions as discussed below. The interested reader may find a discussion on electronic energy losses elsewhere (e.g., Nastasi et al. 1996).

2.1 The Concept of Ion Range

There tends to be some ambiguity in the terms and conventions used to describe ion range data. There are several distinctions between closely related concepts that should be emphasized. The first source of confusion can stem from the shear number of parameters used to quantify the distance that an energetic ion travels in a solid: i.e., range (R), projected range (R_p), penetration depth (X_s), transverse projected range (R_p^t), spreading range (R_s), radial range (R_r), and projected range straggling (ΔR_p). It is simplest to establish range definitions in terms of the interaction of a single ion with a solid. A sound physical interpretation of these definitions allows their application to actual ion beam processes that enlist the action of many ions. Ultimately, the implantation behavior of a single ion can be extrapolated to reflect the implantation behavior of multiple ions in terms of population dynamics.

Beginning with definitions as applied to a single ion, the range (R) described by Equation 1, is defined as the integrated distance that an ion travels while moving in a solid, and is inversely related to its stopping power (Nastasi et al., 1996; Ziegler et al., 1985; Townsend et al., 1976). The stopping cross-section, S(E), is defined as S(E) = (dE/dx)/N, where N is the atomic density. The stopping cross-section may be thought of as the energy loss rate per scattering center.

$$R = \int_{E_0}^{0} \frac{dE}{dE/dx} = \int_{E_0}^{0} \frac{dE}{NS(E)} \quad (1)$$

Thus, R may be defined by the path length for a single ion as illustrated in the schematic diagram in Figure 3. Examination of figure 3 reveals that R is not the same as the longitudinal projected range for a single ion (R_p). R_p for a single ion is the projection of its R onto its incident trajectory vector (sometimes denoted as I_o). Figure 3 is the generic case for arbitrary incidence angle. Only when $\theta = 0°$ (i.e., when the beam is normal to the surface), does R_p equal the implant depth (X_s) as measured perpendicular to the target surface (Nastasi et al., 1996; Gibbons et al., 1975). It should be noted that the statistical R_p, as applied to a collection of ions, is the quantity most frequently used to describe depth for an ion implant. Where R_p for a distribution of ions in a target material is most commonly defined by convention, as the distance measured along the incident ion trajectory at which the highest concentration of implanted ions will be found. It should be noted that the statistical value, X_s, is a more pertinent value of interest for understanding sidewall implantation (e.g., in a FIB prepared TEM specimen).

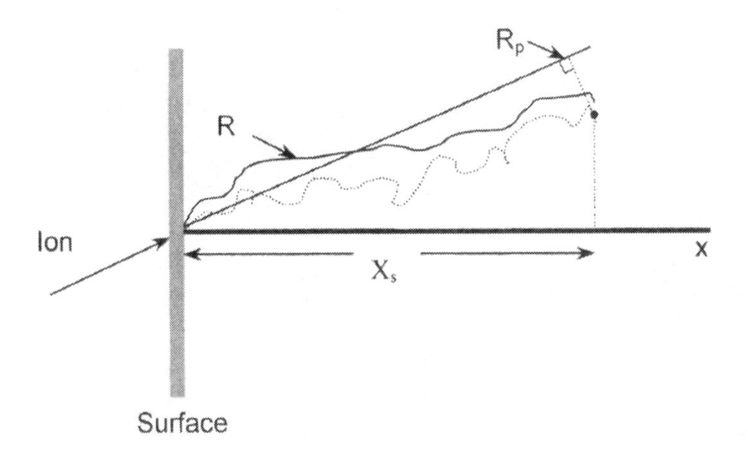

Figure 2-3. A 2D schematic diagram of the path of a single ion that has entered the target at an angle not equal to the surface normal. Note that $R_p = X_s$ only when the incident angle is 0°. (adapted from Mayer, et al. 1970)

A summary of the parameters of interest for the range of travel of an incident ion for the more general 3D case of a single energetic particle entering the solid at (0,0,0) and coming to rest at (x_s, y_s, z_s) is shown in the schematic diagram in figure 4 may be described as follows (Nastasi et al., 1996):

- R is the *range* as defined above
- R_p is the *projected range* for a single ion as defined above
- X_s is the projected range as measured along a vector normal to the surface as shown in figure 3.
- R_r is the *radial range* which is the distance from (0,0,0) to (x_s, y_s, z_s). Note: R_p is the projection of R_r on the direction vector of I_o.
- R is the *spreading range* which is the distance between (0,0,0) and the projection of R_r on the surface (the yz plane)
- R_p^t is the *transverse projected range* which is the vector connecting R_r and R_p.

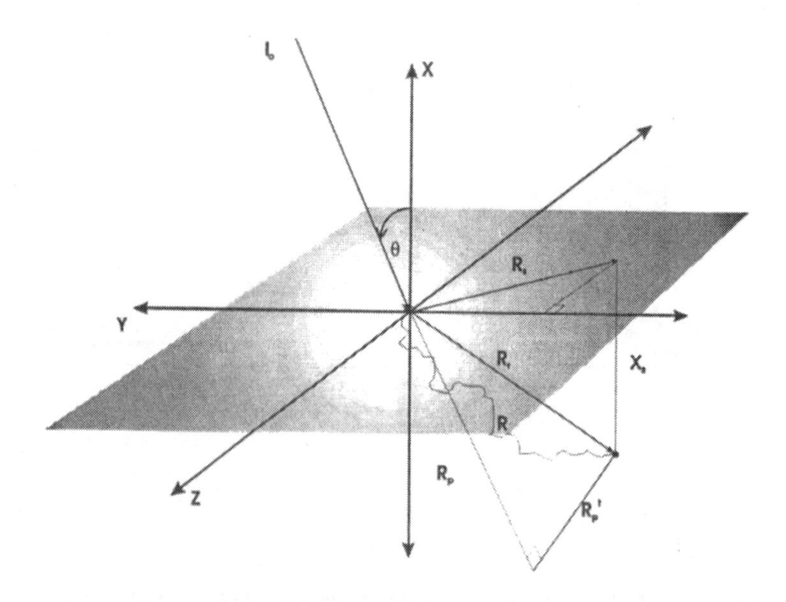

Figure 2-4. A 3D representation of a path taken by an ion that enters a solid at angle θ from the surface normal. (Adapted from Eckstein, 1991)

Although the range definitions above were presented in terms of the penetration behavior of a single ion, ion beam processes require the collective effect of a large number of ions. For the energy ranges associated with FIB applications, stopping of an energetic projectile is a random process, and the range distribution of a sufficiently large population of ions is statistical in nature. The probability function describing implant depth distributions is approximately Gaussian when dealing with relatively low implanted ion concentrations (i.e., less than a few atomic percent) and in the absence of crystallographic channeling effects (Nastasi et al., 1996). The mean of the distribution is the projected range, R_p, and the standard deviation is the projected range straggle, ΔR_p. In a normal Gaussian range distribution of implanted ions, the the largest number of ions will be located at R_p and ΔR_p is $R_p/2.35$ (Nastasi et al., 1996). It should be noted that the range distribution tends to deviate from a Gaussian profile in the presence of either crystallographic orientation effects or high ion doses. Ion channeling causes the depth distribution to be skewed because channeled ions penetrate to depths several times greater than R_p

2.2 Modeling Ion-Solid Interactions in Amorphous Materials

2.2.1 The Collision Cascade

When a target atom is knocked from its position, it can contribute to the collision cascade, (i.e., the moving sea of particles) within a solid under ion bombardment. Sputtering occurs if sufficient momentum is transferred from the collision cascade to a surface or near surface particle. The main parameters that govern the energy loss rate of the incident ion are its energy (E_o), the atomic masses (M_1 and M_2), and atomic numbers (Z_1 and Z_2), of the ion and the target atoms, respectively. The nature of the cascade is dependent on the ratio of the target to ion masses (M_2/M_1) and the incident ion energy. The classification of collision cascades is divided into three regimes. Regime I is called the single knock-on regime, and occurs when either $M_1 << M_2$ or E_o is low. In this regime, the recoil atoms do not receive enough energy to generate a cascade and sputtering is minimal. Regime II is the linear cascade regime where E_o is moderate and $M_1 \approx M_2$. In this regime recoil atoms receive enough energy to generate a cascade, but the density of moving atoms is dilute enough to disregard both multiple collisions and collisions between moving atoms. Linear assumptions that lead to the binary collision approximation are valid in regime II. The linear collision cascade model of regime II is where the FIB generally operates. Regime III is called the spike regime where $M_1 >> M_2$ and/or E_o is large. The result is that the majority of the atoms within the spike volume move during the collision cascade. Regime III is seldom reached during conventional FIB operation (Townsend et al., 1976).

2.2.2 Modeling Energy Loss in Ion-Solid Interactions

The dominant mechanism of energy loss in the energy range used in the FIB process (e.g., 5-50 keV) involves elastic interactions between the ion and a screened nucleus (i.e., nuclear stopping). This is generally modeled using a two-body billiard ball collision model. The "collision" is the distance of closest approach governed by interatomic potentials between the incident ion and the target atom. The influence that the nuclear charge of one atom can exert on another atom is modulated by the shielding efficiency of the orbital electrons. Thus, physical phenomena tend to exhibit periodic fluctuations based on electronic structure and atomic radii of the atoms under consideration.

With each collision, the incident ion losses energy and changes direction by an angle, Θ. Using conservation of momentum in a center of mass coordinate system, the recoil energy of the struck atom is the energy transferred to target atom, T, as shown in equation (2).

$$T = \frac{4M_1M_2}{(M_1 + M_2)^2} E_o Sin^2 \frac{\Theta}{2} = \frac{4E_cM_c}{M_2} Sin^2 \frac{\Theta}{2} \tag{2}$$

In equation (2), $E_c = (E_oM_2)/(M_1+M_2)$, and $M_c = (M_1M_2)/(M_1+M_2)$. The final angle of scatter, Θ, may be expressed in terms of the initial center of mass energy, E_c, the potential, V(r), and an impact parameter, p, as shown by equation (3) (Ziegler et al. 1984), where r_{min} is the distance of closest approach during the collision.

$$\Theta = \int_{r_{min}}^{\infty} \frac{pdr}{r^2 \sqrt{\left[1 - \frac{V(r)}{E_c} - \frac{p^2}{r^2}\right]}} \tag{3}$$

By taking the initial seed value, Θ = π, and iteratively integrating over the entire collision path, the final angle of scatter for the projectile (Θ) can be evaluated in terms of the initial center of mass energy E_c, the interatomic potential V(r), and the impact parameter p. The impact or scattering parameter, p, is basically the effective interaction distance for "collision" between two atoms. Since the nuclei do not actually touch, the impact parameter is defined as a circle of area πp^2 around the target nucleus. Any incident projectile that passes anywhere within this circle will be deflected by some angle greater than θ_c. The actual interaction distance during a collision depends on the energy of the collision. A useful frame of reference for describing collisional interaction distances is bounded by the Bohr radius (a_o = 0.053 nm) and the equilibrium separation, (r_o ~0.25 nm).

Since there is no single potential function that is appropriate for all pairs of ions and all energies, empirical parameters are usually fitted to pre-existing examples of interatomic potentials. Equation 4 describes the range of ions in terms of energy loss (or stopping power). The manner in which energetic particles interact with a lattice of stationary atoms can be described by the way the potential energy of a two-particle system varies with the distance between their centers. Ziegler, Biersack, and Littmark (ZBL) (1984) optimized a function originally developed by Lindhard, Scharff and Schiott (LSS) (1963). The result was a generalized analytical expression called the "Universal Screening Function" used to model interatomic potentials as given by equation 4 (Ziegler et al., 1984).

$$V(r) = \frac{Z_1 Z_2 e^2}{aR}\Phi(R) \tag{4}$$

In equation (4), $\Phi(R)$ is the universal screening function, R = r/a the reduced interatomic separation, Z_1 and Z_2 are the atomic numbers of the each of the two interacting species, and V(r) is the functional form of the interaction potential between the two atoms. The stopping power, S(E), is the average energy transferred when summed over all impact parameters as given in equation (5):

$$S(E) = \int_0^{\infty} T(E,p)2\pi p dp = 2\pi\gamma E_o \int_0^{p_{max}} \sin^2\frac{\Theta}{2} p dp \tag{5}$$

where $\gamma \equiv \frac{4M_1M_2}{(M_1+M_2)^2}$.

Thus, both the conservation of momentum and the interatomic potential are taken into account when the nuclear stopping power of an incident ion in a target material is considered.

2.2.3 Using TRIM for Monte Carlo Modeling of Ion - Solid Interactions in Amorphous Materials

Monte Carlo (MC) simulation of ion-solid interactions does have certain advantages over analytical formulations such as molecular dynamics (MD) calculations. MC methods allow more rigorous treatment of elastic scattering, and explicit consideration of surfaces and interfaces. Additionally, MC models allow energy and angular distributions to be readily determined. MC simulation methods rely on the statistical or random nature of the processes that are modeled. Energy transfer models of this sort are based on the linear superposition of sequential events or the random walk. The validity of the linear cascade approximation holds for cascade Regime II, when the number of moving atoms is small with respect to the total number of atoms contained within the collision volume.

As mentioned above, the nature of the collision cascade depends on M_1/M_2, and the incident ion energy. For a typical FIB application using a 5-50 keV Ga^+ ion beam, E_o is moderate, $M_1 \approx M_2$, and thus the conditions to produce a cascade characterized by linear collision dynamics are satisfied. Therefore, elastic energy losses can be assumed to be the result of a series of uncorrelated binary collisions and, the resulting phenomena can be readily modeled with MC computer simulation methods.

"Transport of Ions in Matter" (TRIM) is a sub-routine of a group of programs called "Stopping and Range of Ions in Matter" (SRIM) created by Ziegler et al. (1985, and Ziegler, 2003). TRIM can be effectively used to physically model the final 3D spatial distributions of ions in either simple or complex target materials. TRIM can generate data regarding the final 3D distributions of ions, as well as all kinetic data associated with the energy loss of the ion to the solid. For example, target damage, average sputtering yield per incident ion, ionization, and phonon production can all be quantitatively modeled using TRIM.

It should be noted that TRIM treats all targets as amorphous, thereby discarding the potential contribution of channeling or other orientation dependent phenomena to range distributions. This is an important consideration when attempting to correlate damage cascades with apparent damage layers that may develop during FIB milling of crystalline materials. TRIM was used to determine ranges for 100 Ga ions at different energies and incident angles for periodic table of elements (for solids) and are listed in the appendix.

With the strengths and the limitations of the SRIM package in mind, the code may be used to produce reasonable models of the variables that govern the FIB ion bombardment process. TRIM was used to physically model final 3D spatial distributions for 25 keV Ga^+ at 0^0 in elements Z = 3-84. The range data was generated using the "Ion Distribution and Quick Calculation of Damage" option (Ziegler, 2003). This option allows the calculation to run more expeditiously by eliminating details about target damage or sputtering yields, while maintaining accurate results for the final distribution of ions in the solid.

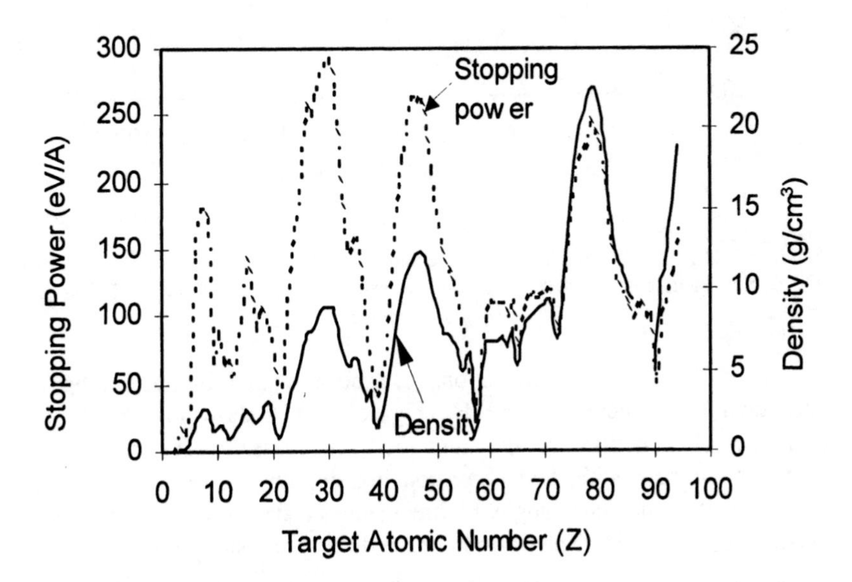

Figure 2-5. TRIM calculations showing the relationship between the total stopping power and the mass density for target elements Z= 1-92 at a Ga+ energy of 25 keV and 0° incident angle.

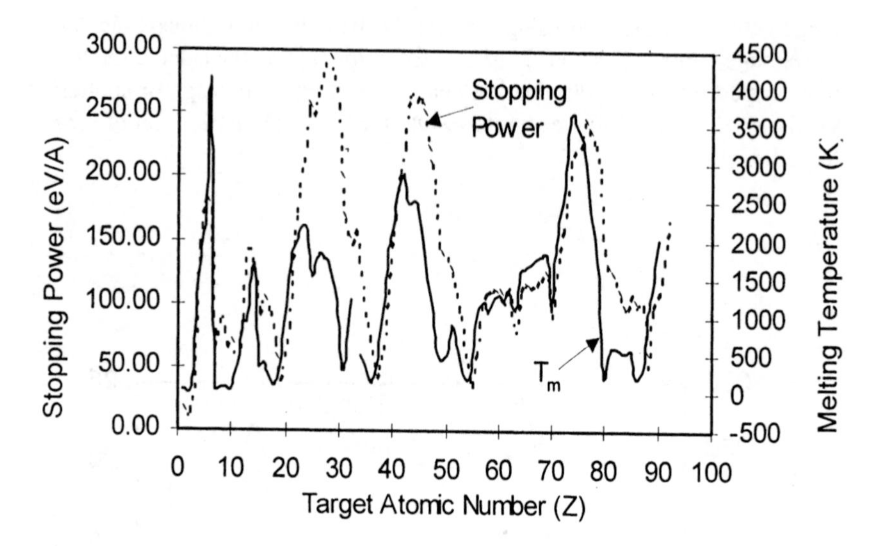

Figure 2-6. TRIM calculations showing the relationship between the total stopping power and the absolute melting temperature for target elements Z = 1-92 for a Ga+ beam at 25 keV and 0° incident angle.

The utility of such calculations can be extended by examining the periodic trends associated with S(E). Figures 5 and 6 show how the physical properties (i.e., melting temperature (T_m) and mass density (ρ_m)) of the elements compare with their corresponding stopping powers for a 25 keV Ga^+ beam at 0° incidence angle. Figure 5 shows a strong positive correlation between ρ_m and S(E). The position of the density and stopping power peaks are coincident for all the elements modeled. This illustrates that S(E) for each element follows the trends associated with its particular group when moving from left to right across the periodic table. In contrast, the magnitude of the individual S(E) peaks exhibit a nearly random distribution in height. This indicates the absence of periodic influence generally associated with increasing mass of the elements when moving from top to bottom within a group. There is an apparent anomaly observed in the peak heights among the elements Z = 58-71, which corresponds to the lanthanide series of elements. The diminished values observed for the physical properties among the lanthanide group are caused by population of the 4f shell in the absence of the covalent bonding contribution of the 5d shell. It is the covalent character of the d shells that imparts the exceptionally high strength of the interatomic bonding forces observed among the transition metals.

Figure 6 shows how stopping power compares with melting temperature for a 25 keV Ga^+ beam at 0° incidence angle. Similar to the example shown in Figure 5, there is also a strong positive correlation between T_m and S(E).

Collectively, the computer modeled data plotted in Figures 5 and 6 illustrates that stopping powers and associated range distributions strongly conform to the periodic trends associated with the electronic configuration of the elements across the periodic table, and to a lesser extent, to the trends associated with mass variations down a group. Thus, predictions with respect to anticipated implantation depth for a given target material may be made based on the position of the element in the periodic table.

2.3 Steady State Ion Implantation Conditions and the Concept of Ion Dose

Established terminology conventions used to describe the characteristics of an ion beam such as ion flux, fluence, dose, beam current, and current density have been used interchangeably and in some cases ambiguously in the literature. Many of the distinctions between the terms are subtle at best and warrant careful clarification. ***Flux*** is defined as the time rate of flow of energy; the radiant or luminous power in a beam. In the case of an ion beam flux is measured as the number of particles flowing through a given area per unit time and has units of ions/cm^2/s. Flux is a rate and remains constant for a given set of parameters such as beam current/aperture setting. (ASTM E1620-97). ***Fluence*** is, the sum of energies, the number of particles or photons incident during a given time interval on a small sphere centered at a given point in space divided by the cross-sectional area of that sphere. Fluence has units of ions/cm^2. Fluence is identical with the time integral of the particle flux density and thus, cumulatively increases as a function of the duration of the time interval that the beam is active (ANSI N1.1-1976). ***Dose*** is a general term denoting the quantity of radiation, energy, or particles absorbed by a medium. In the case of an ion beam, dose has units of ions/cm^2. Dose is the analogue of fluence with the distinction that fluence is the number of ions that pass through a defined area prior to impacting the target and dose is the number of ions that are impacted and absorbed into the target through a similarly defined area. Dose, like fluence, cumulatively increases as a function of the duration of the time interval that the beam is active. (ICRP 60-1990) Similar to flux, the ***beam current*** is also a measure of the time rate flow of energy or how many ions are delivered per unit time. The beam current is measured in amperes, which is equivalent to units of charge per unit time or Coulombs(C)/sec. In the case of a singly charged ion like Ga^+, the beam current can be described in terms of ions/s and is closely related to the flux, which is ions/cm^2/s, where the area is the cross-sectional area of the beam. ***Current density*** is a measure of the energy intensity or number of ions in a given area at any instant in time. Current density has units of C/cm^2 or in the case of a singly charged ion like Ga^+ this is equivalent to $ions/cm^2$.

Although these conventions have origins in the ion implant community, it should be duly noted that certain aspects of the ion-solid interactions inherent to FIB processes are divergent from those that are characteristic of ion implantation. The major manifestation of these differences is seen in the dose i.e., the number of ions that are absorbed and retained in the target. FIB instruments typically operate in an energy range of 5-50 keV, whereas ion implantation is carried out over a broad array of energies that can range from tens of kilo-electron volts to several million electron volts. The energy of the incident beam and the atomic masses of the incident ions and target atoms govern the energy loss mechanisms operative in slowing and stopping the incident ions.

Ion implantation is performed in the energy regime where electronic stopping tends to dominate and sputtering is minimal. This is in contrast to FIB processes where the energy regime and masses are optimal for nuclear stopping which leads to efficient sputtering. Sputtering affects the implant profile as well as limiting the concentration of impurity atoms that can be implanted. As the target material is being bombarded with moderate energy ions, some of the incident beam ions are implanted and retained in the target while the surface that is exposed to the ion beam is simultaneously and constantly receding due to sputtering. FIB sputtering ultimately creates a steady-state condition between impurity implantation and impurity removal by sputtering. Once this steady state condition is reached, the concentration of impurity atoms reaches a maximum value. In general, the maximum concentration value (i.e., the steady-state condition) is attained after sputtering away a thickness comparable to the projected range. The concentration limit is proportional to the reciprocal of the sputtering yield, 1/Y. (Liau et al., 1980) Thus, a material that sputters very rapidly will have a lower peak impurity concentration than one with a correspondingly lower sputter yield.

In addition to the sputtering yield, the characteristics of the implanted impurity atoms also affect the implant profile and peak concentration. As impurity atoms are added to the solid, the attributes of the material are often altered. (Liau et al., 1980) This can lead to modification of the overall sputtering yield due to a change in the SBE and preferential sputtering where generally the ejection of light elements is favored. The preferential sputtering parameter, r, is essentially a ratio of the sputtering rate of the target to the sputtering rate of the impurity atoms. Since sputtering occurs from within the first few surface layers, preferential sputtering can lead to significant variation in the maximum surface concentration of implanted atoms. Note that in contrast to FIB milling or other sputtering phenomena, ion implantation processes operate at very high beam energies and do not result in significant sputtering. Thus, during e.g., semiconductor ion implantation, an increase in ion implantation time (i.e., dose) corresponds to an increase in concentration, since steady state implantation conditions are not met. Ion implantation however, is a non-equilibrium process with the capability of producing materials with compositions unattainable by other conventional means. Thus, in FIB milling, an increase in dose does not alter the steady state equilibrium process of ion implantation, but rather, just increases the time in which the FIB milled surface will recede. Thus, in FIB milling, knowledge of dose may be useful for determining the depth of a receding surface, e.g., the depth of a FIB milled box cut.

In the absence of crystalline orientation effects and at low ion doses, the projected ion range distributions may be characterized by a simple Gaussian curve. However, as the dose increases (for a given beam energy), sputtering occurs and the depth distribution of implanted atoms typically has a maximum at the surface and falls off over a distance comparable to the initial ion range. This phenomenon is shown schematically in figure 7. Figure 7a shows the normalized ion concentration as a function of arbitrary distance units where no sputtering is observed (e.g., similar to that obtained via a simple TRIM calculation). Figure 7b shows an implanted ion concentration distribution as a function of distance once steady state sputtering conditions have been met.

It is therefore important to understand that simple generated TRIM collision cascades are statistical distributions of ion collisions and hence, does not account for steady state conditions between implantation and sputtering. The steady state surface composition for a single element target may be described by the following equation (Nastasi et al., 1996):

$$\frac{N_A}{N_B} = \frac{r}{(Y-1)} \tag{6}$$

In equation (6), N_A and N_B are the concentrations (per unit volume) of the implanted atom and the target atom respectively, r is a preferential sputtering factor (e.g., the ratio of the sputtering yield for a B atom to the sputtering yield for an A atom and is generally in the range of 0.5 to 2), and Y is the sputtering yield (Sigmund, 1991). At high doses, influences of preferential sputtering, segregation, atomic mixing, and chemical effects become important.

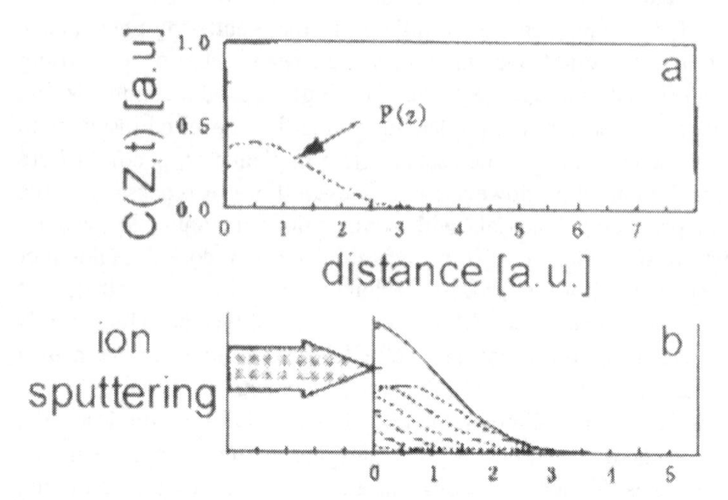

Figure 2-7. Schematic diagram of the implanted ion distribution once steady state sputtering is reached (figure obtained courtesy of Dr. T. Ishitani).

Implanted Ga concentrations of FIB prepared cross-sections in Si were estimated by Ishitani et al. (1998) to be 3.9 at % for r = 1, and 4.3 at % for r = 0.889, for a 30 keV Ga^+ beam at an incident angle of 87.5°. Since TEM specimens contain two surfaces that are FIB milled, they showed that a 50 nm thick TEM specimen would contain 1.3 at % Ga, while a 100 nm thick TEM specimen would contain 0.7 at % Ga.

3. SPUTTERING

The sputtering yield, Y, is defined as the number of ejected particles per incident ion. Sputtering can be considered as a statistical phenomena caused by surface erosion on an atomic scale. In this section we will discuss physical sputtering, whereby a transfer of kinetic energy from the incident ion to target atoms result in the ejection of surface and near surface atoms, also referred to as knock-on sputtering. Sputtering yields for typical FIB energies vary between $\sim 10^{-1} < Y < 10^{2}$ depending on target and incident angle (Andersen and Bay, 1981). Chemical sputtering will be discussed in the chapter on gas enhanced etching, where the physical sputtering rate may be enhanced by chemical reactions which produce an unstable compound with a diminished SBE at the target surface. The mean number of atoms ejected per incident ion with an energy exceeding some arbitrary minimum energy E_o may be expressed by equation (7) as

$$Y(E_o) \cong \frac{1}{4}\Gamma_m \frac{\alpha N S_n(E)\Delta x_o}{E_o} \tag{7}$$

where α is a dimensionless function incorporating the incidence angle, the mass ratio M_2/M_1, and the ion energy E, and Δx_o is the depth interval for which the atoms set in motion have an energy > E_o. (Sigmund, 1981). TRIM was used to determine the sputter yield for 100 Ga ions at different energies and incident angles for periodic table of elements (for solids). These values are listed in the appendix.

3.1 Ejection Direction During Sputtering

Sputtered particles generally possess an energy between ~ 2-5 eV. The emission of sputtered particles generally follows a cosine distribution for normal incidence ion bombardment (Behrisch, 1981). As the angle of incidence increases, the maximum emission of sputtered particles shifts away from the incoming ion beam as shown in figure 8.

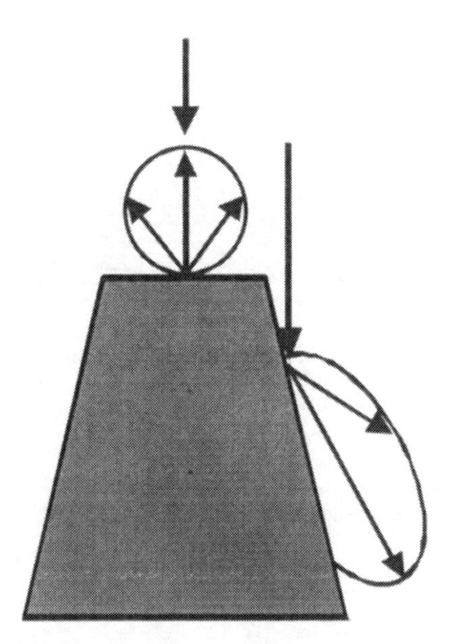

Figure 2-8. Schematic diagram of emission of sputtered particles with incident beam angle.

3.2 Backsputtering

Backsputtering occurs when an incident ion is scattered either directly, or after some number of multiple collisions, out of the target (Sigmund, 1981). (This phenomenon is analogous to backscattering in electron-solid interactions). As shown in figure 9 below, the backsputtering yield of the incident ion increases with angle of incidence. In addition, comparing the backsputtering yields between Si and Cu in figure 9 shows an increase in backsputtering with an increase in mass ratio M_2/M_1. Comparing figure 9 with the discussion on sputtering above, we see that in general, materials with a higher sputtering yield, have a correspondingly higher backsputtering yield of incident ions. Thus, more particles are available for the possibility of redeposition as will be discussed below. TRIM was used to determine the backsputter yield for 100 Ga ions at different energies and incident angles for periodic table of elements (for solids) and are listed in the appendix.

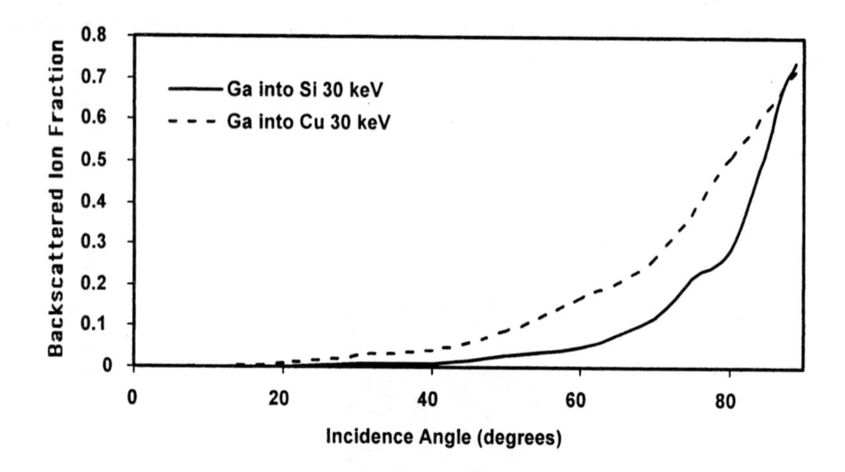

Figure 2-9. TRIM calculations showing backsputtering yield for Cu and Si as a function of incidence angle for 100 Ga ions at 30 keV.

4. REDEPOSITION

FIBs are most often used to create features of high aspect ratio (i.e., deep narrow trenches). Sputtered material and backsputtered ions may therefore deposit on surfaces that are in close proximity to the active milling site (e.g., the sidewalls of a deep narrow trench). Thus, surface degradation due to redeposition of sputtered material must also be considered during FIB milling. Controlling or at least predicting the manner in which redeposition of sputtered material will occur can be significant for the successful and rapid production of high quality TEM and other specimens by FIB techniques.

A hole milled with an FIB tends to be wide at the top surface and tapers down to a point at the bottom of the hole. The formation of this classic "V-shape" has been attributed to the redeposition of sputtered material that occurs when milling at high beam currents (Tartuani et al., 1993; Ishitani et al., 1994; Thayer, 1993; Yamaguchi et al., 1985; FEI, 1993; Walker, 1993). As the hole is made deeper, the effects of redeposition become increasingly severe until the rate of redeposition equals the rate of sputtering, thus limiting the aspect ratio of the milled hole. This effect of redeposition can be counteracted by the local introduction of a reactive gas species (i.e., Cl_2, I_2, and XeF_2) to the milling area (Thayer, 1993; FEI, 1993; Walker, 1993). The gas reacts with the sputtered material allowing it to be volatilized and removed by the vacuum system. Although reactive gas enhanced etching is an attractive solution to the problems resulting from redeposition, the gas has the potential to react with the sample material. The V-shaping that occurs in TEM specimens may be overcome by altering the angle of incidence with respect to the specimen surface for subsequent milling operations. The angle of incidence needed to prevent V-shaping is material dependent (e.g., it depends on the material sputtering rate). For example, while Si-based samples are generally tilted into the beam at 1-2°, it has been shown that a faster sputtering Zn sample must be tilted +/- 14° to create parallel sidewalls for TEM analysis (Prenitzer et al., 1998).

Redeposition is a function of a number of physically and chemically controlled variables, some of which include:

- The kinetic energy of the atoms leaving the surface
- The sticking coefficient of the target material
- The geometry of the feature being milled
- The sputtering yield (Y) of the target material

When an atom leaves a target material as a sputtered particle, it is ejected with a finite kinetic energy. A sputtered particle can, therefore, be considered a projectile capable of producing secondary interactions with local targets that lie in its trajectory. The direction and velocity of the ejected particle will be altered if it collides with another particle or the surface. Depending on the energy of impact and the sticking coefficient of the material, the sputtered atom may be redeposited on the surface that it strikes. The sticking coefficient is a statistical measure of a material's affinity to adhere to a surface, with a value of 1 equal to a 100% probability for sticking. It has been observed that there is very little difference in sticking probabilities for different materials for the low energy range exhibited by FIB sputtered particles (Andersen and Bay, 1981). However, it is shown below that the geometry of the feature to be milled and the sputtering yield play critical roles in the amount observed redeposition effects.

Figures 10a-c are SEM images of three trenches observed at a 70° tilt. The trenches were all milled at normal incidence in (100) Si. A beam current of 1000 pA was used to deliver a variable but known fluence per unit area to the three trenches. The doses were varied by controlling the scan time of the beam. The trench in figure 10a received a fluence of 1.5×10^{12} Ga^+ ions in 4 minutes, the trench in figure 10b received a fluence of 3.0×10^{12} Ga^+ ions in 8 minutes, and the trench in figure 10c received a fluence of 6.0×10^{12} Ga^+ ions in 16 minutes. The series of doses allows the progression of milling to be followed in time. The images provide evidence suggesting an increase in redeposition as the aspect ratio is increased. The shallowest trench has a fairly smooth appearance on the sidewalls and bottom. The sidewalls are still relatively vertical although they are somewhat rounded at the top. The second trench is beginning to show some roughening of the sidewalls and corners which is consistent with the appearance of redeposited material. Also, the sidewalls are beginning to show a greater deviation from the vertical direction. The third case clearly shows the effects of redeposition. The sidewalls have roughened considerably and the shape conforms to a definitive "V".

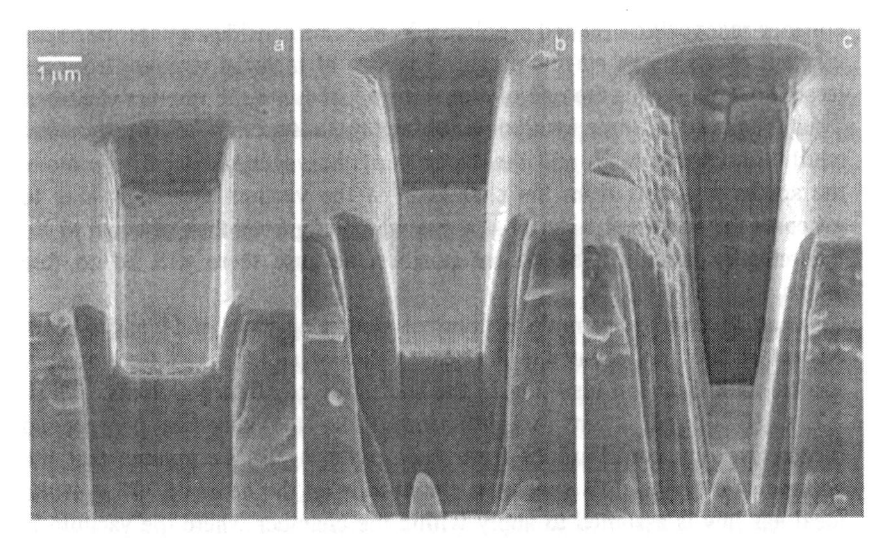

Figure 2-10. Rectangular FIB trenches in (100) Si milled at normal incidence by applying: (a) single fluence, (b) double fluence, and (c) triple fluence of Ga^+ ions at 25 keV.

An increased pressure mechanism for redeposition was proposed by Prenitzer (1999) where it was observed that both the degree to which the sputtered material is restricted from escaping the confines of the trench and the rate at which material is removed from the target have a profound effect on the observed degree of redeposition. Confining trench geometry such as large aspect ratios as well as factors that enhance the kinetics of material removal will tend to increase problems associated with redeposition.

It is important to make a distinction between sputtering rate (Y_t) and sputtering yield, Y. Y can be interpreted as the average number of target atoms ejected from the sample per incident ion. Y is an event dependent measure of the material removal. Y_t is the number of atoms being sputtered from the target per unit time. Therefore, Y_t is the actual time dependent, kinetic material removal parameter. A reasonable estimate of Y_t can be made by multiplying Y by the beam current, (i.e., the rate of delivery of ions to the target surface).

The mechanism that is proposed below to explain the "Classic V Shape" can be instructive in illustrating how the observed limits on aspect ratios attainable by FIB milling are influenced by both kinetic factors and trench geometry. When milling a trench, the rate of material removal can be considered to be in dynamic equilibrium with the rate of redeposition. For redeposition to occur atoms must be ejected from the target material with enough kinetic energy to carry them to a proximal surface with which they may collide and stick. In order to show how factors that enhance Y_t also

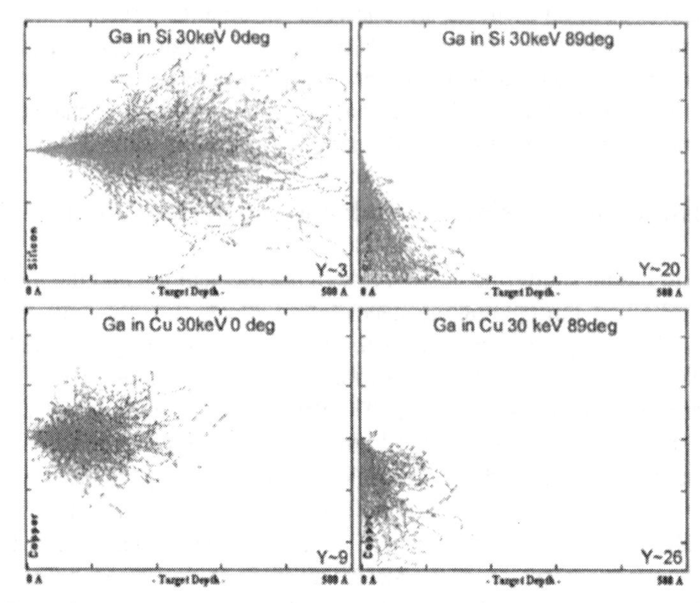

Figure 2-11. TRIM ion trajectories for 500 30 keV Ga^+ ions in Si and Cu and 0° and 89°.

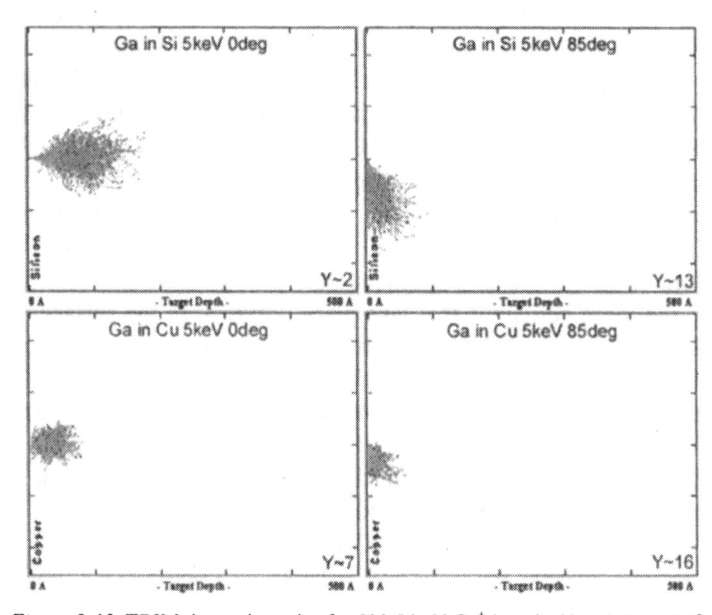

Figure 2-12. TRIM ion trajectories for 500 5 keV Ga^+ ions in Si and Cu and 0° and 85°.

correspondingly higher than that for Si at each incident angle shown. It should be noted that the relative difference in sputtering yield between the two materials is disproportionate to the relative difference in their melting temperatures. Furthermore, the difference in sputtering yield between the two materials decreases as the angle of incidence increases. This indicates that although the dominant mechanism controlling sputter yield is bond strength or SBE, the position of the collision cascade is also significant and may be more of an influence at greater incident angles.

Figure 13 shows how the sputter yield varies as a function of energy for Cu and Si at two different incident angles. The energy range modeled between 5-30 keV is within the nuclear stopping range for this system. The graph shows that the sputtering yield increases steadily with an increase in accelerating voltage at an incident angle of 89°. At the higher incident angle the collision cascade is confined in a region close to the surface. Thus, an increase in accelerating voltage would statistically result in an increase in the number of surface collisions with enough energy to overcome the SBE and consequently increase the sputtering yield. When the beam is at normal incidence with respect to the target surface an increase in accelerating voltage would also statistically result in an increase in the number of collisions cascade atoms with enough energy to overcome the SBE; however, at normal incidence the predominant effect of an increase in the accelerating voltage would be an increased projected range. Thus, the collision cascade would be located deeper under the surface of the target, subsequently mediating the effect of the increased energy of the cascade atoms. The result is a less dramatic increase in sputter yield with an increase in accelerating voltage at normal incidence. This is particularly evident in the Si target where the relatively low stopping power is not as effective at confining the collision cascade relative to Cu.

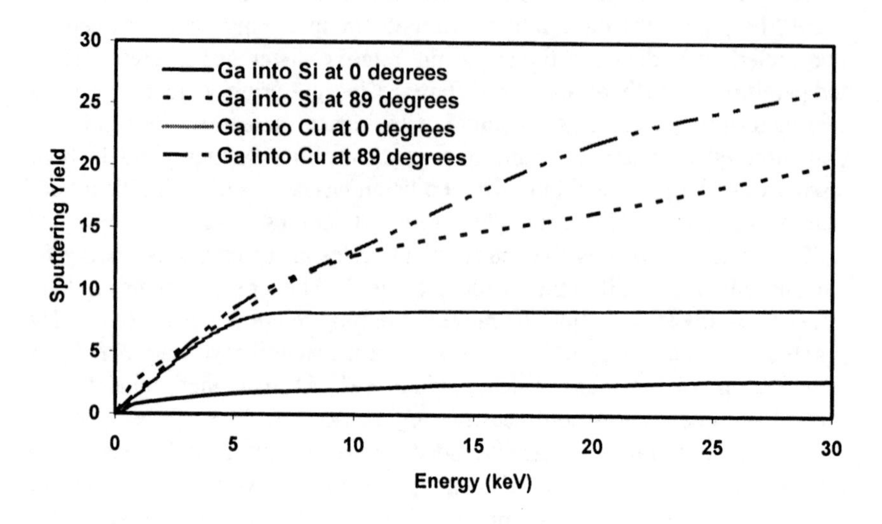

Figure 2-13. TRIM sputtering yields for 30 keV Ga+ ions into Si and Cu.

4.2 Dependence of Sputtering Yield on Target Material and Incident Angle

Figure 14 shows the effects of incidence angle on the sputtering yield for Si and Cu at 30 keV. Note that Y increases to a maximum and then approaches zero for increasing angle of incidence. This maximum sputtering yield occurs between ~ 75-85° depending in the material. It is interesting to note that early uses of broad beam Ar ion milling for TEM specimen preparation were generally performed at these angles of largest sputtering yield. The work pioneered by Barna et al. (1990) soon showed that these large sputtering yield angles were the worst possible angles in which to prepare thin specimens due to the large topography (i.e., "pillows") that developed on these surfaces. Barna's work showed that the best TEM specimens were obtained when ion milled at glancing angles with respect to the sample surface and at the lowest energy possible, thereby, producing a flat and polished surface. These are the same glancing angles (i.e., ~ 89°) that are now routinely used in FIB specimen preparation. The broad beam Ar ion mill is now routinely used after FIB milling to replace FIB damage with a smaller amount of Ar milling damage (see discussion on FIB damage elsewhere in this volume).

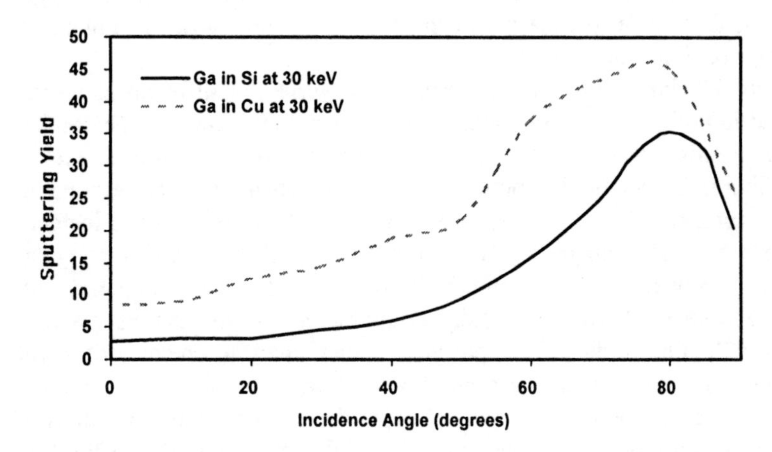

Fi

Figure 2-14. Sputtering yield as a function of incidence angle for 30 keV Ga^+ in Si and Cu.

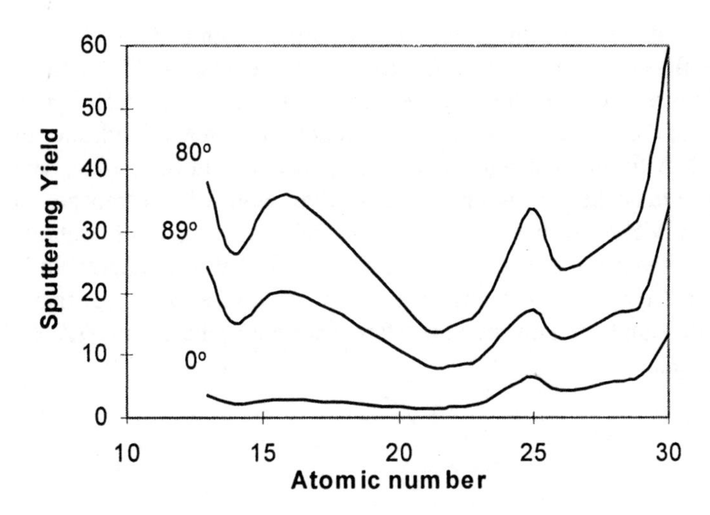

Figure 2-15. Material dependence of sputtering yield for materials Z = 13-30 at various angles of incidence from normal for 25 keV Ga+. (Data modeled with TRIM'97).

The material dependence of sputtering yield (Y(Z)) for a given set of milling parameters is illustrated in figure 15. The results of detailed calculations with full damage cascades modeled with TRIM for target materials, Z = 13-30 bombarded with 25 keV Ga^+ are shown in Figure 15 (and refer back to figures 11,12,13).

Figure 15 shows how the sputtering yield varies for different elements. The position of the peaks and valleys with respect to the x-axis indicates that periodic trends, associated with electronic structure of the elements, exert a controlling influence on the sputtering yield. The positions of corresponding peaks among the three data sets are invariant with respect to the incident milling angle. This suggests that Y(Z) remains consistent while the angular orientation between the target and the beam is varied. Thus, a material with a high relative Y(Z) will mill rapidly at any incident angle. As discussed in section 4.2, this indicates a physical rather than a chemical control mechanism for the angular variance of Y. Although the shapes of the three curves mimic one other, they are not coincident. The relative position of each curve with respect to the y-axis indicates the sputtering efficiency associated with that particular incident angle $Y(\theta)$ as also shown in figure 13. There is also some deviation between the magnitudes of corresponding peaks with incident angle. This is also consistent with the data shown in figure 13.

Figures 16a-d are SEM images of trenches milled with a 1000 pA beam at the same fluence into Zn, Cu, Al, and Si, respectively. The images provide a clear example of how different materials behave uniquely under the influence of the ion beam. Relative trench depths are a good indicator of relative Y(Z) with the incident angle, beam current, milling time, and accelerating voltage held constant. The relative depth of the trenches in Figures 16a-d yields a qualitative measure of the volume of material removed for a given ion fluence. It should be noted that this experiment is not intended to be presented as a rigorous quantification of Y(Z), but rather as an generalization to demonstrate the relative periodic behavior of Y(Z) for various materials.

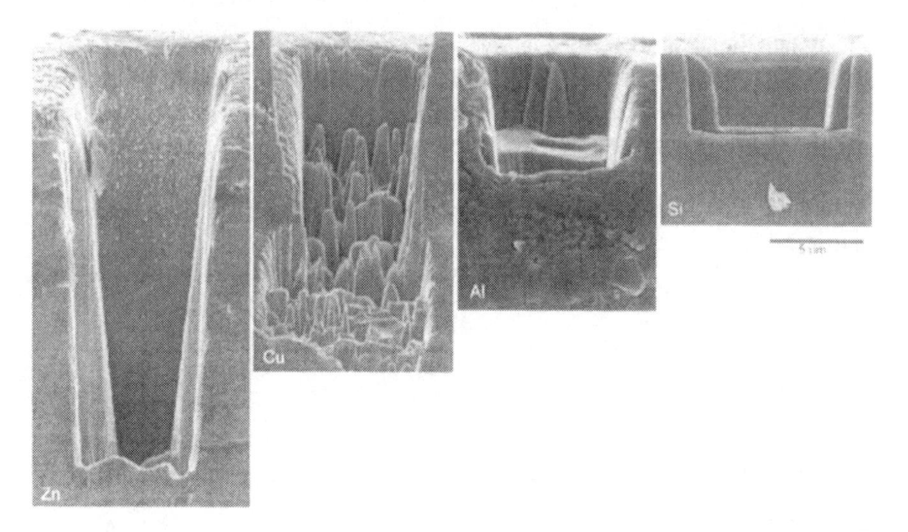

Figure 2-16. SEM images of cross-sectioned FIB trenches milled at 25 keV and constant fluence for Zn, Cu, Al, and Si.

Figure 16a is the trench milled in Zn. The Zn trench is the deepest which is consistent with the theoretical prediction that Zn will mill the fastest of the four materials shown at Y(Zn) = 12.9 atoms/25 keV Ga^+ ion. The Zn also shows the most severe sloping of the sidewalls. This is expected behavior based on the concepts developed to explain redeposition as previously explained. Figure 16b shows a trench milled in Cu. The relative trench depth of Cu is also in agreement with the predicted relative Y(Cu) = 7.0 atoms/Ga^+ ion. Figure 16c shows that the Al does not show roughening under ion bombardment at normal incidence. The trench floor and sidewalls are relatively smooth having a minimum of discontinuities. The relative depth is in accordance with the modeled Y(Al) = 3.5 atoms/Ga^+ ion. Figure 16d is the trench milled into Si. Of the four materials shown, the Si appears to show the least amount of milling induced topography. The Si trench sidewalls and bottom are relatively uniform, and excluding the non-vertical sidewalls, the trench appears virtually free from the effects of redeposition as observed by SEM. The relative trench depth observed in Si also conforms to the predicted relative Y(Si) = 2.1 atoms/ Ga^+ ion.

The periodic fluctuations in sputtering yield demonstrated in the previous section are manifestations of the influence that the interatomic potential exerts on the physical properties of a given material. Thus, range data and sputtering yields predicted by computer modeling, as well as the supporting empirical results show that material dependent properties of a solid are governed by the forces that bind the constituent atoms together.

When the data described above are combined with some other well-established periodic trends, such as T_m as shown in Figure 17, it becomes evident that there are very useful correlations to be formulated. Sputtering yield exhibits an inverse correlation with T_m. This type of relationship is intuitive since T_m is direct indicator of bond strength. The more tightly bound an atom is, the more difficult it will be to eject it as a sputtered particle.

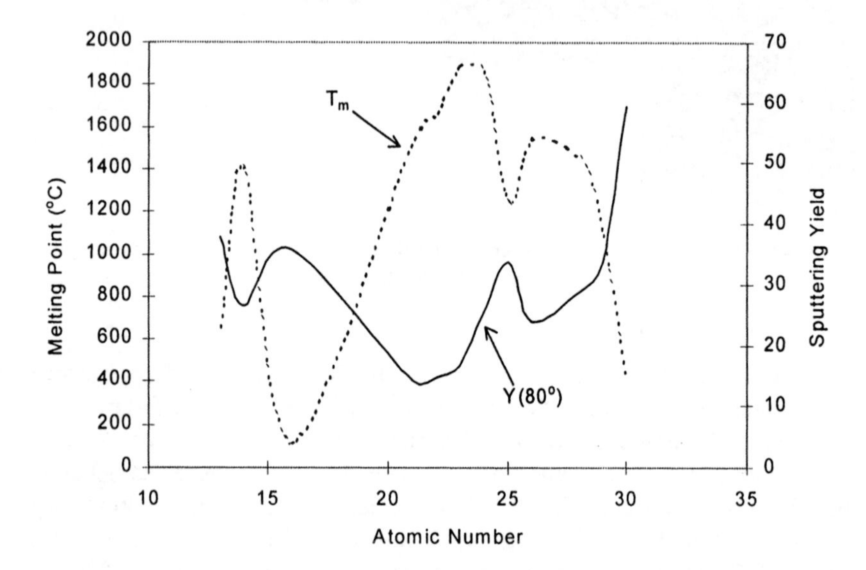

Figure 2-17. The correlation between sputtering yield at 80° angle of incidence and melting temperature for elements, Z = 13 through 30 and a 25 keV Ga+ beam (Data modeled with TRIM'97).

It has been established that material dependent milling induced behavior is influenced by the same interatomic forces that govern other periodic behaviors of the elements. Therefore, phenomena such as sputtering yields and range distributions can themselves be regarded as periodic properties of the elements. This allows the familiar and available periodic table to be used as a tool for predicting milling behavior. Predictions of how an unfamiliar material may respond to Ga^+ milling has many practical implications for FIB applications.

4.3 Ion Channeling

In the discussion of range, stopping power, and sputtering behavior above, it was assumed that the target was disordered, i.e., amorphous. It was observed that the factors that affect sputtering include the atomic number (i.e., mass), energy, and angle of incidence of the ion beam, the mass of the target, and the SBE of the target. The phenomenon of ion channeling occurs in crystalline materials. Channeling is a process whereby ions may penetrate greater distances along low index directions compared to non-channeling directions or amorphous materials. Channeling thereby increases the range of ions, moving the collision cascade further from the surface. Since channeling influences the ion range (i.e., the shape of the collision cascade), it also has dramatic affects on image contrast, damage depths, and sputtering yields.

Ion channeling is responsible for the varying contrast in secondary electron images for polycrystalline samples (Franklin et al., 1988; Kola et al., 1993). Channeling contrast results because the secondary electron yield varies as a function of crystallographic orientation within the sample. Thus, a single crystal region (i.e., a grain in a polycrystalline sample) will appear darker when it is aligned (or nearly aligned) to a low index direction due to a decrease in the number of emitted secondary electrons. Figure 18 below shows an example of the dramatic channeling contrast observed for a polycrystalline Cu specimen. There are many other examples of images showing channeling contrast in this volume.

Ion channeling will also decrease the sputtering yield of a material, the physics and mathematics of which may be found elsewhere (Lindhard, 1964; Onderdelinden, 1966; Sprague et al., 1987; W. Palmer et al., 1990; Hosler et al., 1993). Figure 19 shows secondary electron FIB image of FIB milled trenches in a Cu bicrystal showing differences in sputtering yield with channeling effects (Kempshall et al. 2001). The interesting feature to note in Figure 19 is the correlation between the ion channeling contrast and the milling characteristics. When a grain is oriented to the [100] channeling direction, the milling characteristics of the Cu improve as evident by the flat trench bottoms and clean trench walls. Conversely, the trench milled in the grain that is not aligned with the [100] direction has poor milling characteristics as evident by the rough trench bottom and the sloped trench wall, which is consistent with redeposition effects via an increase in sputtering rate as described above. Furthermore, it is clearly evident by the differing depths in the 24 μm x 2 μm trench that the non-channeled side of the trench mills quicker than the channeled side of the trench. Since the ion channeling contrast has been shown to be directly proportional to the sputtering yield, the differences in milling characteristics can be accounted for by looking at the mechanism that affects both the contrast and the sputtering yield, specifically, channeling. Figure 19 also shows how channeling affects both image contrast (i.e., the generation of secondary electrons) as well as sputtering yield. Figure 20 shows a secondary electron FIB image of the UCF pegasus logo FIB milled into adjacent grains of a 10°/[100] twist boundary. Note that the right grain is aligned to the [100] direction and shows dark contrast, while the left grain oriented 10° from [100] shows bright contrast. The faster sputtering left grain also shows enhanced redeposition artifacts on the sidewalls of the cuts, again showing the relationship between sputtering rate and redeposition artifacts.

Figure 2-18. An ion induced secondary electron image illustrating channeling contrast in a FIB polished polycrystalline Cu sample (sample courtesy of S. Merchant).

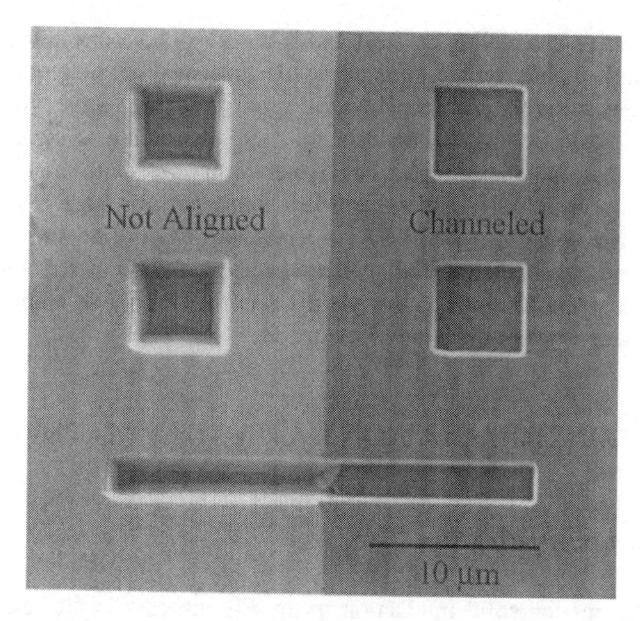

Figure 2-19. Secondary electron FIB image of milled trenches in a Cu bicrystal indicating differences in sputtering yield with channeling effects.

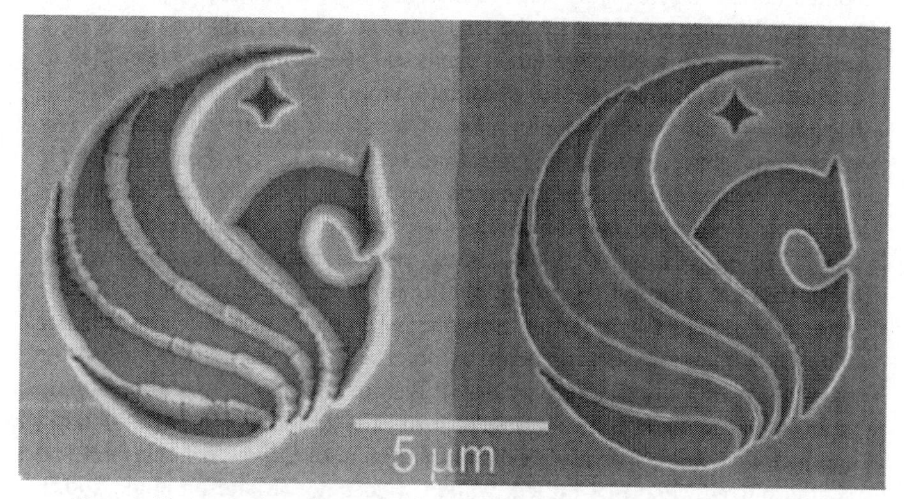

Figure 2-20. A secondary electron FIB image of the UCF pegasus logo FIB milled into adjacent grains separated by a 10°/[100] twist boundary.

The relative channeling contrast that is observed from one material to the next depends mainly on the interatomic planar distances, the atomic density of the target material, and the Thomas-Fermi screening length. In the case of using 30keV Ga^+ as the incident ion, ion channeling is significant for materials that have close packed crystal structures with higher atomic densities like Al, Cu, Ni, and Au and less significant for materials that have lower atomic densities like Si. As a result, the amount of ion channeling contrast, and hence, differential sputtering observed for a material like Cu is significant, while Si tends to not yield a readily observable amount of ion channeling contrast or differential sputtering.

5. ION IRRADIATION DAMAGE IN MATERIALS

5.1 Amorphization

An inherent ion-solid interaction in the FIB process is the result of ion implantation into the target surface. The degree of ion implantation depends on ion energy, angle of incidence, ion species, and target material as discussed above. Sample preparation by FIB uses the process of ion bombardment to selectively remove material. Atoms that are displaced from their equilibrium positions by the impingement of energetic ions generate a collision cascade within the target material. Sputtering occurs if sufficient momentum is transferred to a surface atom. One consequence of ion implantation can be the development of a surface amorphous phase. The amorphous phase induced in crystalline materials by ion bombardment is typically metastable, and its formation depends on unit cell size, complexity of chemical ordering, and the width of an intermetallic phase field (Nastasi, 24). The restoration of the collision cascade induced disorder requires correlated and cooperative motion of alloying atoms. The more complex the material unit cell, the larger the amorphous layer will be. Likewise, smaller unit-celled materials are difficult to amorphize. Additionally, alloys or materials with a broad phase field will remain crystalline, since the atomic packing arrangement is less stringent than line compounds or stochiometric intermetallics. Consistent with the discussion above, we have observed that Si amorphizes when FIB milled, while Cu does not (Matteson et al., 2002). It should be noted that Cu has been observed to form a Cu_3Ga phase when FIB milled at certain crystallographic directions (see chapter by Phaneuf in this volume).

5.2 Local Heating in FIB Milled Surfaces?

Averbeck et al. (1994) have used molecular dynamics simulations to investigate the collision cascade during ion bombardment in crystalline materials. An example of a 10 keV Au particle impinging on a (100) Au surface is shown in figure 21. At the onset of particle bombardment (figure 21a), displaced atoms are observed within the collision cascade. After ~ 1ps (figure 21b), the atoms in the collision cascade are in such disorder that they may be considered to be a liquid. This disorder continues through to ~ 4ps (figure 21c). However, between ~ 10 ps (figure 21d) to ~ 20 ps (figure 21e), the atoms in the collision cascade begin to relax or "self anneal" back into their equilibrium lattice positions. The entire process is over within ~ 30 ps (figure 21f) and the atomic region that included the collision cascade is shown to consist of crystalline point defects. Thus, for the simple centered-close-packed Au structure, ion bombardment results in the generation of crystalline defects. In addition, local heating by the FIB process is restricted to the region defined by the collision cascade. Since the small FIB beam must be rastered over an area in order for milling to occur (contrary to broad beam ion milling where a sample region may experience constant bombardment by the impinging ions), where the beam dwell times are on the order of micro-seconds, heating effects for most materials may be deemed to be negligible. In our lab, we have only observed one case in a proprietary polymeric material where beam damage was so extensive that a thin TEM membrane could not be prepared. *Thus, any specimen "heating" due to FIB milling will be confined to the collision cascade which is on the order of a few tens of nanometers or less for most materials.*

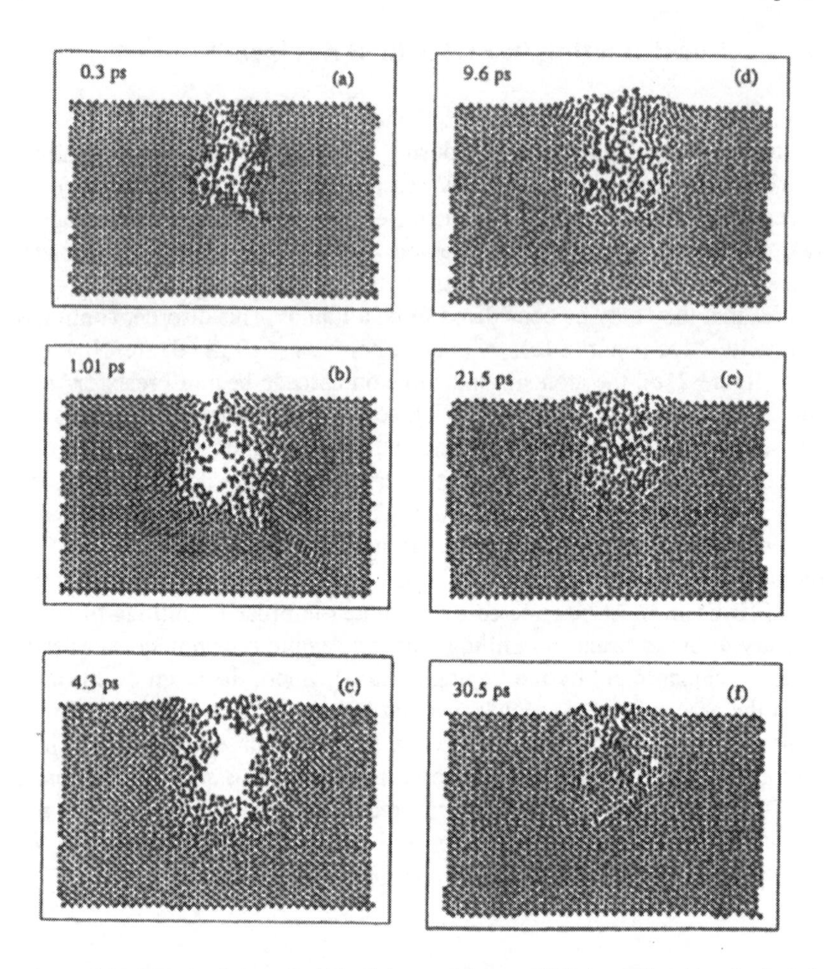

Figure 2-21. MD simulation of a 10 keV Au particle on a (100) Au surface (Averbeck et al., 1994, used with permission, Journal of Applied Physics, © American Institute of Physics)

5.3 Amorphization vs. Redeposition

There seems to be some disparity in the literature as to how much Ga is actually implanted in materials, and how much amorphization damage is expected to occur in certain materials. There is a distinct difference between damage due to inherent ion-solid interactions and redeposition artifacts (Rajsiri et al., 2002). FIB milled side-walls in Si were cross-sectioned using the FIB in-situ lift-out technique as described elsewhere in this volume. As shown by figure 22, the damage along the side-wall actually consists of two distinct regions. The bright contrast region adjacent to the single crystal Si shows ~ 20 nm of amorphization damage. Between the amorphization region and a protective Cr layer is a region of darker contrast (~ 50 nm). XEDS spectra obtained from the dark region shows large amounts of Ga (~37 wt % or ~ 15 at %) in Si. As shown by figure 22, ~70% of the incident Ga^+ ions may be backsputtered during FIB milling at the large incident angles used FIB milling sidewalls. This dark region in the bright field TEM image is consistent with mass/thickness contrast and thus clearly indicates a redeposition layer containing sputtered Si and Ga plus backsputtered Ga.

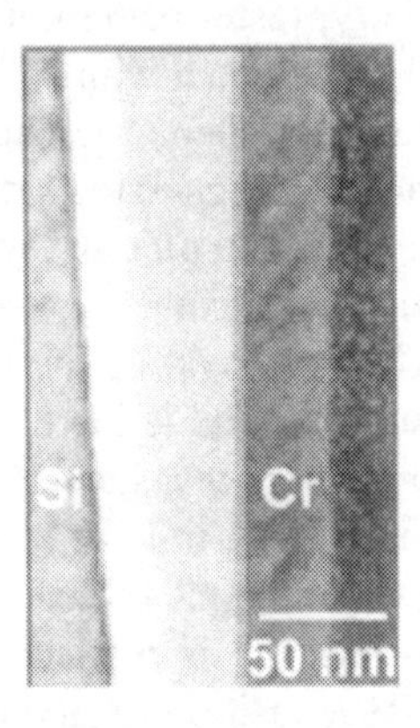

Figure 2-22. Cross-section TEM image of the side-wall of Si FIB milled with 30 keV Ga^+ ions at 1000pA showing an amorphization region and a redeposition region.

If a simple case is considered where 100 Ga^+ ions bombard a Si surface at 89° and 30 keV, then 70 ions will be backsputtered, and the remaining 30 ions will produce ~ 600 sputtered Si atoms. Assuming that no backsputtered ions will contribute to the Si sputter yield, and neglecting that steady state sputtering also removes implanted Ga ions, then ~ 10 at % Ga will be available in the sputtered material. Thus, the ~ 15 at % Ga that is observed in the redeposition region is consistent with the amount of Ga that is also available in the sputtered particles in this simple comparison.

The amorphization layer yielded ~ 0.4 wt % Ga (0.16 at %) within this region. XEDS results obtained from the Si just below the amorphization region showed ~0.1 wt % Ga. This implies that the amount of Ga implanted into the crystalline Si is on the order of the resolution of the XEDS technique. These results are also less than that predicted by Ishitani et al. discussed in section 1.3 above, although the differences may be explained due to differences in incident angles. The importance of these results is that any significant amount of Ga observed in FIB milled Si (i.e., > 1 wt %) is due to redeposition artifacts and not due to inherent ion-solid interactions.

6. SUMMARY

The basics of ion-solid interactions particular to FIB milling have been presented. The concepts and methods to model ion range, sputtering, and ion irradiation damage have been reviewed. It was shown that materials subjected to bombardment by energetic ions tend to exhibit microstructural and topographical disruption. The response of a given target material to the ion beam is strongly dependent on factors such as beam current, incident ion energy, trench/feature geometry, raster pattern, and milling angle. Many of the artifacts that are typically associated with FIB prepared TEM specimens are due to redeposition and not to inherent ion-solid interactions. Problems associated with redeposition (i.e., exaggerated sidewall sloping, formation of undesired topography, or the inclusion of high levels of Ga^+ in the specimen) are controllable. Redeposition is enhanced by factors that increase the rate of sputtering and/or restrict the removal of the sputtered material from the active milling site. The sputtering rate is increased by increasing the beam current, and/or changing parameters that increase the sputtering yield (e.g., milling at an incident angle of ~80°, milling a material with an inherently high sputtering yield, or crystallographic effects). The removal of sputtered material from the active milling site is particularly impeded when milling high aspect ratio trenches. Redeposition related artifacts may be minimized by understanding and selecting appropriate combinations of the milling parameters (e.g., reducing the beam current when milling a material with an inherently high sputtering yield).

Material inherent properties such as sputtering yield or stopping power are dictated by the same interatomic forces that govern other periodic behaviors of the elements. Therefore, phenomena such as sputtering yield and range distributions can themselves be regarded as periodic properties of the elements. This allows the familiar periodic table to be used as a tool for predicting milling behavior.

The objective of the FIB LO method is to rapidly produce a high quality electron transparent membrane to be imaged in the TEM. Because a TEM specimen must be representative of the bulk microstructure, any modifications that the specimen might incur during the preparation process must be well characterized. An understanding of the fundamental ion/solid interactions that govern the milling process as outlined above insures the accuracy of the information obtained from an FIB prepared TEM specimen. Such an understanding of ion/solid interactions can also be used to predict milling behavior of novel materials, improve overall specimen quality and increase the success rates for FIB EM specimen preparation.

ACKNOWLEDGEMENTS

The authors would like to thank the I4/UCF/Cirent/Agere Partnership, NSF DMR #9703281, AMPAC, Florida Space Grant Consortium, Florida Solar Energy Center, and the DOE OIT program for financial support. Special thanks to Jeff Bindell for his undying support of the UCF/Cirent/Agere partnership.

REFERENCES

American Society for Testing and Materials. E1620-97, "Standard Terminology Relating to Liquid Particles and Atomization," in Annual Book of ASTM Standards v. 14.02 General Test Methods; Forensic Sciences; Terminology; Conformity Assessment; Statistical Methods, ASTM International, West Conshohockon, PA, v.14.02 (1997)

Andersen HH and Bay HL, "Sputtering Yield Measurements," in Sputtering by Particle Bombardment I, Physical Sputtering of Single-Element Solids, ed. R Behrisch, Springer Verlag, Berlin, 145-218 (1981).

ANSI N1.1-1976, American National Standards Glossary of Terms in Nuclear Science and Technology, American National Standards Institute New York, NY (1976)

Averbeck RS, Ghaly M, "A Model for Surface Damage in Ion-Irradiated Solids," Journal of Applied Physics, 76, 6, 3908 (1994).

Barna A, Barna, PB, Zalar A, "Ion Beam induced roughness and its effect in AES depth profiling of multilayer Ni/Cr thin films," Surface and Interface Analysis, 12 1-2, 144-150 (1988).

Barna A and Barna PB, Zalar, A, "Analysis of the development of large area surface topography during ion etching," Vacuum, 40, 1-2, 115-120 (1990).

Behrisch R, "Introduction and Overview," in Sputtering by Particle Bombardment I, Physical Sputtering of Single-Element Solids, ed. R Behrisch, Springer Verlag, Berlin, 1-8. (1981).

FEI Focused Ion Beam Application Note, "High Aspect Ratio Hole Drilling Using FIB Enhanced Etch Process," FEI Company 7451 NE Evergreen Parkway; Hillsboro, OR 97124, (1993).

Franklin RE, Kirk ECC, Cleaver JRA, and Ahmed H, Journal of Materials Science Letters 7, 39 (1988).

Gibbons JF, Johnson WS, Mylroie SW, Projected Range Statistics: Semiconductors and Related Materials, Dowden, Hutchinson, and Ross, Inc. Stroudsburg, PA, pp. 3-27. (1975).

Hosler W and Palmer W, Surface and Interface Analysis, 20, 609 (1993).

ICRP 60-1990, "Recommendations of the International Commission on Radiological Protection," in Annals of the ICRP Vol. 21/1-3 Edited by ICRP (1991)

Ishitani T, Tsuboi H, Yaguchi T, and Koike H, "Transmission Electron Microscope Sample Preparation Using a Focused Ion Beam," J. Electron Microsc., Vol 43, pp. 322-26 (1994).

Kempshall BW, Schwarz SM, Prenitzer BI, Giannuzzi LA, and Stevie FA, "Ion Channeling Effects on the FIB Milling of Copper," Journal of Vacuum Science & Technology B, 19(3), 749-754 (May/Jun 2001).

Kola RR, Celler GK, and Harriot LR, Materials Research Society Symposium Proceedings 279, 593 (1993).

Liau, ZL and Mayer, JW, "Ion Bombardment Effects on Material Composition," in Treatise on materials Science and Technology , ed. J K Hirvonen, Academic Press, New York, NY 17-50 (1980)

Lindhard J, Scharff M, and Schiott HE, "Range Concepts on Heavy Ion Ranges" (Notes on Atomic Collisions II), Mat. Fys. Medd. Dan. Vid. Selsk. 33, No. 14, p15. (1963).

Lindhard J, Physics Letters 12, 126 (1964).

Matteson TL, Kempshall BW, Schwarz SW, Houge EC, and Giannuzzi LA, "EBSP Investigation of Focused Ion Beam Surfaces," Journal of Electronic Materials, 31(1), 33-39 (2002).

Nastasi M, Mayer JW and Hirvonen JK, Ion-Solid Interactions: Fundamentals and Applications, Cambridge University Press, Great Britain, (1996).

Onderdelinden D, Applied Physics Letters 8, 189 (1966).

Orloff J, Utlaut M, and Swanson L, High Resolution Focused Ion Beams: FIB and its Applications, Kluwer Academic/Plenum Publishers, NY, (2003).

Palmer W, Wangemann K, Kampermann S, and Hosler W, Nuclear Instruments and Methods in Physics Research B 51, 34 (1990).

Prenitzer BI, Giannuzzi LA, Newman K, Brown SR, Irwin RB, Shofner TL, and Stevie FA, "Transmission Electron Microscope Specimen Preparation of Zn Powders Using the Focused Ion Beam Lift-Out Technique," Met. Trans. A, Vol. 29A, pp. 2399-2405. (1998).

Prenitzer, BI, "Investigation of Variables Affecting Focused Ion Beam Milling as Applied to Specimen Preparation for Electron Microscopy: A Correlation Between Monte Carlo Based Simulation and Empirical Observation, Ph.D. Dissertation, University of Central Florida, (1999).

Rajsiri S, Kempshall BW, Schwarz SM, and Giannuzzi LA, "FIB Damage in Silicon: Amorphization or Redeposition?," Microsc. and Microanal. 8 (Suppl. 2), Microscopy Society of America, 50-51 (2002).

Sigmund, P., Sputtering by Ion Bombardment:Theoretical Concepts, in Sputtering by Particle Bombardment I: Physical Sputtering of Single Element-Solids, R. Behrisch, ed., Topics in Applied Physics vol. 47 (Springer-Verlag, Berlin), (1991).

Sprague JA, Malmberg PR, Reynolds GW, Lambert JM, Treado PA, and Vincenz AM, Nuclear Instruments and Methods in Physics Research B24/25, 572 (1987).

Tartuani M, Takai Y, Shimizu R, Uda K, and TakahashiH, "Development of a Focused Ion Beam Apparatus for Preparing Cross-Sectional Electron Microscope Specimens," Tech. Rep. Osaka Univ., Vol. 43. No. 2143, pp. 167-73. (1993).

Thayer ML, "Enhanced Focused Ion Beam Milling Applications," ISTFA '93, Proc. Int. Symp. Test. Failure Anal., 19th, pp. 425-29 (1993).

Townsend PD, Kelly R, and Hartly NEW, Ion Implantation, Sputtering and Their Applications, Academic Press, pp. 137-42 (1976).

Walker JF, "Focused Ion Beam Applications Using Enhanced Etch," FEI Company 7451 NE Evergreen Parkway; Hillsboro, OR 97124 (1993).

Yamaguchi H, Shimase A, Haraaichi S, and Miyauchi T, "Characteristics of Silicon Removal by Fine Gallium Ion Beam," J. Vac. Sci. Technol. B3(1), pp. 71-4 (1985).

Zeigler JF, Biersack JP, and Littmark U, The Stopping Range of Ions in Solids, Pergamon Press, New York, (1985).

Ziegler, J.F., SRIM 2003, www.srim.org

Chapter 3

FOCUSED ION BEAM GASES FOR DEPOSITION AND ENHANCED ETCH

F. A. Stevie, D. P. Griffis, and P. E. Russell
North Carolina State University, Analytical Instrumentation Facility, Raleigh, NC 27695

Abstract: The utility of a focused ion beam (FIB) to provide material removal for micromachining via sputtering with resolution of better than 10 nm has led to many important applications. In addition, FIB combined with a gas source containing the chemical precursors for deposition of materials or for enhanced, selective material removal provides the capabilities for a much wider range of micromachining applications. This chapter introduces FIB material deposition and chemically enhanced material removal processes, lists some of the FIB chemical precursors in common use and discusses the parameters for their use, and presents several examples.

Keywords: Deposition, enhanced etch, gas sources

1. PROCESS

The material deposition and enhanced material removal processes are similar in that the focused ion beam interacts with a gas phase chemical precursor that is adsorbed on the surface of the sample. Both the material removal and deposition processes utilize essentially the same instrumentation. The chemical precursors are obtained from a gas, liquid, or solid source that can be heated if required to produce the desired vapor pressure. A valve admits the chemical precursor gas to the sample chamber through a narrow delivery tube. The tube or needle, which directs the gas toward the surface of the sample, is similar in dimensions to a hypodermic needle (~0.5 mm) and can be manipulated for proper positioning with respect to the specimen surface. This experimental arrangement is shown schematically in Fig. 1.

The precursor gas density desired at the sample surface is obtained via regulation of the precursor gas flow and by the proximity of the end of the needle with respect to the sample surface (approximately 100 μm). (Xu, 1993). The rate of precursor flow is controlled by suitable adjustment of the temperature of the chemical precursor to control precursor vapor pressure combined with either a leak valve or a limiting aperture to control flow. Additionally, the use of a smaller diameter needle can be used increase the local precursor pressure at the sample surface, but the use of too small a needle can lead to clogging. Multiple precursors can be accessed through several individual precursor sources/needles or by using a manifold and a single needle.

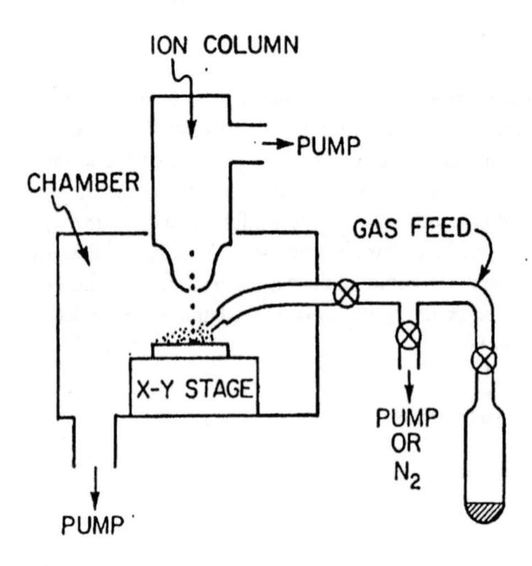

Figure 3.1. Schematic of FIB instrument with gas introduction system for material deposition or enhanced material removal. From Ratta, 1993, with permission of author and American Institute of Physics.

Both the FIB induced deposition and material removal processes involve adsorption of the chemical precursor onto the sample surface followed by deposition of energy by the focused ion beam. This deposition of energy by the ion beam results in a chemical reaction involving the precursor, and, particularly when it is desired to control material removal rate, with the sample surface. The FIB induced chemical reaction can result in deposition of material or provide material removal selectivity. For material deposition, a precursor is selected that will decompose into non-volatile products having desired materials properties depending on the application. For material removal control, chemical precursors are selected either to retard or enhance the sputtering rate of the sample. To retard the removal rate of a material, a precursor is selected that reacts with the substrate to form reaction products that are non volatile and have a reduced sputtering rate. To enhance the material removal rate, a precursor is selected that will react with the substrate to produce volatile products or products which have a higher sputtering rate than the native sample material.

Fig. 2 shows the spatial relationship of the gas source, the Ga^+ focused ion beam, the sample surface, and the volatilized and sputtered species.

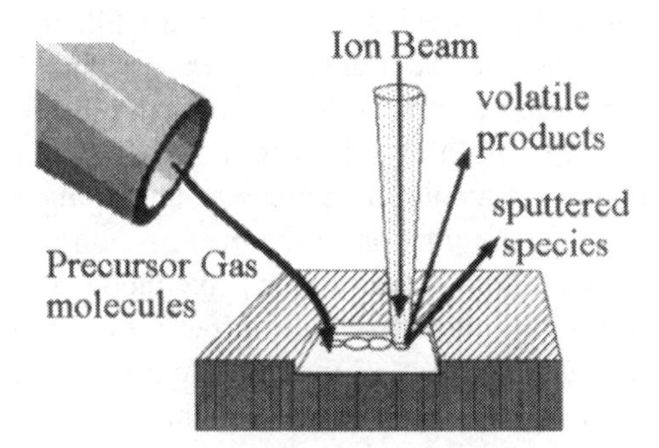

Figure 3.2. Schematic drawing of deposition/controlled material removal process. The enhanced etch process is shown. If adsorbed gas decomposes to non-volatile products, then deposition will take place.

In order to optimize the micromachining process, whether for material deposition or chemically enhanced material removal, a detailed understanding of the micromachining process (Harriot, 1993; Bannerjee, 1993) is required. The micromachining process can be broken down into two separate components, precursor deposition and focused ion beam/precursor/sample interaction, which must both be controlled with respect to the other to obtain optimum micromachining.

First consider non-chemically enhanced FIB material removal. For this process, the focused ion beam is digitally steered in a repeating raster pattern as shown in Figure 3. To form the raster, the digitally steered focused ion beam is stepped in a serpentine pattern over the area to be micromachined. When the farthest extreme of the raster scan is reached, the FIB is quickly stepped back to the beginning of the pattern and the pattern is repeated. Generally, the raster scan is repeated many times. The size of the beam steps in the raster pattern is usually set to equal one half of the FIB beam diameter in order to provide sufficient beam overlap ensuring that the rastered area is uniformly exposed to the FIB ions, i.e., the FIB ion dose (ions/cm^2) is uniform. Thus the diameter of the focused ion beam and the size of the rastered area dictate the number of beam steps or the number of pixels in the raster. The number of pixels and the time per pixel determine the time required to complete one raster which is called the raster refresh time.

For chemically enhanced FIB material removal, each pixel of the rastered area must adsorb or otherwise have available an optimal amount of the chemical precursor. If there is too little precursor with which the FIB ions/sample surface can interact, the optimal (highest) material removal rate will not be achieved. If there is too much precursor adsorbed, the FIB must remove the excess precursor before sample material can be removed and the material removal rate is again reduced as compared to the optimal rate. The amount of precursor available for reaction is determined by three parameters: 1) the precursor flux or the amount of precursor impinging on the rastered area per unit time; 2) the combination of the precursor sticking coefficient and the evaporation rate which determine the likelihood of an impinging precursor molecule remaining on the sample surface; and 3) the raster refresh time which determines the amount of time for the precursor to collect on the surface before being impinged upon by the FIB ions. Assuming that optimum coverage of a pixel by the precursor has been achieved (most often a single monolayer of the precursor), optimum utilization of the adsorbed precursor molecules depends on the proper selection of the focused ion beam current or flux and the pixel dwell time which in combination determine the pixel dose or the number of FIB ions impinging on a particular pixel diameter during the pixel dwell time. If too few FIB ions impinge on the pixel during one pixel dwell time, all precursor molecules coving that pixel area will not be reacted and thus will not be utilized for material removal rate enhancement. If too may FIB ions impinge on the pixel, the material removal rate enhancing precursors will be consumed before the pixel dwell time elapses and the removal rate will revert to the material removal rate resulting from physical sputtering only.

While theoretically possible, it is difficult to calculate the optimum parameters for chemically enhanced FIB material removal (Harriot, 1993). In practice, the optimum micromachining parameters can be experimentally determined by systematically varying FIB ion current, pixel dwell time, raster refresh time or precursor flux while holding the remaining parameters constant (e.g., Harriot, 1993; Stark, 1995). In these experiments, the material removal yield, i.e., the volume of sample material removed per incoming FIB ion, is determined as the selected experimental parameter is varied and the other parameters held constant. The combination of results is then analyzed to determine the set of experimental parameters that provide the optimal removal rate.

While the discussion above has focused on enhanced material removal, the micromachining parameters discussed above also apply to FIB induced material deposition and to chemically enhanced selective material removal whether through retardation or enhancement of material removal rate. All FIB processes based on chemical precursors are dependent on the combination of FIB ion current, pixel dwell time, raster refresh time and precursor flux chosen for a particular process. Finally, while most micromachining is performed using the raster scanning method described above, there are specialized cases when more exotic beam scanning strategies, both with and without the use of chemical precursors, are used (Thaus, 1996; Casey, 2002).

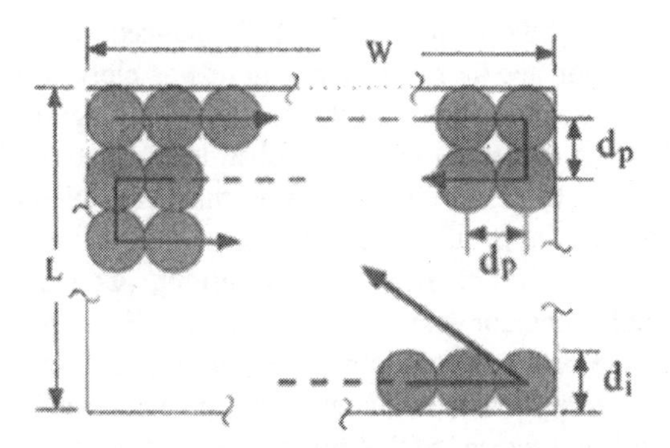

Figure 3.3. Drawing showing exposure parameters. The beam is repeatedly scanned over the milled area. (Area = LxW, beam step = d_p, and beam diameter = d_i) From Stark, 1995, with permission of author and American Institute of Physics.

2. DEPOSITION

FIB induced material deposition provides the ability to deposit useful materials on almost any solid surface with very high (nanometer) spatial precision. FIB material deposition is currently commonly employed for deposition of conductors and insulators for IC circuit edit and for deposition of material to mask or protect the sample surface during FIB micromachining for cross section analysis, for TEM/STEM sample preparation and other micromachining applications (Bannerjee,1993; Giannuzzi, 1999; Yaguchi, 1999). The high precision and resolution of FIB material deposition also provides the ability to construct novel microstructures which many believe will have increasing application in the future (discussed below).

For FIB induced material deposition to occur, a precursor must have two properties. The precursor must have a sufficient sticking probability to stick to a surface of interest in sufficient quantity, and it must, when bombarded by an energetic ion beam, decompose more rapidly than it is sputtered away by the ion beam. A commonly utilized example of FIB material deposition employs the precursor $W(CO)_6$ which decomposes to W (with volatile C and O containing compounds given off) for deposition of W (see discussion below). Although the deposited material also contains significant Ga and C, the deposited W containing material is sufficiently conductive to be technologically useful by providing the capability to perform circuit rewiring for FIB IC edits.

To optimize the precursor flux available for FIB induced material deposition, the precursor injector needle must be positioned as close to the area as possible and must be oriented such that the precursor impinging on the area is maximized. On non planar samples, any obstruction that would interfere with precursor flow to the deposition site must be avoided. Following precursor injector needle positioning, for optimal FIB deposition rate and material composition, a combination of FIB current, pixel dwell time, raster refresh time and precursor flux must be chosen. For a given precursor flux and focused ion beam current, a pixel dwell time and refresh time must be defined to allow sufficient build up of the precursor at each pixel before the rastered ion beam returns to that pixel. The pixel dwell time must be sufficiently long to decompose as much of the available precursor as possible as completely as possible, but short enough to remove the minimum amount of material via sputtering. If the dwell time/refresh time/ion current is not optimum, the deposition rate will be slowed. In addition, the deposited materials properties will not be optimum due to incomplete decomposition resulting in incorporation of unwanted impurities. In the extreme, sample sputtering will occur and a crater will form where the deposition was to take place. It must be remembered that these parameters will vary with the area over which deposition is desired. For example, if FIB induced deposition conditions are optimized for a 10μm by 10μm area and those same micromachining parameters are used for a 1μm by 1μm area, it is highly likely that a hole will be sputtered in the sample. The reduction of the area by 100 will reduce the refresh time by 100 and insufficient precursor will deposit between exposures of a given pixel. The precursor will be quickly decomposed and then be sputtered away along with significant amounts of the sample. Unintentional sputtering of a sample as a result of an improper poor choice of deposition conditions is presented in the left micrograph in Figure 4 (Mohr, 1993). The right micrograph in Figure 4 shows the result after more appropriate deposition parameters are selected.

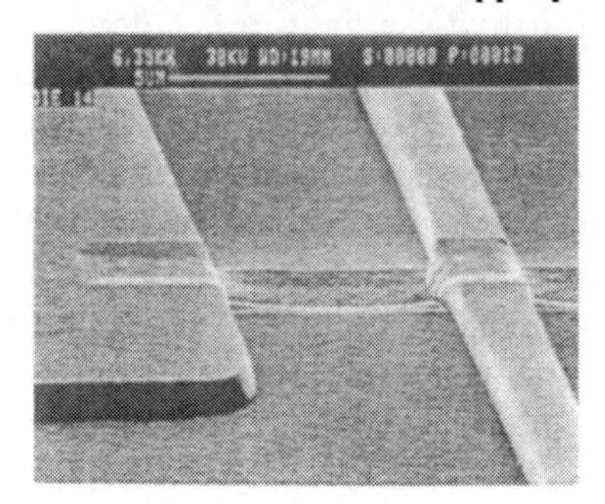
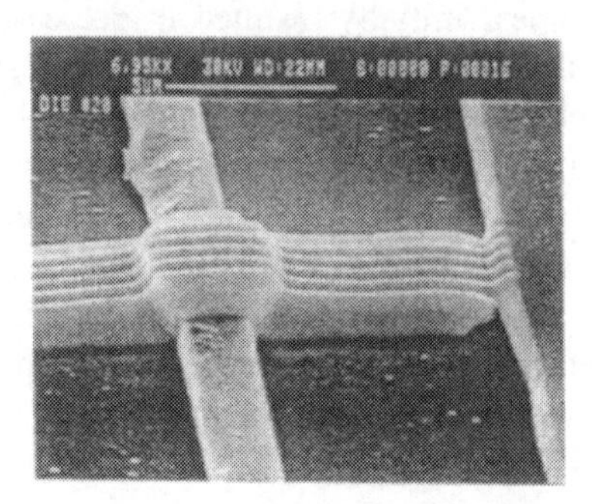

Figure 3.4. Examples of effect of choice of deposition parameters on quality of deposition. The left figure shows a poor deposition for tungsten deposited with 15 µs dwell and 0.16 µm step size, because the sputtering rate exceeds the deposition rate. The figure on the right shows that tungsten deposited with 0.3 µs dwell and 0.6 µm step size provides a good deposition. From Mohr, 1993, with permission of ASM International.

Of the many precursors that have been successfully used for FIB deposition (See Table 1), those whose FIB induced decomposition results in metal deposition (e.g. W, Pt, Au, Al, and Cu) are the most commonly used (Overwijk, 1993, Mohr, 1993, Ro, 1994). The microstructure, composition and resistivity of the deposited metals varies. FIB deposited Pt and W films are amorphous, but FIB deposition of Cu and Au produces polycrystalline films. W and Pt are by far the most commonly deposited metals (See Pt example in Figure 5) and are routinely and mostly interchangeably used for semiconductor circuit edits and as a surface protection for SEM or TEM/STEM specimen preparation.

All deposited films contain not only the desired metals but also incorporate impurities from the incompletely decomposed precursor and contain Ga from the focused ion beam (See Table 2). The percentage of contamination can vary significantly depending on the deposition conditions, but in all cases, the amount of gallium and carbon is a significant amount of the total. Depositions made using high pixel dose have lower resistivity and fewer contaminants compared with those deposited using low pixel dose although the material deposition rate is lowered under the high pixel dose conditions (Li, 1994; Telari, 2002). While the presence of these impurities increases resistivity e.g. 250µΩcm for the FIB deposited W film versus 5.5µΩcm for pure W (Langfischer, 2002), the resistivity is sufficiently low for IC circuit editing except when exceptionally long conductors must be written. It has also been shown that resistivity varies with the thickness of the FIB deposited metal film. For example, during the initial stages of deposition, resistivity for a W film was relatively high. While increasing thickness initially resulted in decreased resistivity, a limit in this decrease was reached. In one study, the resistivity of a W film decreased with thickness until a film thickness of about 0.2μm was reached (Langfischer, 2002). For IC device edits, the practical length limit for an edit is ~1000 μm at which point signal delays can become significant because of the total resistance of the line. This length is also a practical limit in terms of deposition time. Longer conductors are usually deposited using other methods such as plating or by deposition by decomposing an adsorbed precursor using an ultraviolet beam (Van Doorselaer, 1993; Van Doorselaer, 1994).

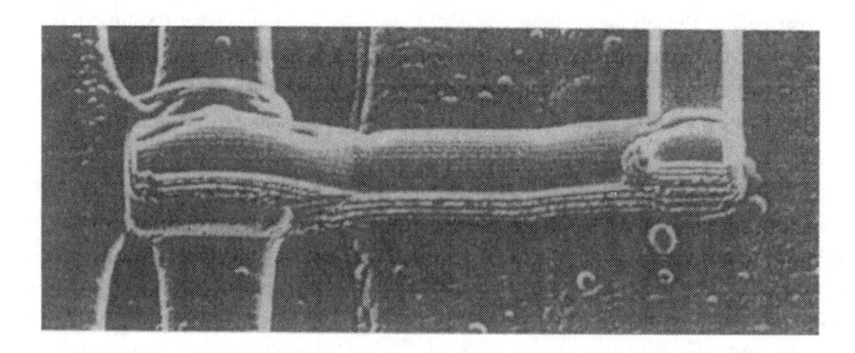

Figure 3.5. A FIB induced platinum used to connect two metallization lines on an integrated circuit. The Pt deposition resulting from each pass of the ion beam and the layer accumulation can be discerned. The width of the deposited line on the right leading up to the top of the page is 2μm (Jim Cargo, Agere Systems)

Table 1. Common deposition gases and source temperature

Element	Precursor gas	
W	Tungsten Hexacarbonyl, $W(CO)_6$	55° C
Pt	Methylcyclopentadienyl platinum trimethyl, $(CH_3)_3(CH_3C_5H_4)Pt$	30° C (melting point)
Al	Trimethylamine alane (TMAA), $(CH_3)_3NAlH_3$	25° C, 1 torr vapor pressure
SiO_2	O_2 and tetraethoxysilane (TEOS)	
SiO_2	O_2 and tetramethoxysilane (TMOS), $Si(OCH_3)_4$	
C	Phenanthrene	

Table 2 Composition and resistivity of deposited metals

Metal	Composition (metal:C:Ga:O)	Resistivity (μohm-cm)
W	75:10:10:5	150-225
Pt	45:24:28:3	70-700

While deposition of metal films is the most common application of FIB induced deposition, the deposition of insulating materials is also of great technological importance. The FIB induced deposition of silicon oxide is used commonly used for complex IC device edits where space is limited and micromachining through more than one set of alternating conductor and insulator layers is required In order to avoid unintended connection of FIB deposited metal layers with sample conducting layers, a layer of insulator must be often be FIB deposited. There are many examples for insulator deposition presented in the literature. (Young, 1995; Campbell, 1997; Edinger, 1998).

Other applications of FIB induced material deposition include lithography mask repair and the fabrication of three dimensional structures. The almost limitless range of applications for a focused ion beam system with deposition capability is exemplified by the capability of producing three dimensional objects with nanoscale resolution. Perhaps the best example of this capability is the work by Matsui (Matsui, 2000; Hoshino, 2003; Morita, 2003). Using phenanthrene as the precursor and innovative software for writing three dimensional structures, both fanciful (See figure 6) and potentially technological useful structures have been fabricated. A 30 keV, 0.4 pA Ga^+ beam with diameter of 7 nm was used to decompose the phenanthrene to diamond like carbon to fabricate the structures shown in Figure 6 a and b. Although the spaceship in Figure 6a is quite fanciful, it ably demonstrates the level of complexity that can be attained using this technique. The wire structures in figure 6 b show promise for more practical application. A 4μm length of 110nm diameter diamond like "wire" of the same composition of those used in Figure 6b had a resistance of 600MΩ indicating a resistivity of approximately 100Ωcm.

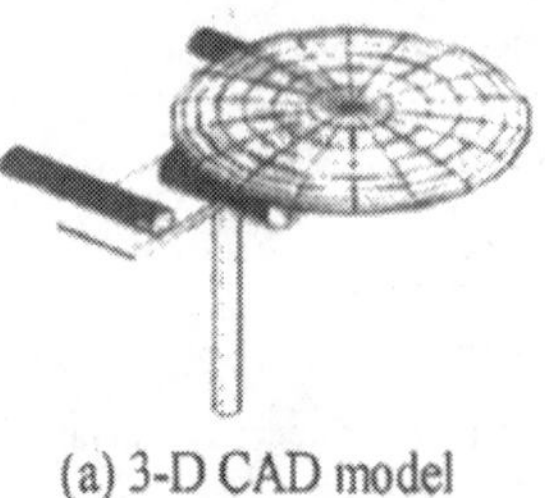

(a) 3-D CAD model

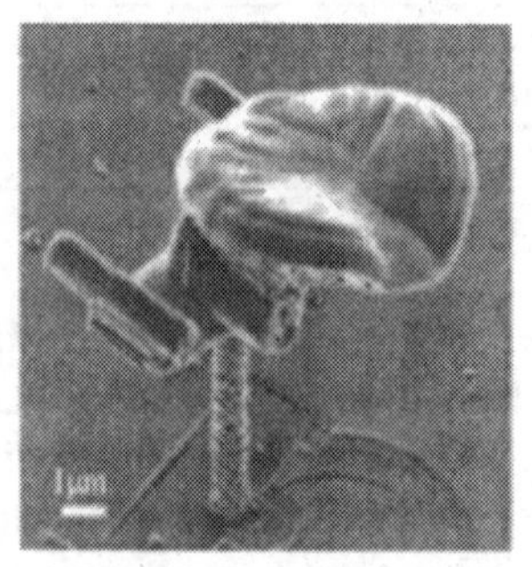

(b) SIM image (tilt 45deg)

Figure 3.6. 3D space ship fabricated of diamond produced by FIB induced decomposition of phenanthrene precursor. From Hoshino, 2003, with permission of author and American Institute of Physics.

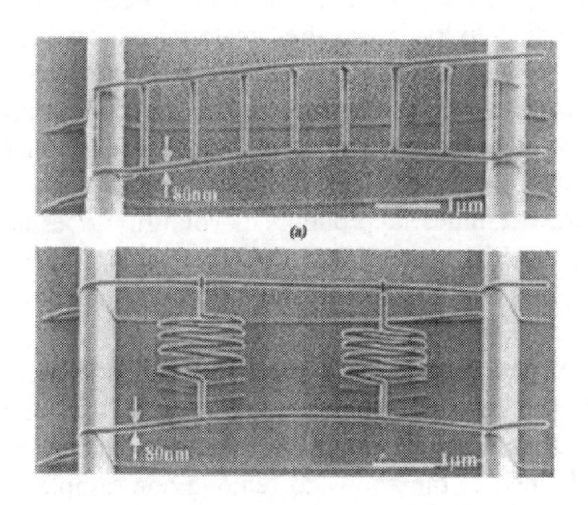

Figure 3.7. SEM micrographs of (a) free space wiring with bridge shape, (b) free space wiring with parallel resistances. Both are fabricated of diamond produced by FIB induced decomposition of phenanthrene precursor. From Morita, 2003, with permission of author and American Institute of Physics

3. ENHANCED ETCH

The reliance on physical sputtering alone for material removal using FIB has significant disadvantages. Material removal rates are often an order of magnitude slower than those obtained using a suitable removal rate enhancing chemical precursor gas. In addition, the lack of precursor enhanced volatility of the material physically sputtered from the sample can result in unwanted redeposition of the material. This redeposited material can result in short circuits after IC edit and often limits the obtainable aspect ratio of a micromachined feature to 10. Physical sputtering alone can also produce unwanted selectivity and thus unwanted topography due to differential sputtering rates, particularly in polycrystalline materials. The topography resulting from uneven material removal is often sufficiently severe to make unenhanced FIB micromachining (FIBM) of materials exhibiting these sputtering properties impractical. The need for rapid, controlled material removal is such that most practical FIBM tasks rely on chemical precursors for material removal enhancement.

The ideal precursor gas for chemically enhanced focused ion beam micromachining (CE-FIBM) of a particular material must provide the desired material selectivity, most often material removal rate enhancement, desired while avoiding any deleterious effects. The ideal precursor has a high sticking coefficient for the material to be micromachined and only chemically reacts with the material when bombarded by the Ga^+ focused ion beam. The ideal precursor does not spontaneously chemically react to produce unwanted residues, byproducts or volatility; does not react with the components of the FIB system; and does not require specialized chemical handling techniques due to corrosiveness, toxicity or environmental issues (Thaus, 1996).

The precursor gas chosen for a specific CE-FIBM process must, when bombarded by the focused ion beam, react with the material for which material removal rate enhancement or selectivity enhancement is desired. Selectivity is defined as the ability to remove one sample constituent while leaving other sample constituents relatively unaffected. In cases where a single material is being micromachined, i.e., no other sample constituents present, an increase in material removal rate is all that is required. In this case, a precursor is chosen that will increase the material removal rate of a desired sample constituent. In complex integrated circuit samples having multiple layers of many materials particularly where micromachining end point detection is required, it is often also advantageous if the precursor at least does not also accelerate or ideally retards the material removal rate of other sample materials. In cases where a material removal rate enhancing precursor cannot be found for a material, a precursor is chosen that will retard the removal rate of other sample materials (Gonzalez, 2001; Gonzalez, 2002).

The selectivity produced by the CE-FIBM process is a function of the sticking coefficient and reactivity of the precursor with the material(s) being micromachined and the micromachining parameters discussed in the Process section of this chapter. Assuming that a precursor with a sufficient sticking coefficient and reactivity has been chosen, the optimum CE-FIBM process parameters can then be determined through careful, systematic variation of these parameters as previously described. As always, a refresh time must be chosen that allows sufficient accumulation (often a single monolayer is ideal) of the precursor on the sample surface and a pixel dwell time must be chosen such that the Ga^+ dose during the chosen dwell time is just sufficient to react all of the adsorbed precursor. For example, for CE-FIBM of Si using

Cl, refresh times for the pattern are generally at least 1msec, and dwell times no more than 0.5μs. Ion flux will be limited to about $5x10^{18}$ ions/cm^2/sec.

The most commonly used precursor gases are halogens or halogen containing molecules including halogenated organic compounds because these gases form volatile reaction products with many materials including those of interest to the semiconductor industry (Abramo, 1994; Young 1993). Water provides tremendous selectivity for carbon based materials by dramatically increasing the material removal rate of carbonaceous materials and retarding the removal rate of many other materials (Stark, 1995). Table 3 shows the material removal rate enhancement for some materials in common use by the semiconductor industry. The numbers indicate the change in the material removal rate versus physical sputtering alone when a CE-FIBM is employed.

Table 3 Etch rate enhancement of CE-FIBM vs. FIBM

Gas	Si	SiO_2	Al	W	GaAs	InP	PMMA
Cl_2	7-10	1	7-10	1	50	4	-
Br_2	5-6	1	8-16	1	-	-	-
I_2	5-10	1	5-15	-	-	11-13	2
XeF_2	7-12	7-10	1	7-10	-	-	4
H_2O	0.05-0.1	0.1-0.15	0.02-.05	-	-	-	18

Examples of applications utilizing CE-FIBM include semiconductor device modification (Van Doorselaer, 1993), semiconductor device analysis (Casey, 1994), and other applications such as diamond micromachining (Russell, 1998).

Semiconductor integrated circuit (IC) editing for current technology devices would not be possible without CE-FIBM. Device editing usually requires material to be removed, both for exposing and often for cutting selected conductors (device metallization) that are used to distribute electrical signals and power. Due to the small feature size of modern IC devices, very high aspect ratio (depth to width ratios of greater than 10) micromachining is often required to expose structures of interest without damaging neighboring device structures. High aspect ratio cuts cannot be produced unless the material removed is volatile and does not deposit on the walls of the cut. In addition, since the intent of circuit editing is usually to edit or alter the electrical conducting metallization of the IC by cutting and/or rerouting conductors, any redeposition of micromachined material which may result in unwanted conduction paths or interfere with the electrical connection of FIB deposited conductors to circuit metallization cannot be tolerated. The required high aspect ratio cuts and in many cases the circuit rewrite are not feasible without the ability of CE-FIBM to convert micromachined material into volatile species (Abramo, 1994).

An example of the material selectivity that can be obtained using CE-FIBM is provided by CE-FIBM of an IC that has PMMA passivation and Al metallization (Stark, 1995). Figure 8 shows a 30μm x 30μm feature cut into such a device using water CE-FIBM. CE-FIBM parameters used were 70mTorr H_2O pressure at the sample surface, a 1nA 0.16μm diameter 25keV Ga^+ ion beam, a pixel dwell time of 0.5μs and a raster refresh time of 10ms. Pixel overlap was 50%. This feature was micromachined in 11 minutes. Using physical sputtering alone to remove the PMMA would have required 225 minutes and the Al metallization would be completely removed. The water CE-FIBM provides a selectivity for PMMA versus Al of over 500 which allows the rapid removal of the PMMA while preserving the Al metallization.

Figure 3.8. Use of water enhanced etch to remove polyimide passivation layer over a 30μm x 30μm region and expose intact two levels of aluminum metal. From (Stark, 1995, with permission of author and American Institute of Physics.

With the advent of wide usage of Cu metallization on state of the art integrated circuits, the ability to analyze and edit integrated circuits with copper (Cu) metallization is critical. Unfortunately, CE-FIBM techniques developed for Al metallization are not directly transferable. Commonly used halogen precursors are not applicable since these precursors often react spontaneously with the Cu producing unwanted often conductive residues (Casey, 2002). If no chemical assistance is employed, FIB physical sputtering alone results in very non homogeneous Cu removal due to differential sputtering resulting from the varying crystalline orientation of the crystallographically textured Cu metallization films (Phillips, 2000) as illustrated in Figure 9a.

Various CE-FIBM methods have been developed which provide uniform Cu removal. One method (Casey, 2002) developed for thick Cu films (>0.5μm) employs a novel FIB scan strategy along with simultaneous introduction of $W(CO)_6$ which combine to greatly reduce the channeling of the Ga^+ in polycrystalline Cu films by increasing scattering. Since material removal is slower in the channeling direction, the reduction in channeling results in more uniform Cu removal. For thin Cu films (<0.5μm), non uniform Cu removal due to differential sputtering of Cu grains is significantly less of a problem due to the limited thickness of the Cu film. In this case, H_2O and O_2 CE-FIBM are employed to provide the Cu/SiO_2 selectivity necessary to completely clear the Cu without undue damage to the underlying SiO_2 dielectric. A CE-FIBM approach with demonstrated applicability for textured Cu films over a wide range of Cu thickness (demonstrated for up to 3μm Cu thickness) has also been developed that utilizes 1-Propanol (Gonzalez, 2002) (See Figure 9b). While CE-FIBM selectivity for Cu versus SiO_2 will continue to be important, there is much development currently underway to switch from SiO_2 to low K dielectrics for some applications. It should also be noted that 1-propanol provides the best selectivity for Cu versus SiLK, one of the low K dielectrics currently in use. Figure 10 shows dielectric selectivity for some chemical precursors that have been studied for use in CE-FIBM of Cu.

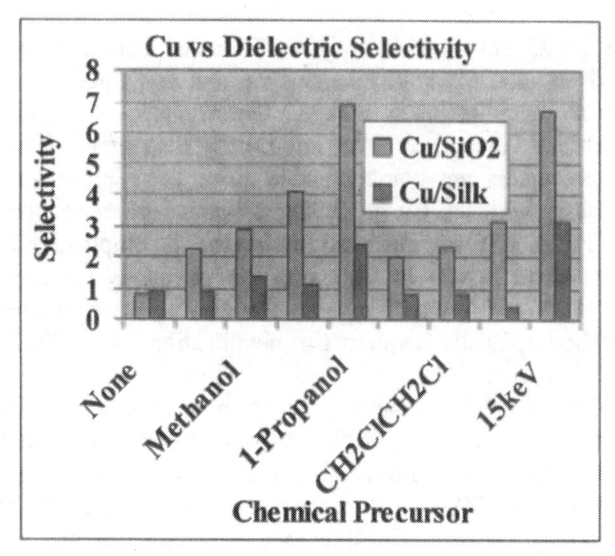

Figure 3.10. CE-FIBM selectivities obtained for SiO_2 and SiLK using various precursors. From Gonzalez, 2002, with permission of author and American Institute of Physics.

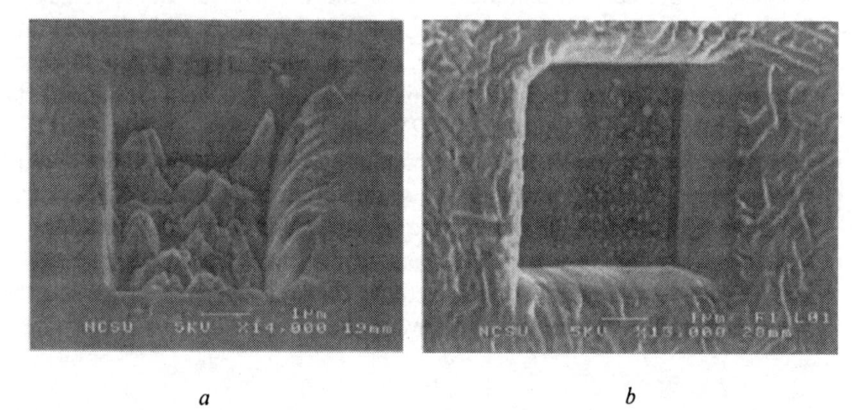

a *b*

Figure 3.9. SEM micrographs of Cu films FIB micromachined (a) with no chemical enhancement and (b) with 1-Propanol CE-FIBM. From Gonzalez, 2002, with permission of author and American Institute of Physics.

Machining at an edge is faster than over a flat region because significantly more surface area receives energy from the collision cascade sufficient to eject atomic or molecular species. Addition of enhanced etch to edge machining improves the removal rate. It should be noted that the refresh time for edge machining will be shorter than for a flat region because of larger surface area. The orientation of the gas needle with respect to the edge is also important. Best results are obtained with the needle facing the edge where material is removed. (Thaus, 1996)

The selectivity of CE-FIBM can also facilitate device analysis via SEM. The selectivity of CE-FIBM provides the means to duplicate etch methods used for decoration of specimens prepared for SEM analysis. For example, the selectivity of Cl_2 CE-FIBM results in Si and Al material removal rates that are almost an order of magnitude greater than that of SiO_2. An SEM micrograph of a cross sectioned sample decorated to delineate the sample structure using Cl_2 CE-FIBM is shown in Figure 11 (Weiland, 2000). Alternatively, XeF_2 CE-FIBM which provides Si and SiO_2 material removal rates that are almost an order of magnitude greater than that of Al could have been used. If a dual beam FIB/SEM instrument is available, the analysis can be performed without removing the sample from the instrument. Otherwise, the sample must be transferred to an SEM.

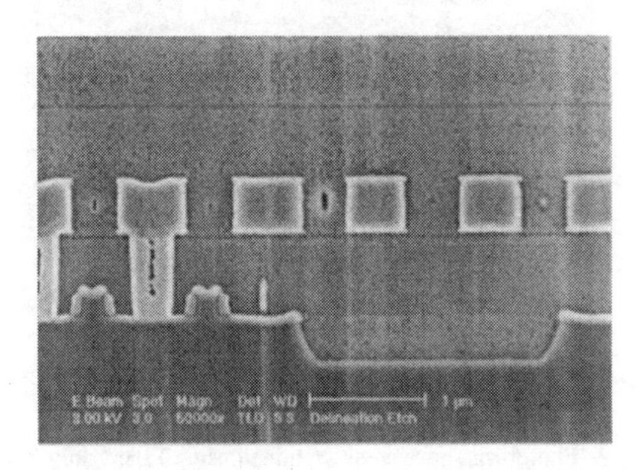

Figure 3.11. SEM image of FIB prepared cross section. Identification of layers improved with delineation etch. From Weiland, 2000, with permission of ASM International.

Finally, CE-FIBM combined with geometrical micromachining enhancement can further enhance micromachining capability. This combination has been exploited for micromachining permalloy (C_2Cl_4 CE-FIBM; Thaus, 1996), for sharpening of diamond cutting tools (H_2O CE-FIBM; Russell, 1998; Adams, 2003) and for measurement of photoresist line widths without sample cleaving (H_2O CE-FIBM; Stark, 1996). Geometrical micromachining enhancement takes advantage of the greater sputtering rates available at very high impact angle, i.e., close to parallel to the surface to be micromachined. The addition of chemical enhancement further improves material removal rate and minimizes redeposition. Figure 12 illustrates the advantage of this technique by comparing photoresist feature cross sections obtained employing sample cleaving, physical sputtering only and H_2O CE-FIBM. It is clear that CE-FIBM cross sectional micromachining prevents the redeposition of substrate material onto the edge of the structures, allowing clean cross sections with minimal artifacts to be produced.

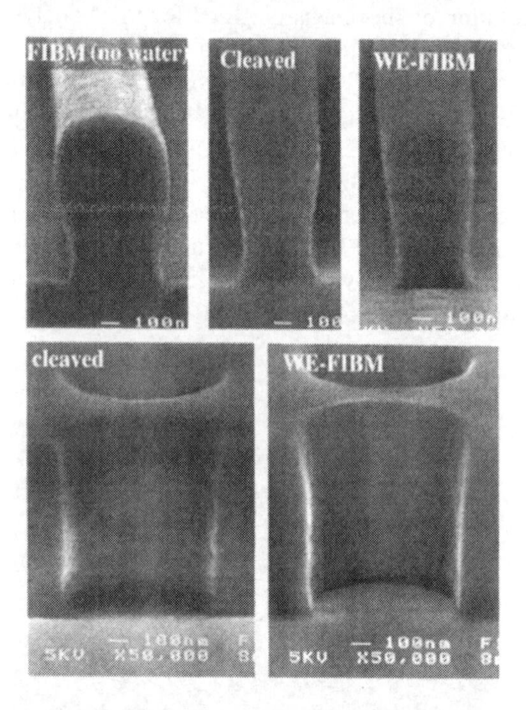

Figure 3.12. Photoresist lines cross sectioned using FIBM only (top right), by sample cleaving (top center) and using H_2O CE-FIBM (top right). Holes in photoresist cross sectioned by cleaving (bottom left) and using H_2O CE-FIBM (bottom right). From Stark, 1996, with permission of author and American Institute of Physics.

4. SUMMARY COMMENTS

The utilization of chemical precursors for FIB induced deposition and for enhancement of micromachining material selectivity, whether through enhancement or judicious retardation of material removal rates, greatly broadens the applicability of FIB to a wide variety of technological problems. The development of new circuit edit methods as IC technology advances is testament to the fact that research into FIB deposition and material removal methods not a stagnant area of development. Continued investigation of new FIB precursors insures that the range of applications for FIB will continue to be expanded.

REFERENCES

Abramo M, Hahn L, and Moszkowicz L, Proceedings of 20th International Symposium for Testing and Failure Analysis, ASM International, p. 439 (1994).

Adams DP, Vasile MJ, Mayer TM, and Hodges VC, J. Vac. Sci. Technol. B21, 2334 (2003).

Bannerjee I and Livengood RH, J. Electrochem. Soc. 140, 183 (1993).

Campbell AN, Proceedings of 23rd International Symposium for Testing and Failure Analysis, ASM International, p. 223 (1997).

Casey JD Jr., Doyle AF, Lee RG, Stewart DK, and Zimmermann H,Microelectronic Engineering 24, 43 (1994).

Casey JD Jr., et al., J. Vac. Sci. Technol. B20, 2682 (2002).

Edinger K, Melngailis J, and Orloff J, J. Vac. Sci. Technol. B16, 3311 (1998).

Giannuzzi LA, and Stevie FA, Micron 30, 197 (1999).

Gonzalez JC, Griffis DP, Miau TT, and Russell PE, J. Vac. Sci. Technol. B19, 2539 (2001).

Gonzalez JC, da Silva MIN, Griffis DP, and Russell PE, J. Vac. Sci. Technol. B20, 2700 (2002).

Harriott LR, J. Vac. Sci. Technol. B11, 2012 (1993).

Hoshino T, Watanabe K, Kometani R, Morita T, Kanda K, Haruyama Y, Kaito T, Fujita J Ishida M, Ochiai Y, and Matsui S, J. Vac. Sci. Technol. B21, 2732 (2003).

Ishitani T, et al., J. Vac. Sci. Technol. B9, 2633 (1991).

Langfisher H, Basnar B, Hutter H, and Bertagnolli E, J. Vac. Sci. Technol. A20, 1408 (2002).

Li SX, Toyoshiba LO, Delenia E, and Kazmi S, Proceedings of 20th International Symposium for Testing and Failure Analysis, ASM International, p. 425 (1994).

Matsui S, Kaito T, Fujita J, Kanda K, and Haruyama Y, J. Vac. Sci. Technol. B18, 3181 (2000).

Mohr J, and Oviedo R, Proceedings of 19th International Syposium for Testing and Failure Analysis, ASM International, p. 391 (1993).

Morita T, et al., J. Vac. Sci. Technol. B21, 2737 (2003).

Overwijk MHF, and van den Heuvel FC, Nucl. Inst. Meth. Physics Research B80/81, 1324 (1993).

Phillips JR, Griffis DP, and Russell PE, J. Vac. Sci. Technol. 18, 1061 (2000).

Ratta ADD, Melngailis J, Thompson CV, J. Vac. Sci. Technol. B11, 2195 (1993).

Ro JS, Thompson CV, and Melngailis J, J. Vac. Sci. Technol. B12, 73 (1994).

Russell PE, Stark TJ, Griffis DP, Phillips JR, and Jarausch KF, J. Vac. Sci. Technol. B16, 2494 (1998).

Stark TJ, Shedd GM, Vitarelli J, Griffis DP, and Russell PE, J. Vac. Sci. Technol. B13, 2565 (1995).

Stark TJ, Griffis DP, and Russell PE, J. Vac. Sci. Technol. B14, 3990 (1996).

Telari KA, Rogers BR, Fang H, Shen L, Weller RA, and Braski DN, J. Vac. Sci. Technol. B20, 590 (2002).

Thaus DM, Stark TJ, Griffis DP, and Russell PE, J. Vac. Sci. Technol. B14, 3928 (1996).

Van Doorselaer K, Van den Reeck M, Van den Bempt L, Young R, and Whitney J, Proceedings of 19th International Symposium for Testing and Failure Analysis, ASM International p.405 (1993).

Van Doorselaer K, and Van den Bempt L, Proceedings of 20th International Symposium for Testing and Failure Analysis, ASM International p. 397 (1994).

Weiland R, et al., Proceedings of 26th International Symposium for Testing and Failure Analysis, ASM International, p. 393 (2000).

Xu X and Melngailis J, J. Vac. Sci. Technol. B11, 2436 (1993).

Yaguchi T, Kamino T, Ishitani T, and Urao R, Microscopy and Microanalysis 5, 365 (1999).

Young RJ, Cleaver JRA, and Ahmed H, J. Vac. Sci. Technol. B11, 234 (1993).

Young RJ, and Puretz J, J. Vac. Sci. Technol. B13, 2576 (1995).

Chapter 4

THREE-DIMENSIONAL NANOFABRICATION USING FOCUSED ION BEAMS

Takashi Kaito
Seiko Instruments, Inc.,. Japan

Abstract: Creating nanometer scale structures using a FIB has been a topic of recent interest. In this chapter we report on complicated nanometer level structures, such as a nano-wine glass and a nano-toilet, based on CAD data produced by the FIB CVD process.

Key words: FIB-CVD, Carbon deposition, 3D nanofabrication, 3D CAD data

1. INTRODUCTION

The Focused Ion Beam (FIB) instrument has three functions: imaging, etching, and deposition, and has been used for wide range of applications including the preparation of site-specific cross section TEM specimens. The "deposition" function is achieved by FIB chemical vapor deposition (FIB-CVD) techniques. Seiko Instruments Inc. (SII) has developed the use of FIB deposition techniques (Kaito, 1984, 1986) for a number of specific applications (Kaito and Yamamoto, 1985; Kaito et al., 1986; Kaito and Adachi, 1988).

The three-dimensional nanofabrication technology introduced in this chapter has been jointly developed by Himeji Institute of Technology, NEC, and SII (Matsui, Kaito and Fujita, 2000). Using the FIB-CVD technique allows the formation of 3D structures that cannot be made with existing technologies.

We will first explain the configuration of the FIB equipment and its basic functions. Then we will describe the 3D nanofabrication technology and present several examples of 3D structures, applications, and suggested future applications.

2. CONFIGURATION OF THE FIB EQUIPMENT

The external view of FIB equipment used for smaller sample preparation (model SMI2050) is shown in Figure 1. The FIB consists of the main body, a computer table and a power rack. The equipment is mainly operated with a mouse while observing the PC display. The adjustment of focus, stigmators, contrast and/or brightness can be made via the operation panel. The sample stage is driven by a 5 axis motorized and computer controlled stage (e.g., X, Y, Z, T (tilt), and R (rotate)).

The ion optical system is usually used with an acceleration voltage of 30kV. The aligner electrode placed between the condenser lens and object lens contributes to make the beam alignment easier. The liquid metallic gallium is used as an ion source. A two-stage electrostatic type lens is used to improve both chromatic aberration and spherical aberration resulting in the improvement of beam performance. A wide range of ion beam currents, from 0.1pA to 20nA, can be used by switching movable apertures and the strength of the condenser lens. The beam diameter at the minimum beam current is 5nm, so it allows observation at magnification of up to ~ 100,000x with the scanning ion microscope. The maximum current density is ~ 20 A/cm^2 and high-speed processing and ultrafine processing in the units of a few nA may be realized.

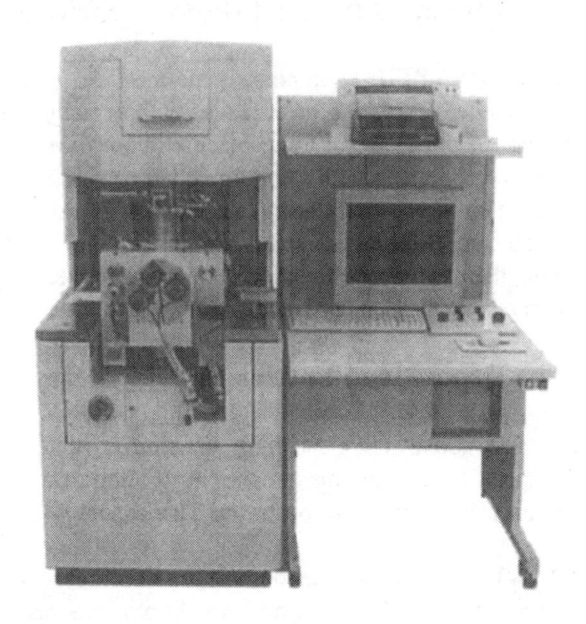

Figure 4-1. Schematic diagram of FIB equipment used for specimen preparation.

3. BASIC FUNCTIONS OF THE FIB

3.1 The Imaging Function (Scanning ion microscope)

The principal function of FIB equipment is that of a scanning ion microscope. The secondary electrons excited by the ion beam is collected and displayed to form an image. This secondary electron image using a scanning ion microscope is called a SIM image and is used for the positioning of processing points, monitoring of processing conditions, and observing cross sections. Comparing images obtained with a scanning electron microscope (SEM), the SIM image can show the channeling contrast more strongly. For example, precipitation of silicon in aluminum wiring can be easily observed. One of the biggest features of SIM image observation is the crystal grain structures of metals. The resolution of a SIM image was only about 100nm in 1985 but today it has improved to 5 nm.

3.2 The Etching Function (Maskless Etching)

Another function of FIB equipment is the maskless etching function. If the FIB is irradiated over one area requiring processing, several tens of nanometer level of processing can be made without using a mask process. Also, contrary to processing with using broad ion beams, FIB processing is a "soft" process that uses small beam current and yields little heating or distortion problems. Moreover, since the processing conditions can be observed using the microscope function mentioned above, very accurate and high precision results may be produced.

3.3 The Deposition Function (Maskless Deposition)

As mentioned above, the deposition function is achieved by the FIB-CVD technique. This function is the most important aspect of 3D nanofabrication using a focused ion beam. This aspect will be explained in detail.

The FIB-CVD method uses the ion beam to decompose a source gas that adsorbs onto a sample surface into nonvolatile products resulting in the net formation of a deposited film. It is also referred to as "beam excitation surface reaction technology." Carbon films and metallic films such as tungsten have been routinely used for depositioin applications. Both films are amorphous films which include impurities from ion beam such as Ga.

3.3.1 Keys for Increasing Film Growth Speed

The most important variable in increasing the film growth speed is to increase the source gas spraying density. The increase in spray density, however, should be limited to the range in which the source gas does not precipitate. By moving the spray nozzle of the source gas as close as possible to the irradiation position on the sample surface, the density of the gas can be increased locally without deteriorating the vacuum level of the sample chamber. The best combination of nozzle size and distance to the sample consists of a nozzle diameter of 500μm or less with the distance between the nozzle tip and the sample of 500μm or less.

Another important consideration is the selection of a suitable gas. The gas selected should include a large number of atoms that easily adsorbs onto the sample surface in a single adsorption layer. For the deposition of C, phenanthrene, $C_{14}H_{10}$, is a gas that yields rapid growth. An example of a gas used to deposit a conducting metal (e.g., W) is tungsten hexacarbonyl $W(CO)_6$.

Comparing the gas structure with the deposition of these two materials, it is observed that 14 C atoms can be used to create films from the former gas, while the latter gas has just one (1) W atom. Thus, it is observed that C films grow faster than W films.

3.3.2 The Upper Limit of Film Growth Speed

The theoretical limits and the experimental values for several kinds of source gases and film growth speed are described below. The theoretical limit described here is a film growth speed calculated based upon the simplest assumptions that the all source gases are adsorbed, decomposed, and accumulated, and that the density of an accumulated film is equal to the bulk density of the atoms. The theoretical limits and the experimental values calculated from these assumptions are shown in the Table 1.

Table 4-1. Source Gas and Growth Speed

Type of Gas	Chemical Formula	Theoretical Growth Limit (μm/min)	Experimental Growth Limit (μm/min)	Experimental Limitation
Phenanthrene	$C_{14}H_{10}$	18.1	11	Controller temperature limit
Pyrene	$C_{16}H_{10}$	2.4	2	Precipitation limit
Tungsten Hexacarbonyl	$W(CO)_6$	6.7	1	Sample chamber vacuum limit

For phenanthrene, the precipitation limit cannot be reached due to the design limit of temperature controller (i.e., the maximum temperature setting of the gas reservoir is low). If there is no limit for temperature control, then the film growth speed could reach the theoretical limit. Although the experimental value of 11 μm/min does not reach the theoretical limit of 18 μm/min, the growth speed of phenanthrene is 5 to 10 times faster than that of other kinds of gases. With phenanthrene gas, the sample chamber pressure is 10^{-5} Pa even near the precipitation limit, so it can be considered as a very efficient source gas.

For pyrene, since its vapor pressure is lower than that of phenanthrene, the precipitation limit of the source gas restricts the growth speed much lower than that of phenanthrene.

For tungsten hexacarbonyl used for metal deposition, in addition to the reason mentioned in section 3-3-1 above, the pressure limit of the sample chamber restricts the growth speed much below the theoretical value. The experimental growth value using tungsten hexacarbonyl compared to phenanthrene is smaller by a factor of 2.

4. THREE-DIMENSIONAL NANOFABRICATION TECHNOLOGY

Details of the 3D nanofabrication technology using FIB-CVD will be explained herein. The feature of this technology is the accumulation of a layer from the bottom up by overlapping each layer with the one beneath it as shown in figure 2. This allows the formation of an over-hang structure, which cannot be formed by etching from the top down. The computer-controlled beam also allows the formation of structures having a high degree of freedom. As indicated from Table 1, using phenanthrene as a source gas results in the formation of an amorphous carbon film at a growth speed of 10 μm/min.

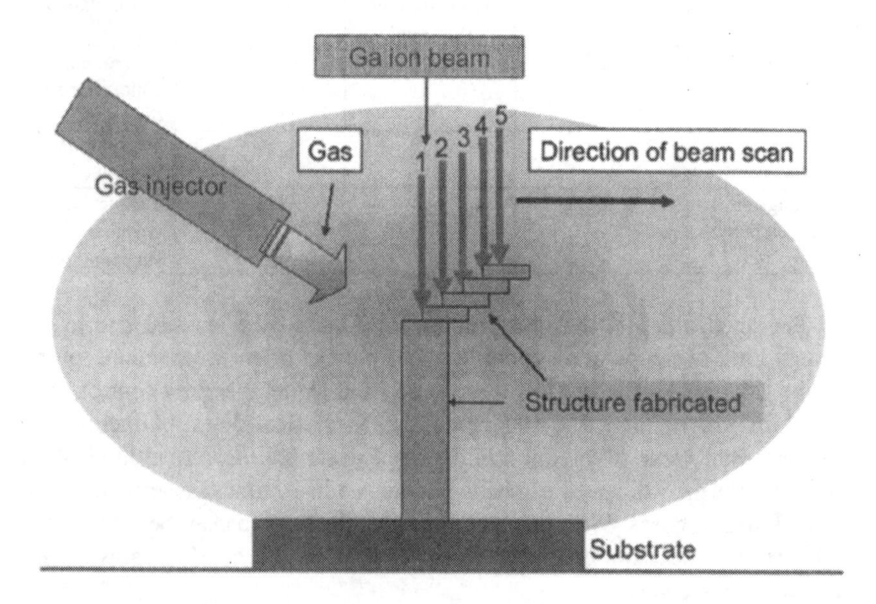

Figure 4-2. A schematic diagram of the premise of the 3D nanofabrication technology.

Several examples of basic structures and the methods of their formation are described below. All of these examples were made by the deposition of amorphous carbon. A beam current of 1pA was used to make these structures. The processing time was less than 10 minutes for all examples.

Figure 3 shows the 3D fabrication of a circular cylinder. This feature is deposited by rotating the ion beam with a fixed radius at a high speed.

Figure 4-3. 3D fabrication of a circular cylinders having out diameter of 1.7 μm and 1.2 μm.

Figure 4 shows the 3D nanofabrication of a pillar feature. This feature was prepared in only 60s by keeping the ion beam spot at a constant location. The upward growth of the film creates the pillar structure that is 100nm in diameter.

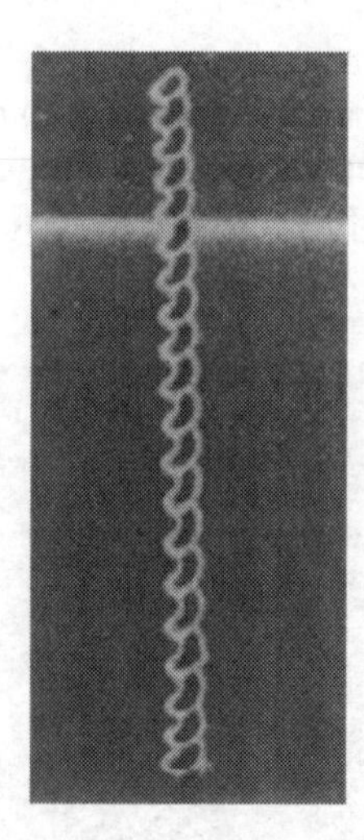

Figure 4-4. 3D nanofabriction of a 100 nm diameter pillar.

Figure 5 shows the 3D fabrication of a spring coil structure. The spring coil was formed by rotating the ion beam at a 13s cycle. The diameter of the coil is 600 nm, while the diameter of the wire is 80 nm.

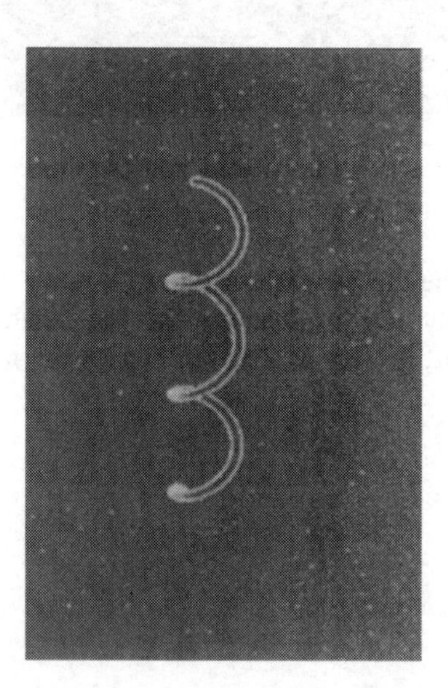

Figure 4-5. 3D fabrication of a spring coil.

Figure 6 shows the 3D deposition of a bellows structure that is 2.75 μm in diameter. Rotating the ion beam at high speed while periodically changing the radius of beam rotation makes the bellows structure.

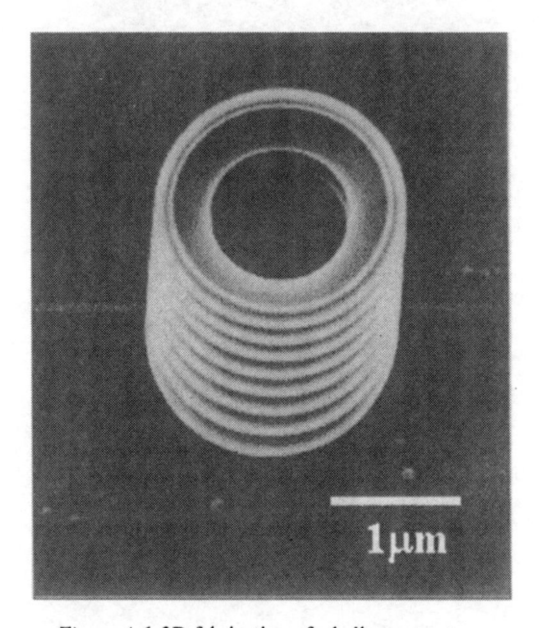

Figure 4-6. 3D fabrication of a bellows structure.

A wine glass having the diameter of less than 3 μm formed using this technique has appeared frequently in the media as well as technical magazines and is widely known by the general public. This nano-wine glass is shown in the Figure 7. The 3 μm diameter dimension of the wine glass is one third that of red blood cell. The wine glass was prepared by rotating the ion beam at a high speed while slowly changing its radius.

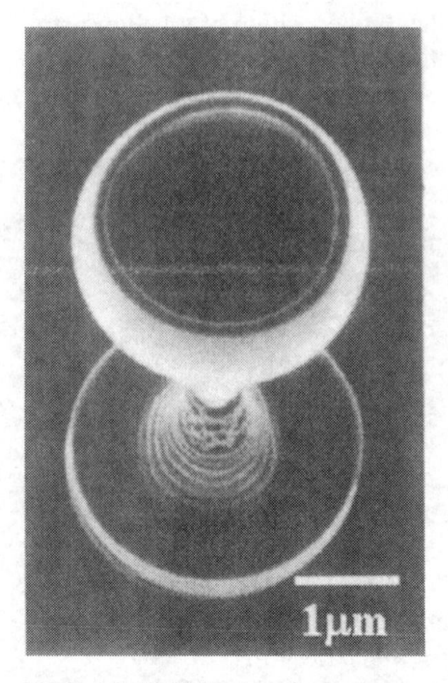

Figure 4-7. 3D fabrication of a wine glass.

These examples demonstrate that the FIB-CVD technique allows for forming three-dimensional structures at a fixed position, which could not be achieved using existing techniques. The SIM2000 series scanning ion microscope has a function of making FIB processing data automatically based upon the 3-D CAD data. This function allows users to form three-dimensional nano structures very easily.

An example of this function is shown in the Figures 8 and 9. First, the user makes a CAD file of any three-dimensional structure as shown in the Figure 8. This CAD data is divided in the Z-axis direction into several pieces of data to make a series of FIB processing data automatically. Then this series of data is loaded to the SMI2000 series scanning ion microscope and deposition begins. The SMI2000 uses the CAD data to automatically form the three-dimensional nano-structure as shown in the Figure 9.

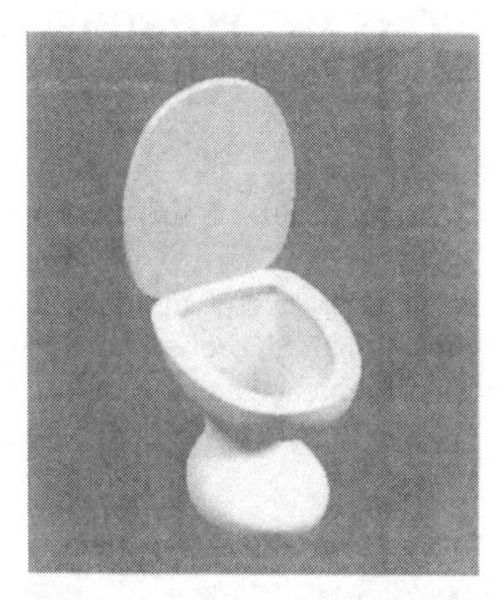

Figure 4-8. 3D CAD drawing of a feature.

Figure 4-9. 3D FIB fabrication performed automatically from the CAD drawing in figure 8.

5. APPLICATIONS AND POSSIBLE APPLICATIONS OF 3D NANOFABRICATION

5.1 Tips for Scanning Probe Microscope

A cylindrical probe for measuring critical dimensions may be formed by cutting the tip of a conventional Si probe and depositing a high-aspect tungsten pillar that is 80 nm in diameter and 1 μm in length. This probe will allow for e.g., the microscopic measurement and resistance measurement of semiconductor devices.

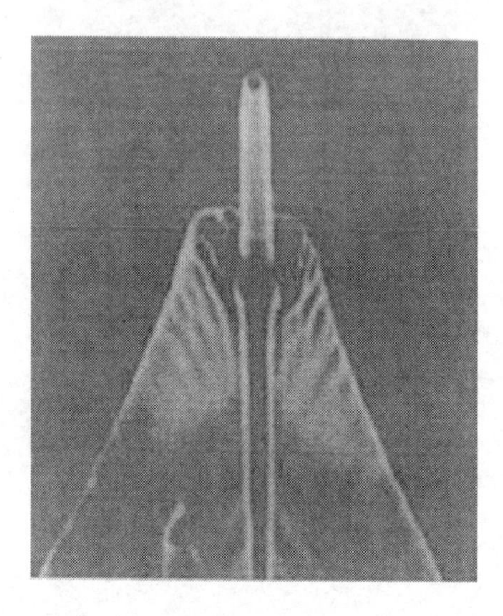

Figure 4-10. An 80 nm diameter tip FIB prepared for a scanning probe microscope.

5.2 Extractor Electrode of a Scanning Atom Probe

To realize a scanning atom probe instrument, an ultrafine extractor electrode is needed. Using a current metal foil electrode as a base, a cone-shaped electrode having an ultrafine open tip formed by FIB-CVD may be added to it. It is now expected that the use of this technology will accelerate the practical application of a scanning atom probe.

5.3 Micro-nano mold for nanoimprint

Much research relating to the preparation of a nano-mold may be performed. It is considered that the use of FIB-CVD and FIB milling methods may allow the formation of a small mold that may easily compare with other methods requiring several kinds of expensive devices.

5.3.1 Processing the tip of micro-manipulator

Applying the following processes to the tip of micro-manipulators using FIB-CVD and FIB milling may allow the preparation of useful tools in the field of biotechnology for cell operation and protein analysis. Some of these features may include:

a) High-aspect ratio needles with the size on the order of a few 10's nm.
b) Nozzles having a hole diameter of a few score 10's nm (e.g., a nano-pipette)
c) Nano-tweezers

6. SUMMARY

In this chapter, the configuration and functions of a FIB having an ion beam diameter as small as 5nm has been described. Three-dimensional nanofabrication techniques and its applications have also been presented.

Existing lithography methods are suitable for mass production of features but requires expensive facilities and semiconductor processing knowledge. On the other hand, 3D FIB-CVD technology requires only FIB instrumentation. This technology can easily form 3D structures with fine control in the nanometer dimension level. It is proposed that there could be many possible applications taking advantages of these features; e.g., making prototypes in small numbers or making tools for research and development.

REFERENCES

Kaito T, Nakagawa Y, Sato M, and Yamamoto M, "Carbon Pattern Film Fabrication by Focused Ion Beam Chemical Vapor Deposition", Proc. of Japan Society of Applied Physics 14p-T-15, 265 (1984 Autumn).

Kaito T and Adachi T, "Property of Tungsten Film fabricated by Focused Ion Beam Chemical Vapor Deposition", Proc. of Japan Society of Applied Physics 4a-Z-9,339 (1986 Spring).

Kaito T and Yamamoto M,"Mask Repair using Focused Ion Beam", Proc. of the 9 th Symp. On ISIAT 207 (1985).

Kaito T,Adachi T, and Kobayashi K,"Circuit Modification and Failure Analysis on IC Chip using Focused Ion Beam Chemical Vapor Deposition and Milling", Proc. of Japan Society of Applied Physics 28a-ZG-10,348 (1986 Autumn).

Kaito T and Adachi T, "Focused Ion Beam System for IC Development and Its Applications", Proc. of the 1st Micro Process Conference 142 (1988).

Matsui S, Kaito T, Fujita J, Komuro M, Kanda K and Haruyama Y, "Three-dimensional nanostructure fabrication by focused-ion-beam chemical vapor deposition", J. Vac. Sci. Technol. B 18, 3181 (2000).

Yagyu T, Watanabe M, Tanaka K, Taniguchi M, Kaito T and Nishikawa O,"Correration between the STM Images and the Open End of the Scanning Electrode of a Scanning Atom Probe", Proc. of Japan Society of Applied Physics 13p-ZW-6,507 (2001 Autumn).

Yasutake M and Kaito T, Proc. of Japan Society of Applied Physics 25a-ZQ-7, 590 (2002 Autumn).

Chapter 5

DEVICE EDITS AND MODIFICATIONS

Kultaransingh (Bobby) N. Hooghan
Agere Systems, 555 Union Blvd. Allentown, Pa 18109-3286

Abstract: The ability to physically modify prototype Integrated Circuits (ICs) helps in reducing time to market and increases the likelihood of having fully functional board level systems before the IC's come out in mass production. This can be achieved using a Focused Ion Beam (FIB) system, which can be used to make device/circuit edits on prototype IC's, in order to correct and subsequently test, design/process flaws, in an iterative manner, before mask changes are made. This includes but is not limited to making metal connections, disconnecting metal lines, make probe pads, speed up/ delay circuits etc.

Key words: Deposition, milling, device edit, device modification, vias, navigation, clean-up.

1. INTRODUCTION

Focused Ion Beam (FIB) systems have come a long way since the early days of mask repair. A stable Gallium ion source and enhanced resolution have really turned these systems into very versatile tools in the semiconductor industry. In addition to its analytical capabilities, one application that stands out, is the ability of the FIB system to carry out device/circuit modifications on prototype chips. A word of caution is warranted at this point before proceeding further. Gallium beam interaction with the substrate is inherently destructive. The damage inflicted is essentially twofold: impact damage due to heavier mass of the Ga^{+} ions, and surface and subsurface charging effects. The damage inflicted under both these categories has been well documented elsewhere in the literature, and suffice to say, one needs to be aware of the damage and minimize it when working on the prototype chips. (Wills, 1998; Campbell, 1998; Benbrick, 1998)

Modifications carried out on prototype chips include, (but are not limited to) making new connections using metal deposition (either W or Pt), breaking connections, and making probe points for tapping signals. These modifications may be to fix design errors, or carry out design modifications presented by customers. In either case, a FIB system is usually able to turn around the modified chips in a matter of hours, compared to mask changes, which could take a few weeks, are quite expensive and with changes based only on simulations, run the potential risk of some kind of mismatch.

Eventually, mask changes have to be made, but testing of new concepts and system level scrutiny can be carried out using the modified chips before the fab turns out new chips. Prototype chips can be worked on iteratively until the right result is achieved.

The basic and underlying theme in working on prototype chips is to provide fully functional chips in a timely manner. In practice, customer needs change continuously, and these changes are often forwarded to the chip manufacturers, typically, at the last minute. In order to keep the customer happy one needs to implement the changes asked for, test the changes on prototype chips and send them back to the customer. The customer in turn puts the modified chips on their board level application, and checks their system out. Traditionally this would take weeks (going through mask and testing cycles). With a FIB system around, design changes can be made on a few prototype chips that can literally save the day. The customer can carry out all their board level tests, and be assured that the chips being manufactured will have the necessary changes incorporated in them.

Just like any other project, a device edit can be broadly divided into two phases namely, (1) planning and (2) execution/implementation. These are discussed in detail below.

1.1 Planning

The planning phase primarily involves the engineer requiring the fix, but efficient planning at this stage ensures smoother execution subsequently. Defining and explaining what needs to be achieved is vital to the success of the device edit, whether it is rerouting a signal, gate inversion, tagging gates/circuits high or low, or making probe points for e-beam/mechanical probing. The planning stage can be broadly divided into the following sections: a) decapsulation b) data collection, and c) deposition layout.

1.1.1 Decapsulation

An important part of planning is to have the devices decapsulated and tested after decapsulation. Devices are often damaged during decapsulation, therefore it is imperative to test the devices after decapsulation to ensure proper functionality, prior to FIB operations.

1.1.2 Data Collection:

This can be further classified into the following: (i) targeting information (ii) accessibility, and (iii) chip orientation

1.1.2.1 Targeting information:

After identifying the fix to be performed, one needs to gather the targeting information necessary to carry out the fix. Depending on the technology (interconnect line width) that one is dealing with, this could range from visual reference plots, to 3-point alignment co-ordinates and/or database information for overlaying the CAD layout. In any case, plots are essential to any technology being dealt with. Plots need to illustrate the fix location, with the level of metal that one is dealing with, along with metals above that level so one can have access without damaging them. Basic 8 1/2" x 11" plots are satisfactory, but larger colored plots are often more useful.

1.1.2.2 Accessibility:

At this point the designer should try to identify the most convenient, optimal and feasible location to carry out the fix. A point to remember is that top level metal is always preferred, but lower level metal layers are acceptable. Alternately one may have to perform FIB fixes from the backside of the chip, depending upon the technology and/or package type. This is especially true when carrying out lower level repairs in today's seven or eight layer metallizations. It augurs well for the design engineer to trace signals and see where the fix can be carried out at the most easily accessible level. Success rates are the highest for top level metal fixes and fall off considerably when dealing with lower levels of metals. This is especially true when connections are to be made on lower level metals. This is because end-point detection gets difficult on lower level metals, and filling up the vias is more challenging. Cuts are not that much of a problem. This is discussed further in 1.2.3.

1.1.2.3 Chip Orientation:

Plots showing the fix location with reference to the entire chip are essential, because some fix locations may be near bond pads along the periphery of the chip. These locations require appropriate orientation of the chip in the system, such that the etch/deposition needles do not touch the chip on extension.

1.1.3 Deposition layout:

At this point, the FIB engineer might look at the length of the depositions involved. One can use a different deposition technique (for lengthier connections) or may use a redundant runner (which is not being used for that particular circuit). The basic idea behind all this is the optimization of beam time, so as to impart minimal damage to the chip being FIB'ed.

It behooves the FIB engineers and the design engineers to sit down together, after all the relevant information is assimilated, and discuss the fixes involved. Go over the plots and make sure the fix is feasible, and that it can be done in the manner requested by the designer. This is a good time to navigate using the plots and try to carry out the fix "on paper." Note: *Designers usually lay the most direct route between two points. The FIB engineer has to make sure that the fixes do not run over active circuitry. This is so because of the inherently destructive nature of the beam as mentioned earlier. Capacitors charge up and can be damaged when directly FIB imaged, and even laying a line across a capacitor causes them to charge up. In addition, transistors undergo parametric changes when imaged using the FIB. Some of the changes may be temporary and reversible, but that entails additional work, after the FIB repair is complete* (Campbell, 1998; Beam-It)

After all the above is said and done, one moves on to the implementation/execution phase of the project. So far the chip has not been put under the beam but has been "worked" with on paper so as to optimize the beam time in this phase. Optimizing beam time is essentially a joint effort between the engineer requesting the design fix, and the FIB engineer carrying it out. By working in tandem, this team can carry out the required fixes successfully and in a timely manner.

1.2 Implementation/Execution

The implementation/execution phase essentially consists of the following elements: a) sample mounting and grounding, b) imaging and navigation, c) milling vias, d) filling vias and making connections, e) disconnecting lines and cleanup.

1.2.1 Sample Mounting and Grounding

This is the first step in getting the chip in the system to be worked on and serves a dual purpose: 1) the chip is mounted such that it does not move in the chamber, and 2) that the chip is well grounded.

1.2.1.1 Sample Mounting

FIB manufacturers do offer numerous sample-mounting contraptions, but typically a sample holder machined out of Aluminium works great. This can be custom machined out of aluminium stock and diameters can vary from 2" to 8" depending on the size of the chamber. Electron Microscopy suppliers also offer smaller sample holders, which fit in most electron microscopes and FIB systems. The choice of sample holders is limited only by the needs, ingenuity or innovation of the FIB engineer.

The samples can be mounted on these sample holders using double-sided carbon tape, enhanced with Al tape, depending on the IC package one is dealing with. Figures 1-4 below show different devices and the way they are mounted on Al sample holders.

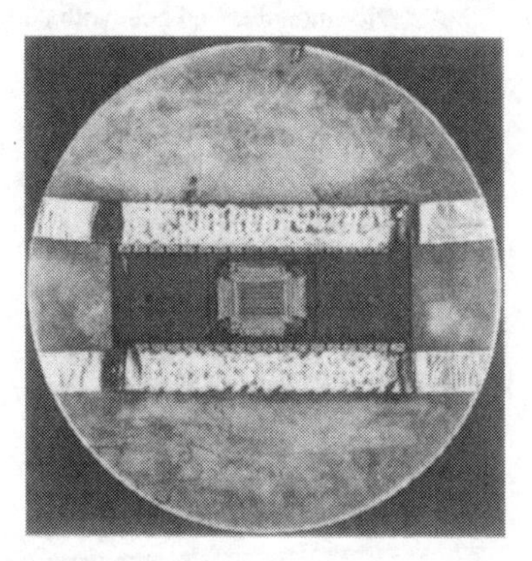

Figure 5.1. A chip mounted with Al tape.

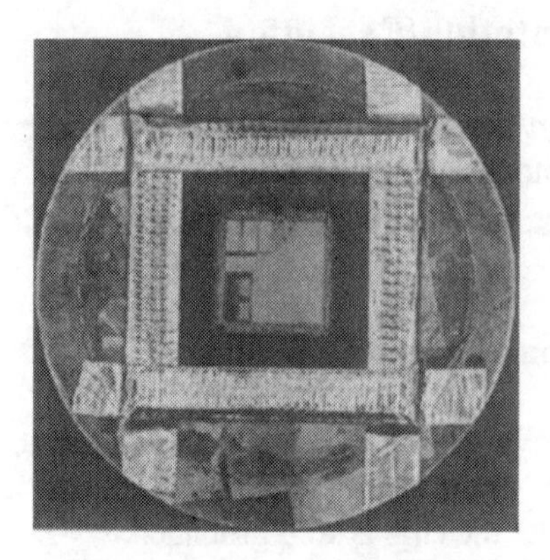

Figure 5. 2. A device mounted with Al tape.

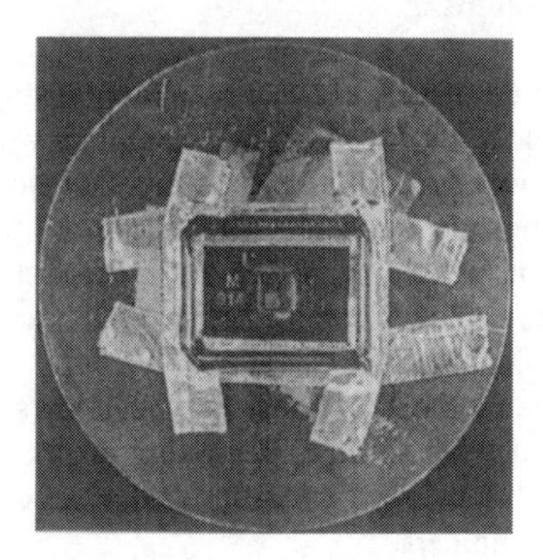

Figure 5.3. A device mounted and held with Al tape.

Figure 5.4. A device held by a clamp.

1.2.1.2 Sample Grounding

Being grounded is vital to the health of the chip, as the beam striking the surface has a positive charge which has to be dissipated as it is being worked on. Hence, one must ensure that the leads touch a conductive surface in order to effectively dissipate the charge generated by the beam. On a well grounded chip, bond pads appear very bright as evident in Figure 5.

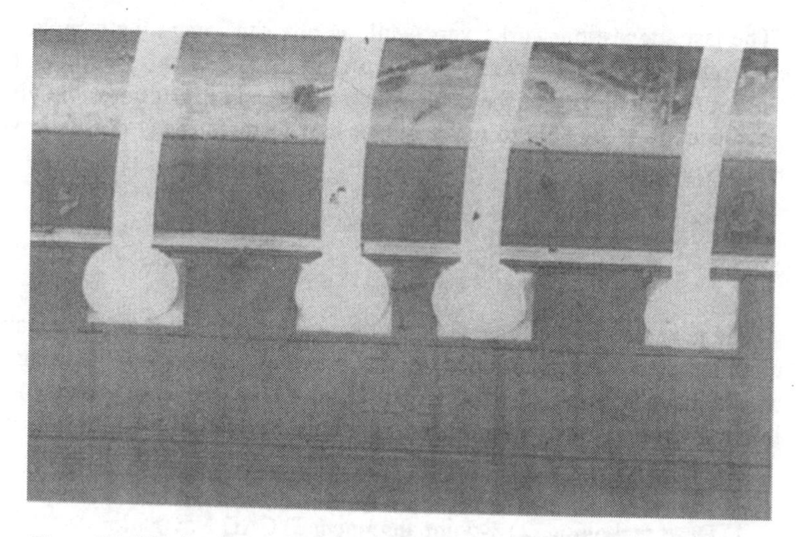

Figure 5.5. Wire bonds showing bright contrast indicating a well ground device.

1.2.2 Imaging and Navigation

1.2.2.1 Imaging:

Imaging essentially is dependent on effective mounting and excellent grounding of the sample. Occasionally, one runs into samples that will charge up, despite mounting in the best possible way. For such samples one can :

a) lower the beam current, though there is a limit to that, since at lower beam currents, the secondary electron yield reduces significantly and hence, making out features on the chip becomes difficult, or

b) apply a local Platinum (Pt) coat. This is done by extending the Pt needle, commence imaging and manually opening the valve in order to let the Pt flow onto the surface and hence dissipate some of the charge, and/or

c) apply a global carbon coat. The only caveat to this being, that the carbon has to be ashed in a plasma etcher after completion of the fix, and one must make sure that the leads are well grounded while the chip is in the asher.

The last suggestion works very well, as one can image the sample using low beam currents (between 30-50 pA), which is a safe working beam current for a majority of the devices being worked on. Besides, the yield is good enough to be able to make out surface features easily during imaging and navigation.

1.2.2.2 Navigation:

The next step in the process is navigating to the location of the repairs or modifications. As mentioned in the previous section on Planning, one should have a good idea of target locations on the chip, based on the information previously assimilated. On-chip navigation broadly falls under three categories:

1) Dead reckoning, 2) 3-point alignment 3) CAD Navigation

1.2.2.2.1 Dead Reckoning:

This is just what the name implies, when you use visual references to reach your target location. Top-level metal plots are essential, and once the location is identified on paper, the rest is relatively easy. One has to make sure the orientation of the chip is correct and then proceed to the target location. Figures 6-9 show a typical example of dead reckoning navigation on a chip. Figure 6 shows the overall chip with the region of interest (RoI) indicated by a square. Figures 7-8 show the RoI zoomed in further and Figure 9 shows the target location. Figure 10 shows the fixes involved.

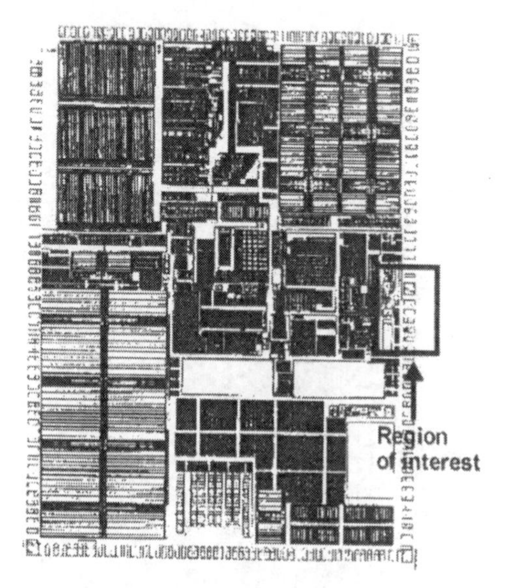

Figure 5.6. Overall chip plot, showing region of interest.

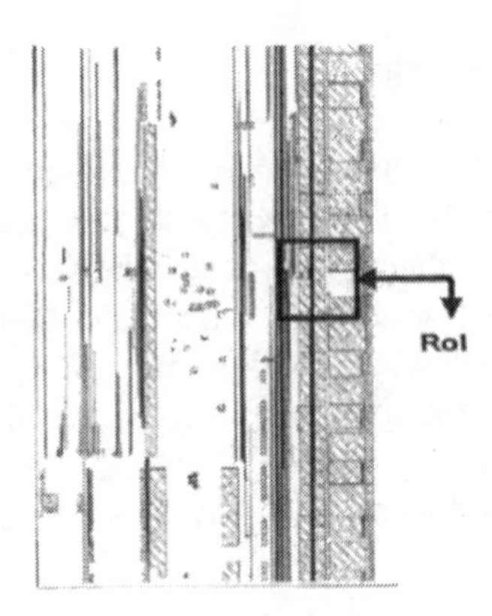

Figure 5.7. RoI zoomed in

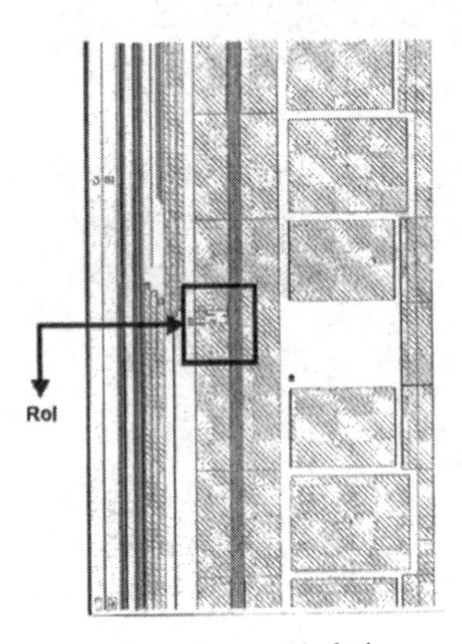

Figure 5.8. RoI zoomed in further

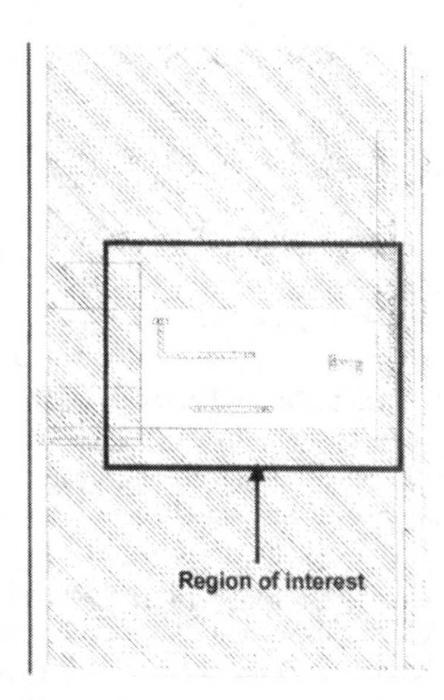

Figure 5.9 RoI zoomed in, showing FIB modification location

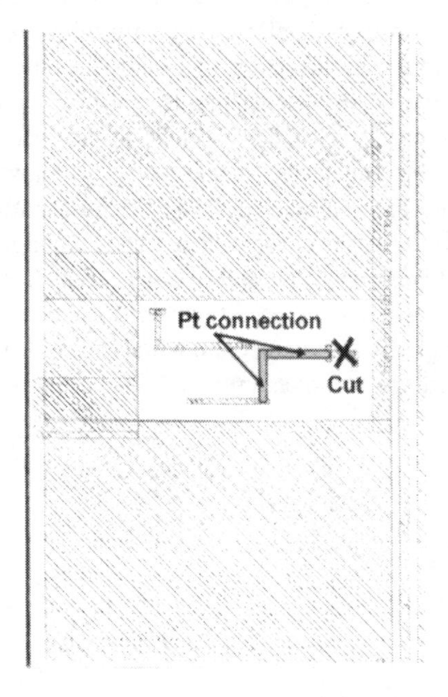

Figure 5.10. Actual modification schematic

1.2.2.2.2 3-Point Alignment:

The next level of progression in navigation is using three-point alignment. This is where the stage of the FIB system is realigned using the co-ordinates from the chip obtained from the chip database. This is usually with the chip center as the origin. Stage alignment software on the FIB system allows one to enter the co-ordinates of the three edges of the chip. Once the alignment is complete the stage is aligned to the user-defined units, hence one can navigate by entering the co-ordinates of the target location. One suggestion for the alignment would be to get the co-ordinates for all four corners of the chip. Enter the three points as required, then enter the remaining point to gauge the accuracy of the alignment. Stage offsets would be noticed and corrected at this point. Also the corners are near the bond pads and charge dissipation is better there. With charge dissipation better near the bond pads, imaging is easier and subsequently, chip damage due to the charged beam is minimized.

The same chip is used to illustrate the 3-pt navigation in the figures below. Figure 11 is an overall plot of the chip with the corners marked. A chip schematic is necessary in order to orient the chip prior to alignment. One can use a company logo or a "pin 1" notification on the package for orientation. In this case, box 1 corresponds to the lower left corner of the chip. The corners are zoomed into in Figure 12, as shown. These are mask alignment marks and are oriented this way before the alignment procedure. They are similar on the other corners also. These marks were used because they were there and were convenient, but other references i.e. top level metal features (in each corner) may be used to achieve the same results. One would have to use the design layout to correlate with features on the chip. Figure 13 shows the target location and fixes involved.

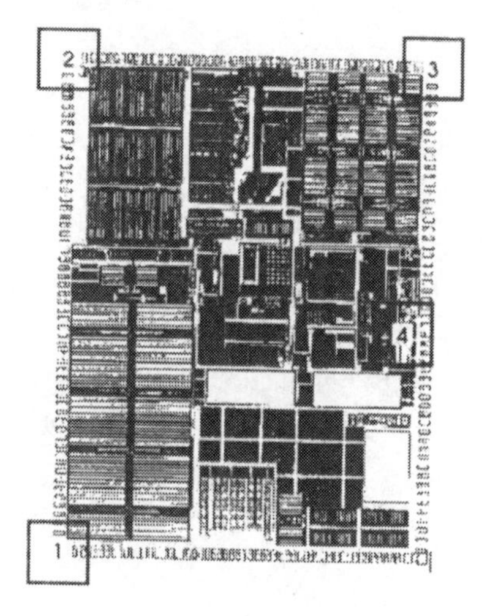

Figure 5.11. Overall chip plot, showing corners for 3-pt alignment

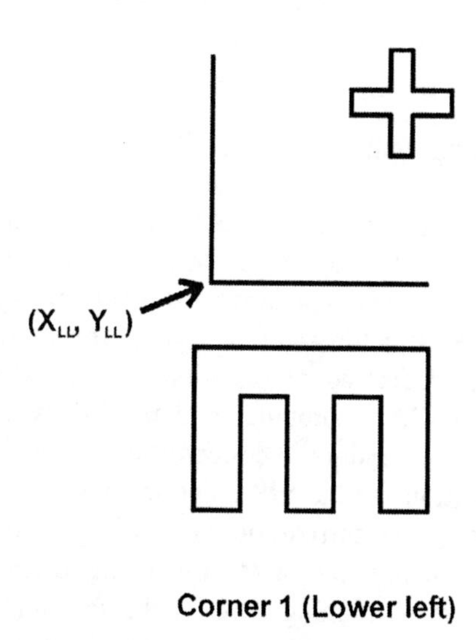

Figure 5.12. Corner 1, zoomed in

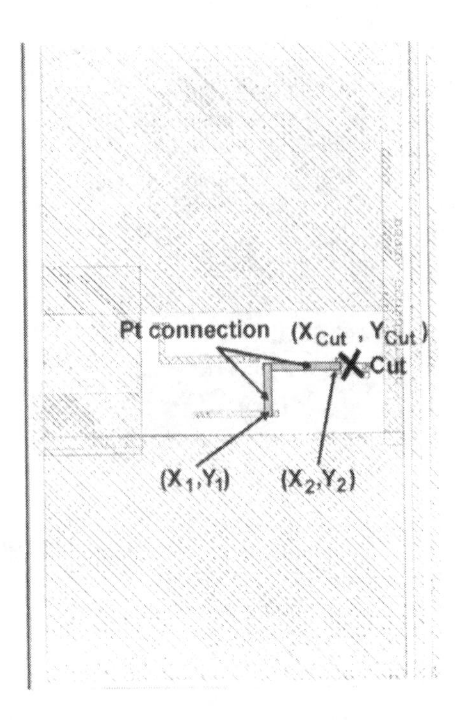

Figure 5.13. Schematic of modification, with connections and cut marked.

1.2.2.2.3 CAD Navigation:

The next natural step up for efficient on-chip navigation is using the database information for the layout of the chip. This can be obtained as a .gds file. This information is then converted using the x-y mask information on the different layer structures for the chip. The conversion is typically done on a separate, dedicated workstation/PC, and there is typically an interface between the CAD computer and the FIB system. Different FIB systems use a variety of vendors to convert the information on the .gds files into tangible information for the FIB stage to facilitate on chip navigation. Once the conversions are carried out, the stage is aligned to the CAD navigation software using the 3-pt alignment mentioned above. One important point worth mentioning here, is the fact that, as one aligns the stage and database one has to make sure that the deposition and etch needles do not interfere with the bond pads or wires on their way in and out on the chip. This is especially critical when the FIB locations are near or around the bond pads. Of course, one should be aware of this information from the earlier planning stage, when the locations are appraised on paper. The database software usually allows the database information to be rotated as needed, so the needles have an easy in and out path.

It would thus suffice to say that one would essentially study the plots, be aware of the FIB locations, check the path of the etch and deposition needles, and then align the CAD information from the database. All this can be checked when the eucentric adjustments are made on the FIB system.

Once the CAD system is aligned to the FIB stage, one can go to the target location either by entering co-ordinates, or by navigating using the CAD software. Once the target area is reached, out come the zoomed-in plots. Also, the different layers can be highlighted on the CAD software. At this point, in today's highly planarized chips, one can hope to find a top-level metal structure, which can be used to fine-tune the alignment. (Van Doorselear, 1993) Typically one uses alignment marks which are similar to the ones mentioned in the previous section, and then proceeds to the target location.

One important point at this juncture is the fact that after the alignment is done, and the target area is arrived at, and the point of contact is determined, one has to typically use an enhanced etch gas for efficient milling. Unfortunately, there is always an image shift when the etch needle is inserted just above the chip. That may be enough to be milling in the wrong place. One way out of this predicament is to define the mill box, as needed, and lightly mill the surface of the chip, so as to make a small indentation on the surface. In this way when the needle is inserted, and image shift occurs, one can correct the location of the mill box using the indentation created as mentioned earlier.

Another unfortunate outcome of planarization is the fact that more often than not, one has to make some "seeker" holes to align with underlying and buried targets. Coupled with the extremely accurate stages provided by the FIB manufacturers, one should be able to reach underlying targets with relative ease. Depending upon how far the seeker holes are one may need to fill them in using FIB induced insulator deposition to avoid unwanted shorting of metals. It is hereby left to the FIB engineer to use whatever means available at his/her disposal to hit underlying metal targets.

1.2.3 Milling vias:

Vias are typically milled using the same beam current used to image the sample. Beam currents between 30-50pA work well for a majority of samples. Vias for top-level metal connections can be just a couple of microns on the side. However, as one needs to make vias on lower level metals, via size needs to increase. This is because of the higher aspect ratio of the holes, and the fact that when one deposits metal in the vias for making contact, there are chances that the metal bridges on the top of the via instead of making contact with the metal at the bottom. Also larger contact areas ensure lower contact resistance. One can increase the area by making via shapes rectangular instead of being square.

Vias were made in the earlier times using just the regular mill using just the Ga beam. Over time, gas chemistries have enhanced the milling rates for specific materials. The enhanced etch gas used to assist the milling of oxides enhances the milling rate about 7-9 times. What the gas does is essentially convert the sputtered material into a volatile compound, which is pumped away by the vacuum system. This ensures that the via hole is made with minimal redeposition, and the beam is on for a shorter time. It is also a good idea to watch a real time monitor (RTM), update box etc. to check for drift, and also to ascertain when one hits metal. Using end point detection is also highly recommended. (Beam-It; Ashmore; Walker)

1.2.4 Filling vias and making connections with platinum:

1.2.4.1 Filling vias:

Filling vias in top-level metal is straightforward. It is just part of the connection. One just has to make sure that the deposition box is smaller than the via dimension. Making fills on lower level vias is a lot trickier. This is so because if one is not careful, improper deposition techniques will lead to "voiding" in the vias. In this case the Platinum bridges across the top of the via and does not make contact to the metal below.

A couple of good rules of thumb regarding via filling are as follows: a) Make the deposition box smaller (in both dimensions, x and y) than the mill box and b) use the 2 1/2 minute rule. This number was empirically ascertained. (Beam-It) *One needs to alter the dimensions of the deposition box (x, y, z) and the beam current so that the time for the deposition is ≥ 2 ½ minutes*. This combination ensures better filling in the vias. (Beam-It) Some FIB systems have material files for via fills. The material files usually alter the dwell and overlap times, besides dropping the beam current.

1.2.4.2 Making connections:

Another suggestion for figuring out the beam current is to calculate the area of the deposition box and multiplying by 2-6 pA/μm^2. These give a ballpark figure for the beam current to be used. Keep in mind that with higher beam currents one needs to use different material files in order to keep ahead of the deposition curve. (Van Doorselear, 1993)

Typical connections run about 10-75 µm long, 1.5-4 µm wide and about a micron to a micron and half thick. (Beam-It; Carlson, 1997) The deposition tracks need to be wider when using higher beam currents (typically > 350pA). (Beam-It; Carlson, 1997) While doing depositions at higher beam currents (>350 pA) it is generally a good idea not to grab too many frames while depositing, so as to minimize over spray. One can check and compensate for drift using a Real Time Monitor, update box etc.

1.2.5 Disconnecting lines and Clean-up:

1.2.5.1 Disconnecting lines:

In order to avoid refilling of cuts with redeposition, lines are cut or disconnected after all depositions are complete, and prior to over spray clean-up (discussed in the following section). Typical beam currents for exposing the lines to be cut are similar to ones used for exposing lines prior to deposition, i.e. 30-50 pA. It is quite tempting at this stage to use a higher beam current in order to get the job done quickly. It is however prudent to maintain control over the cuts and avoid using beam currents greater than 300 pA. One may start exposing underlying metal levels very quickly, when using a higher beam current. Use of insulator and metal enhancing etch gases is highly recommended so as to have cleaner cuts. Knowledge of barrier metal materials is essential so one can use gases to attack the barrier metals and get the cuts done more quickly and cleanly.

1.2.5.2 Clean-up:

Clean up is typically required if there are several modifications in close proximity to each other and only if the Pt deposition over sprays overlap. If the depositions are far enough away and do not overlap, no clean up is necessary.

Cleanups can be carried out in a couple of ways:

I) One can draw individual mill boxes around the deposition paths and mill them in series or parallel. Alternatively, one can use the outline mill boxes to achieve the same result. In both cases, one needs to be extremely careful so as not to mill too much into the top surface. Monitoring the mill on the RTM and rapidly grabbing frames would ensure knowing when to quit. One needs to stop when the Pt over spray is milled through.

II) One can also do this: Zoom into the sample (typically around 30-50 KX) extend the enhanced etch needle, start the flow of gas, and start imaging. Soon one will find the over spray separating out, then one can just move using the joystick and carry on the clean up. (Beam-It; Van Doorselear, 1993)

One important point to be remembered during clean up is to make sure that one does not clean up on lines just deposited. That is quite easy to do and one has to be extremely careful not to do that. Figure 14 gives an example of a clean up following a "complex and messy" FIB modification. The thick white lines are the Pt depositions. The lighter white is the over spray and the black lines in between are the clean ups. The four cut locations are also marked. In this case the depositions needed to be isolated from each other, and were successfully carried out as shown.

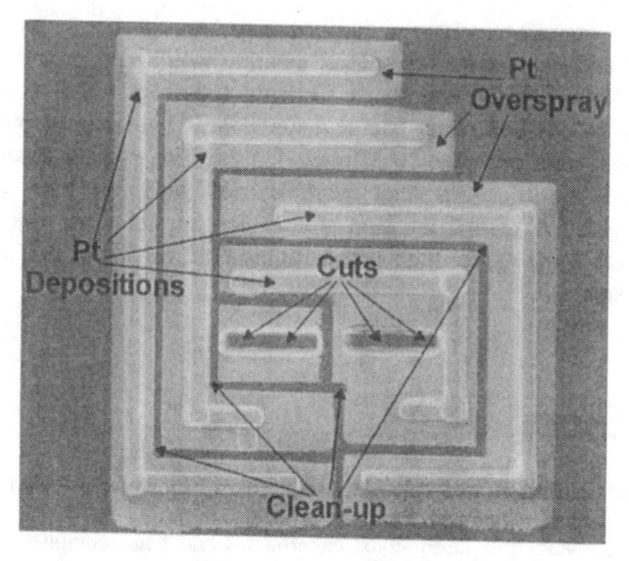

Figure 5.14. Complex modification and clean-up

2. SUMMARY:

For future reference, it may be a good idea to design spare gates at top level metals, and also give power and ground contact points at top level metal. These could be scattered all over the top surface, so FIB engineers would not have very far to FIB certain changes that are needed. These would serve two purposes: 1) Provide easy access to tie signals high or low, and 2) Give additional visual reference points to aid navigating in today's planarized multi-layered circuits.

3. ACKNOWLEDGEMENTS

The author would like to acknowledge all his co-workers at Beam-It Inc., during his tenure there for all the discussions, and a lot of experimentation we were encouraged to do, in order to carry out difficult repairs. Also on the list would be folks from FEI Company, Richard Young and Peter Carleson in particular, for helpful advice and useful discussions over the past several years. Credit also goes to Sho Nakahara, Bob Privette, for their constant support and encouragement, and Judy Moore, Agere Systems, for help with the figures and manuscript. A special mention would also go out to Robert Ashmore, Robert Pajak, and Bruce Griffiths (all from Agere systems) for their contributions and many useful discussions. The original pictures for Figures 6-13 were provided by Clinton Holder, Esq., Agere Systems (and were actual modifications that were carried out), and are gratefully acknowledged. A special mention also to Joe Serpiello and my boss Jim Cargo, Agere Systems, for their help with the final version of the manuscript and comments.

REFERENCES

Ashmore R, Griffiths B, and Pajak R, private discussions of the PAL FIB group, Agere Systems, Allentown, Pa.

Beam-It, Inc., Numerous discussions, experiments and notes compiled during the author's tenure.

Benbrick J, Rolland G, Perdu P, Casari M, Desplats R " Focused Ion Beam Irradiation Induced Damages on CMOS and Bipolar Technologies," Proceedings, ISTFA'98, pp. 49-57 (1998).

Campbell AN and Soden JM, "Voltage Contrast in the FIB as a Failure Analysis Tool", Electronic Failure Analysis desk reference, pp. 129-135 (1998).

Carlson P, "Integrated Circuit Modification", Course notes The Essentials of Focused Ion Beam Technology, University of Maryland, October 5,6, (1997).

Van Doorselear K, Van Den Reeck, Van Den Bempt L, Young R, Whitney J, "How to Make Golden Devices using Lesser Materials", Proceedings ISTFA, pp. 404-414 (1993).

Walker J, Applications note, "Focused Ion Beam Applications Using Enhanced Etch", FEI Europe, LTD

Wills KS, and S. V. Pabisetty SV, "Microsurgery Technology for the Semiconductor Industry", Electronic Failure Analysis, desk reference, pp.227-276 (1998).

Chapter 6

THE USES OF DUAL BEAM FIB IN MICROELECTRONIC FAILURE ANALYSIS

Becky Holdford
Si TRAC , SC Packaging Development, Texas Instruments, Inc.

Abstract: Focused Ion Beam systems have many uses in failure analysis (FA), and dual beam systems are a must for state-of-the-art integrated circuit and MEMS device failure analysis. Finding opens/shorts in circuitry by utilizing passive voltage contrast is a common usage. Using the dual beam for defect review on wafers enables you to find and section defects too small to see with optical microscopes. To cross-section particles and mechanically-weak structures and to generate smear-free cross-sections of structures having layers of differing hardnesses is very difficult to do using mechanical techniques; this becomes routine using the dual beam FIB. The dual beam FIB is a must-have tool for cross-sectioning MEMS devices without inducing catastrophic damage.

Key words: FIB, dual beam FIB, IC, MEMS, MOEMS, passive voltage contrast (PVC), defect review, cross-sectioning, particles, weak structures, smear-free sections, Cu metal

1. SINGLE-BEAM VS. DUAL-BEAM

Focused ion beam (FIB) systems have long been recognized as useful tools for integrated circuit (IC) and microelectromechanical systems (MEMS) device failure analysis. The introduction of dual beam systems has made the tool an order of magnitude more useful and a must-have for state-of-the-art IC/MEMS failure analysis. The drawback to single-beam systems (ion column only) is the imaging. Imaging is done with the ion beam and this has two drawbacks. First, the beam is milling the surface as it is imaging and this causes damage to the surface of interest; the amount of damage varying by beam size and magnification. Second, the resolution of the images generated with the ion beam is lower due to the size of the ions making up the beam. With dual beam systems, the addition of an electron column (e-beam) to the ion column (i-beam) solved both these problems. The electron column (e-beam) handles the imaging without milling the surface and the resolution of the image is comparable with high-end dedicated SEMs. The ion column is free to do the milling but can be used for imaging when the user desires. Most of the work in this chapter was done on an FEI DualBeam 830. The column arrangement of this tool is shown in Figure 1.

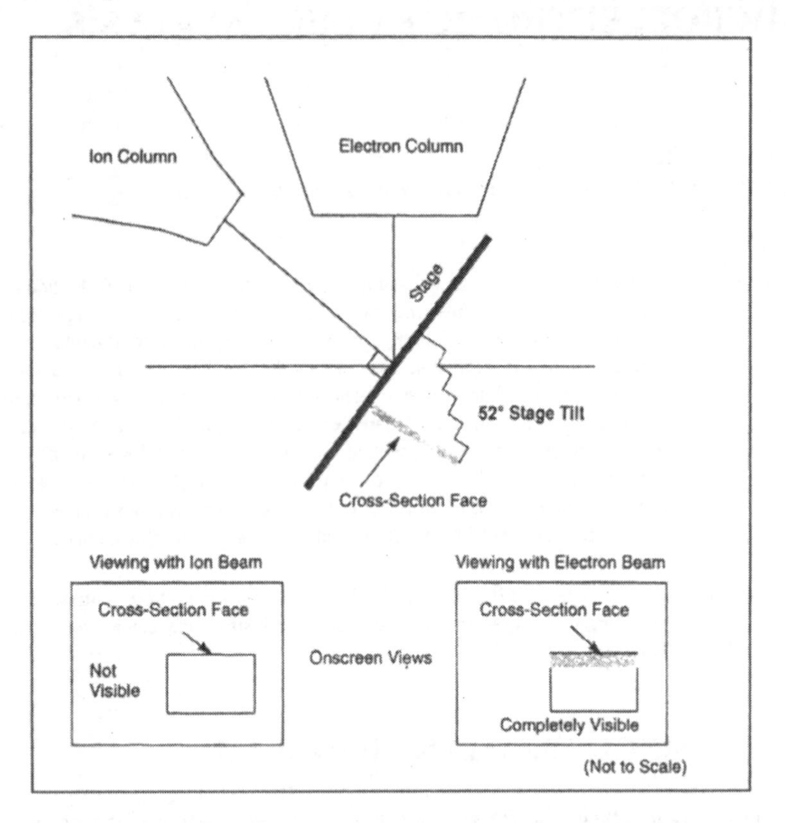

Figure 6.1. Dual Beam Column Arrangement (image courtesy FEI Company)

The columns are arranged so that, at eucentric sample height, the focus point of both beams is coincident. When the stage/sample is tilted normal to the i-beam, the face of interest is at a 52° tilt to the e-beam. The sample can be milled and imaged without moving the sample just by switching imaging modes. This arrangement also gives a working distance for the e-beam of 5 mm while the working distance of the i-beam stays at 16 mm, helping to improve the e-beam resolution. So now there are two columns; what can be done with them with them?

2. PASSIVE VOLTAGE CONTRAST

Passive voltage contrast (PVC) is a charging phenomenon that is very helpful in finding opens and/or shorts in circuitry. (Campbell) PVC uses the charge induced on the surface of a conductor by the e-beam to provide a contrast between structures charged bright (negative) and those charged dark (positive). PVC is especially useful on small repetitive structures, such as via chain test structures. Via chains are sets of metal links on two levels that are connected by vias. These are typically rectangular structures consisting of thousands of links with a probe pad at either end of the chain. When electrical testing shows a particular chain is open, how does one find the bad link/via in those thousands? Ground one of the probe pads, usually by cutting a hole through it to the substrate of the die. Then image the whole (or largest part you can) at a slow scan rate until the charge contrast becomes evident. Imaging with the e-beam enables the contrast to become apparent without damaging the surface of your structure. One can then zero in on the suspect area. Switching to i-beam will make the contrast more obvious (as it's at a higher accelerating voltage) and then mark the bad link and set up to cut the cross-section. Figure 2 shows the location of a bad link in a via chain found by PVC and Figure 3 shows the end result of the cross-section.

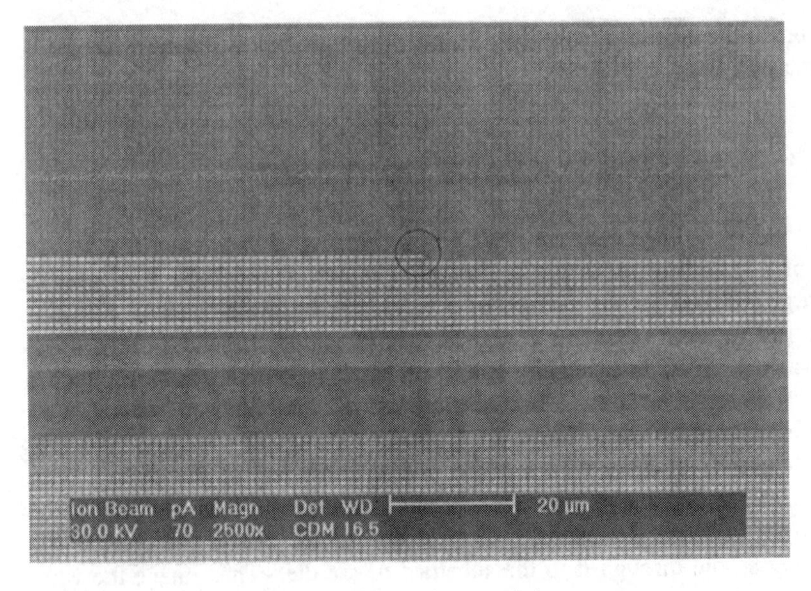

Figure 6.2. PVC of via chain.

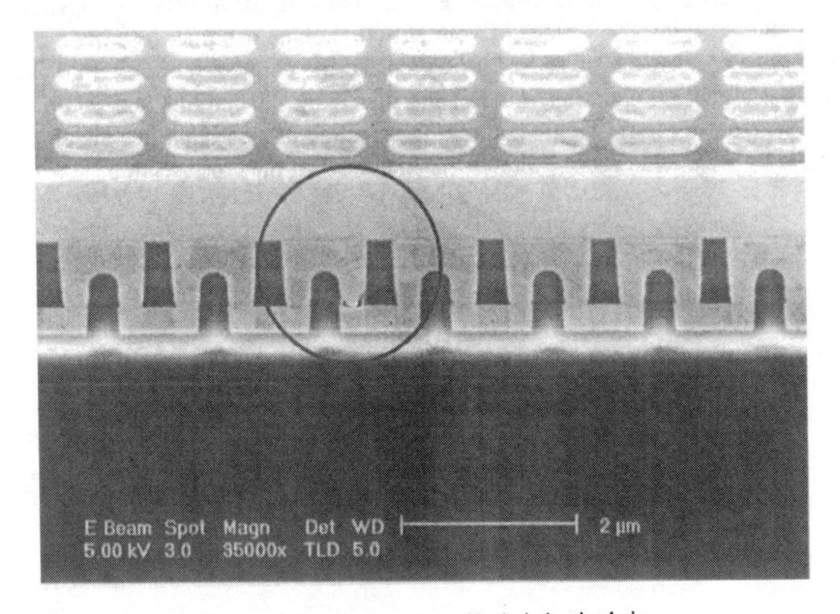

Figure 6.3. Cross-section of bad via in via chain

3. CROSS-SECTIONING DIFFICULT MATERIALS

3.1 Cross-sectioning materials of greatly differing hardness and/or mechanical strength

Getting a smear-free, defect-free cross-section of a stack of materials that differ greatly in hardness can be a trying experience if using mechanical polishing methods. The softer layers will polish faster than the harder layers, resulting in round off of the cross-section face. Rounding off will obscure the interface between the layers. If the layers are not mechanically strong or well attached to each other, the cross-sectioning forces will cause delamination of the layers or destruction of the structure. Using the dual beam FIB will remove most if not all of the headaches associated with making these types of cross-sections. A good example of this type of problem cross-section is a gold ball bond on an aluminum bond pad, as shown in Figure 4.

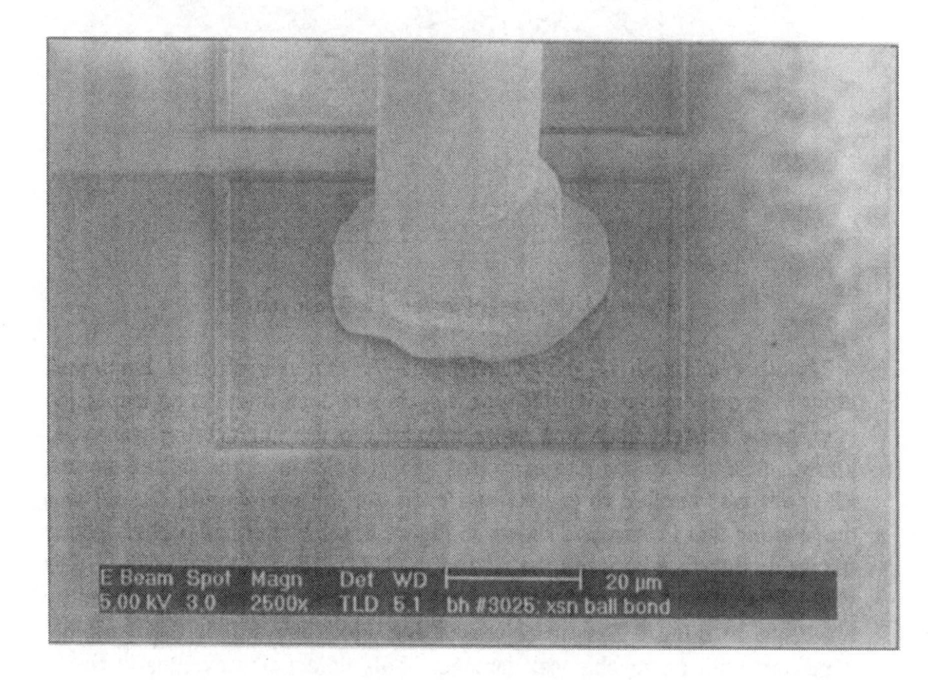

Figure 6.4. Au bond wire on Al bond pad

The gold and the aluminum are much softer than the silicon of the die, also the bond wire and the bond pad will smear and peel off the surface using conventional mechanical sectioning. One could encapsulate the whole device to give support to the wire and bond pad, but then no more testing could be done to that device. In the FIB, this structure can be cross-sectioned almost as easy as cutting a cake, as shown in Figures 5 and 6. Depending on the bond pad structures, the pad and wire can still be electrically connected to the rest of the die.

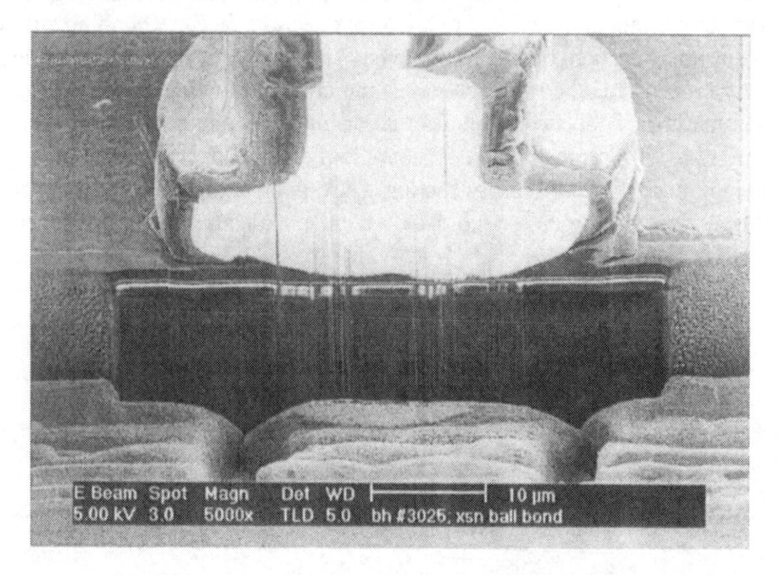

Figure 6.5. Overview of sectioned ball bond and pad

Another example is shown in Figure 7. This is another bond pad, aluminum over copper. Blister-type defects had been found at an inspection step, some blisters open and some intact. Process engineering wanted to know where the metals had separated and at what level the failure started. Platinum was applied to protect and even out the surface and the cut was made along that Pt line. As shown in Figure 8, the blister had peeled open at the oxide interface between the Al and the Cu, pulling out part of the Cu as well. The opened flap of the blister has not moved, although is it eroded by ion-beam imaging. The metal cracks and voids are still is their original condition with no mechanical damage. This defect was eventually traced back to a CMP slurry problem.

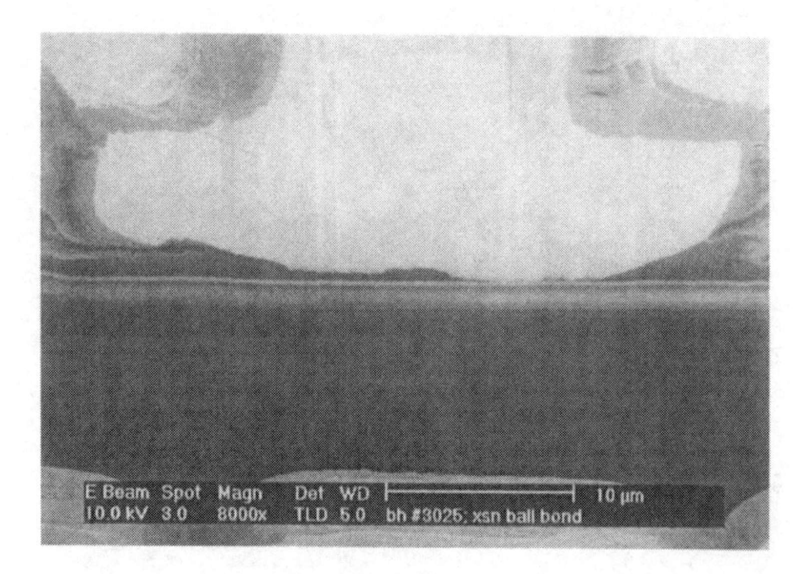

Figure 6.6. Cross-section of ball bond and pad with no smearing or damage

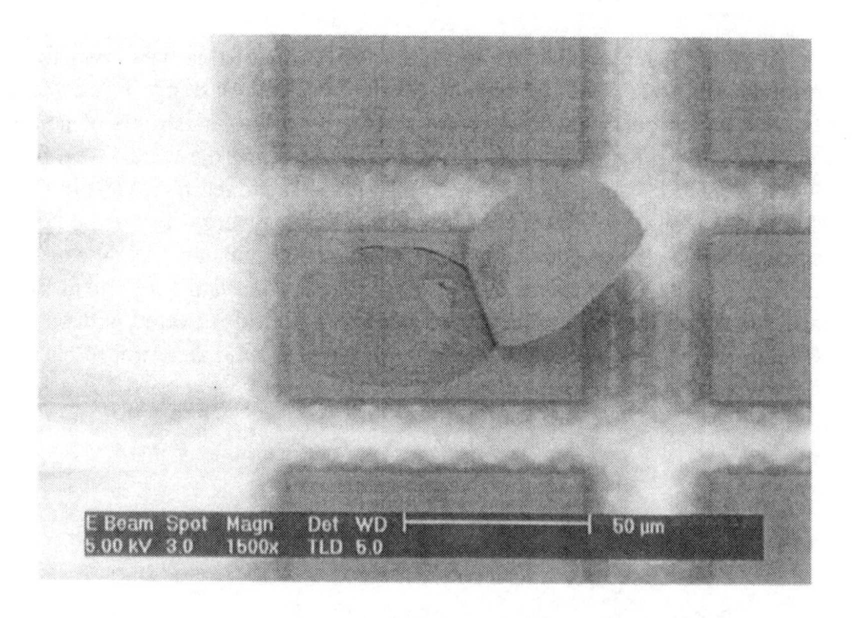

Figure 6.7. Peeled open blister. The flap is aluminum.

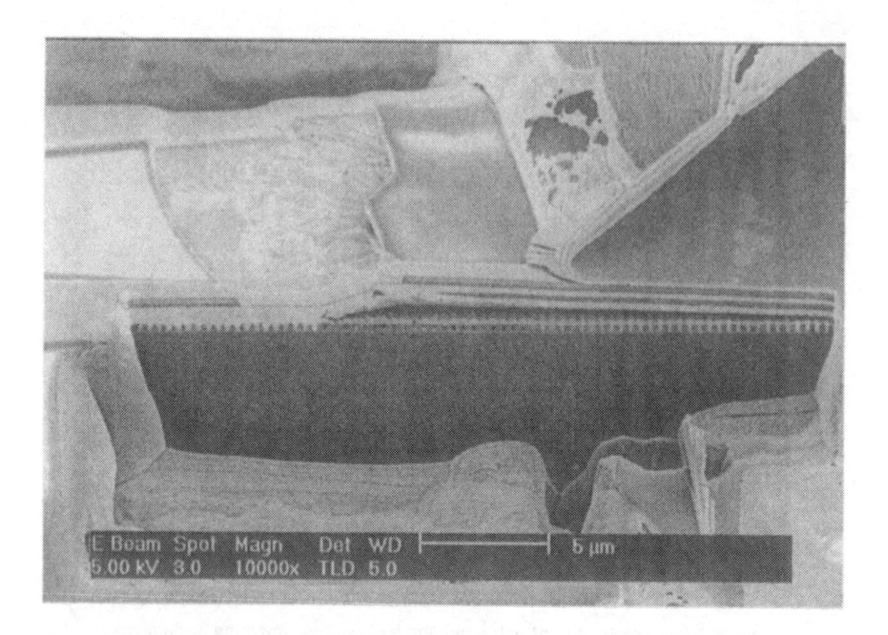

Figure 6.8. Cross-section through blister and flap revealing the damage below.

Another example is shown in Figure 9. Flakes of this type were found prior to a copper CMP (chemo-mechanical polishing) step. The selected flake is precariously attached to the surface by a few angstroms of copper and would fly off at an attempt to cleave it by breaking the wafer. The flake was protected by deposited platinum then cross-sectioned to see the internal structure. With this structure revealed, EDX (energy-dispersive x-ray spectroscopy) can be done to determine elemental makeup. EDX data can then give clues to the origin of the flake. The section and subsequent EDX analysis found it to be a thin flake of aluminum fluoride covered with copper (Figure 10). The aluminum fluoride particles were ultimately traced back to a prior dielectric etch tool.

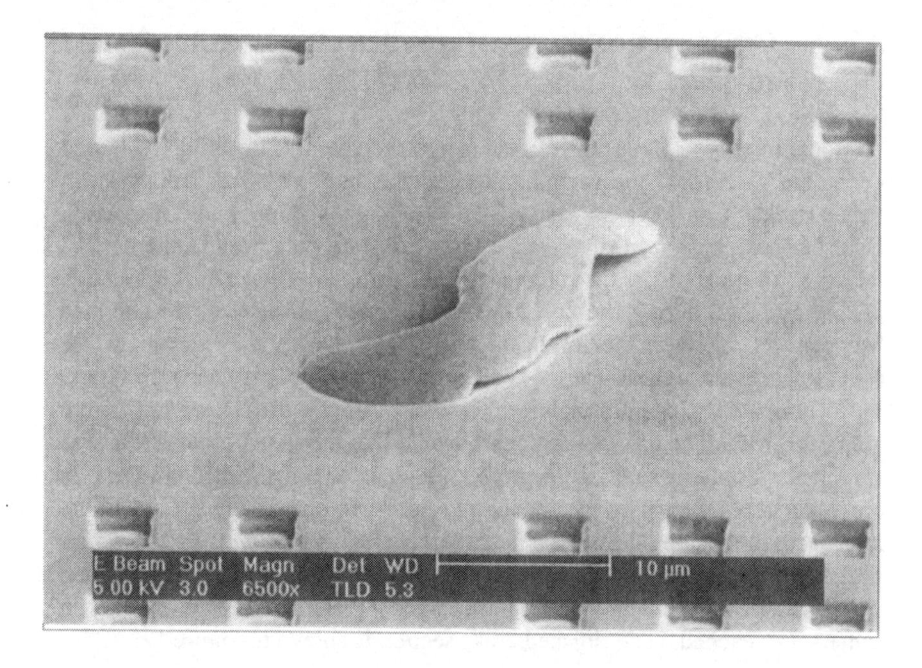

Figure 6.9. Flake on copper surface

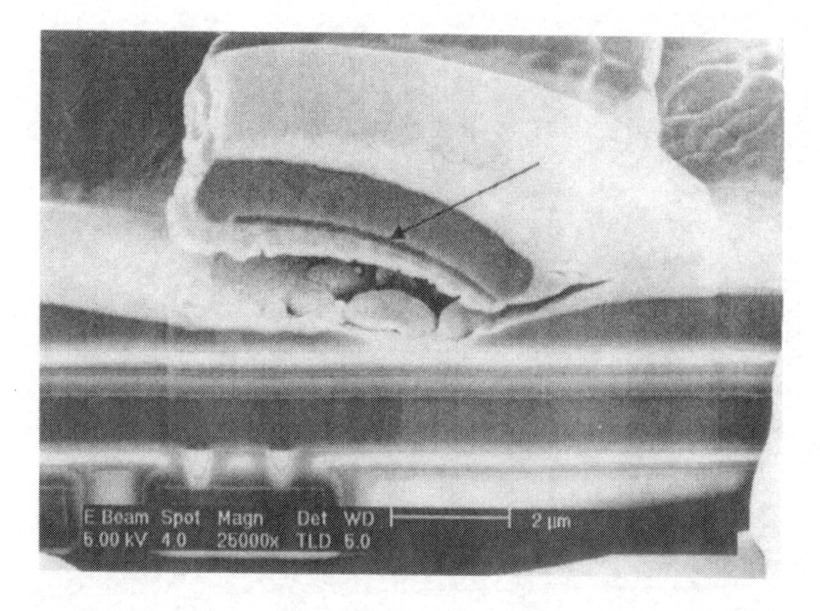

Figure 6.10. Cross-section of flake showing internal structure. The thin dark line is the aluminum fluoride flake.

3.2 Cross-sectioning structures/defects not visible from surface

Sometimes you have an interest in a structure or defect that is buried under the surface of the sample, such as vias under blanket metals or an oxide stack. You have an idea of the location but don't know it exactly. Dual beam FIB is an ideal way to gently excavate your way to the desired location. One can make the rough cut in the general location of the structure of interest, then during the polishing or clean up milling one can alternate between cutting with the ion-beam and imaging with the e-beam to see exactly when the area of interest is reached. An example is shown in Figure 11. This is a via chain structure after being filled with copper but before going to CMP. The process engineers would like to know how well the vias are filled. Copper metallization does not cleave or polish well and the vias are under a blanket of copper metal and not visible from the surface. One can image the vague outline of where the via chain structure is on the wafer, so make the rough cut somewhere in the middle. Then during the polishing step, switch between the ion beam and the e-beam to see when the via centers are reached. The finished cross-section is shown in Figure 12.

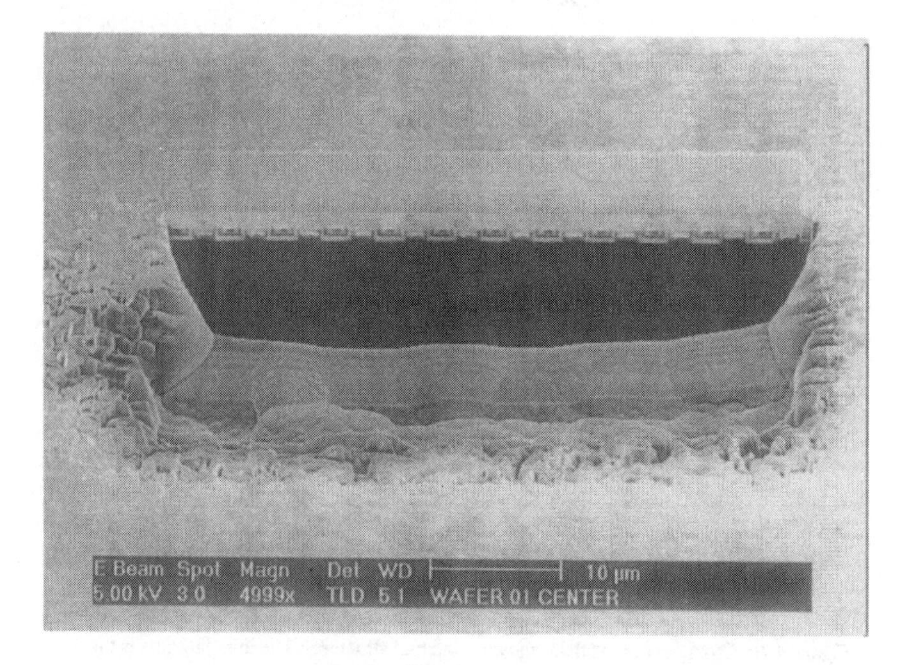

Figure 6.11. Blanket Cu over vias

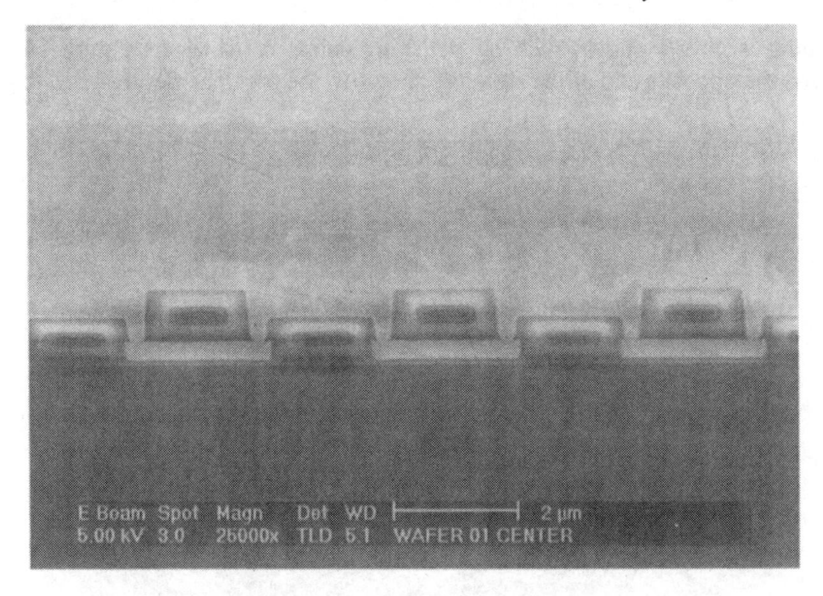

Figure 6.12. Final polish of via chain structure showing center of vias and quality of fill

3.3 Cross-sectioning defects on packaged devices

Getting a cross-section of a defect in packaged device is especially fraught with headaches because typically the package is huge compared to the area of interest. And if the package material is ceramic, the hardness of the ceramic material compared to anything else in the stack-up makes it really difficult to get an acceptable cross-section. The example is Figure 13 is a worst-case scenario. This device is an electromigration (EM) test structure utilizing minimum linewidths and minimum via sizes and had failed electrical testing with opens. These structures are right at the edge of resolution for optical microscopy. They are packaged for testing using silver-filled die bond epoxy in a metal-lidded ceramic DIP (dual in-line package). Mechanical cross-sectioning was not really an option due to the size of the structures, the ceramic package material, and the fact that the die couldn't be gotten out of the package without shattering it. Since the desired structure was not visible from the surface in the FIB, the distance the structure of interest was from visible landmarks was measured optically and transferred into the FIB. Now having a general location for the structure, the rough cut was made. Then a fine polish pattern was set up and milled slowly into the structure until the middle of the line and via was reached. The EM voiding is clearly shown with no polishing debris or damage. Figure 14 shows the opposite end of the structure shown in the previous figure.

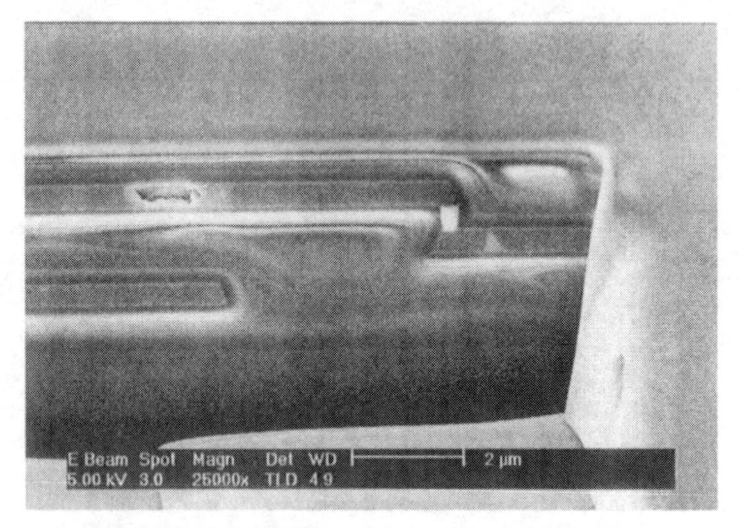

Figure 6.13. One end of EM test structure showing void in line

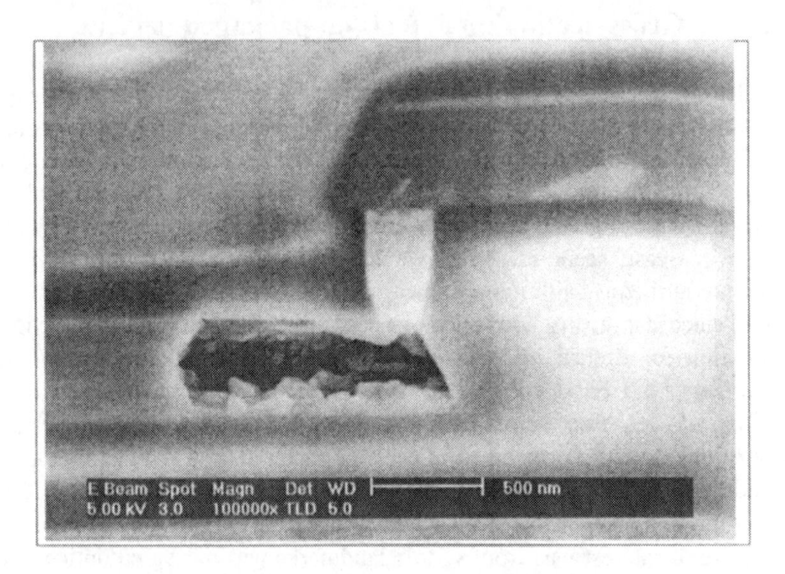

Figure 6.14. Opposite end of structure showing EM void at the via

3.4 Cross-sectioning voids without contamination

As shown in the previous section, dual beam FIB is an excellent tool to obtain cross-sections of voids/cavities with out contaminating the voids/cavities. Mechanical polishing will always fill a hole with polishing debris and polishing media, confusing the results. Shown in Figure 15 is a die ID fuse structure after the fuses were blown, then the slag and debris cleaned up by a wet etch process. The process engineers wanted to see how much metal remained after cleanup. Figure 16 shows a higher magnification view of the left side of the fuse. Note that the FIB cross-sections are very clean with no debris left in the voids to confuse the results.

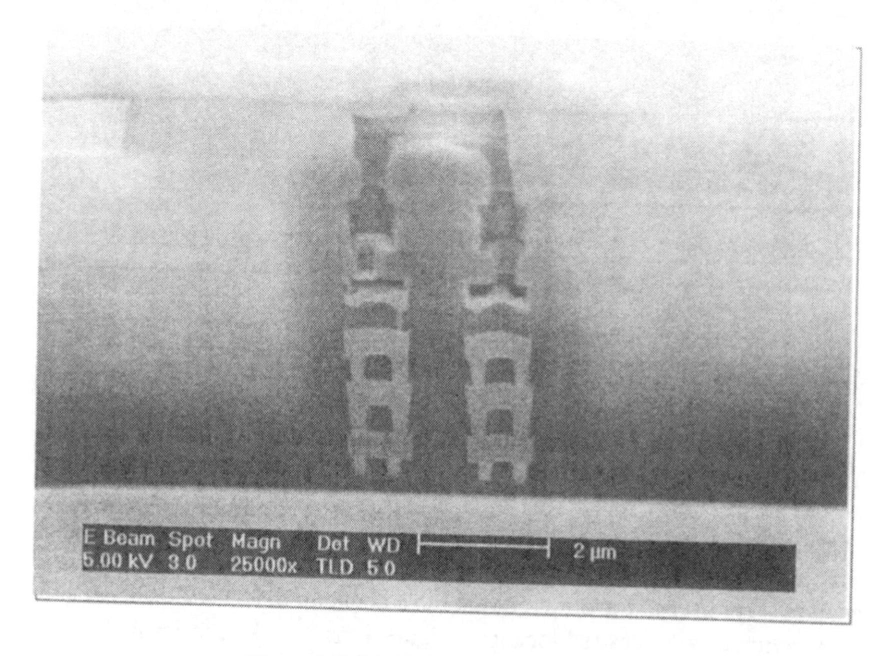

Figure 6.15. Die ID fuse structure after slag etch

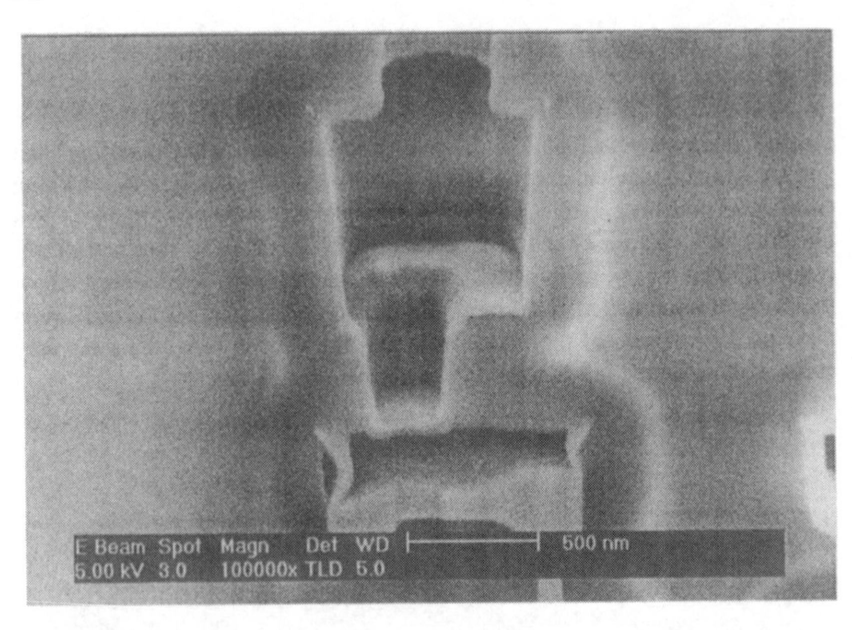

Figure 16. Higher magnification of die ID fuse showing very clean cross-section

4. CROSS-SECTION FOR DEFECT REVIEW

With geometries getting smaller, killer defects are also getting smaller. This fact, coupled with the larger wafer sizes, means defect review tools are a must in the fab. To be able to find and cross-section defects found by these tools is also a must to get the full benefit of the tools. Most full-wafer FIB tools have defect review software available for them. This software enables them to read the output of the most common defect review tools and then navigate to the desired location. Figure 17 shows the end result of such a job. This wafer had many random defects of the type shown by the arrow, which were tiny cracks or slits in the PO (passivation oxide). Optically the metal pattern was very odd. An FIB cross-section showed the defect was caused by a problem that started at metal 4 and was propagated to the surface by the rest of the layers. Sometimes, obviously, the defects found by defect review are large enough to see and cleave manually. But if the metal stack is copper, confusion can result due to the way copper cleaves. Copper is a very ductile metal and will stretch and pull-out badly during a mechanical cleave which can cause voids and/obscure voids and delaminations.

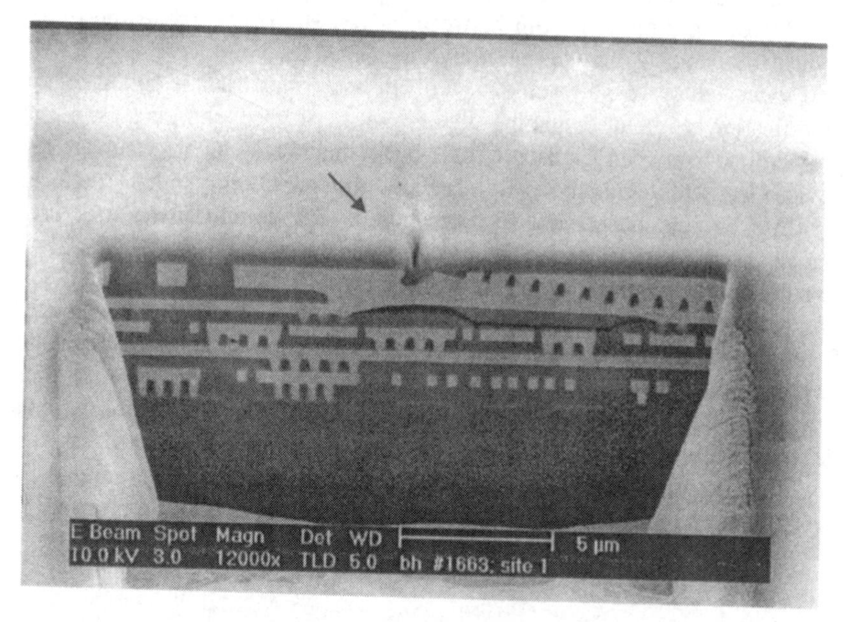

Figure 6.17. End result of a defect review job

5. DEVICE MODIFCATIONS OR EDITS

Probably one of the most popular uses for dual beam FIB is device modification or editing. This is a way of checking a proposed design change without incurring the expense of mask generation and device fabrication. It also saves an enormous amount of time, especially if it becomes obvious the proposed change doesn't work. This type of modification is also very handy in failure analysis for generating test pads for probing and for checking results. An example: if a particular fail mode is supposed to be caused by a short/open at a particular node, repair that node and see if it solves the problem. Most FIBs can be interfaced with pattern navigation software that can read the device layout database and navigate to a desired node or structure. Very complex modifications can be made if the FIB is equipped with some type of insulator deposition. Other authors have shown examples of this work. (Wills; Li, 1998; Casey, 2002; Wilson, 2002) Figure 18 shows the end result of a very simple device edit. The designers wanted two of these nitride-covered test pads opened up and connected together to verify a design change. Holes were cut in the nitride layer, exposing the metal.

Then the holes were connected together by filling them with platinum (Pt) and depositing a platinum run between them. The resistivity of the run is about 10-20 ohms/square per micron of deposited thickness.

Device edits are still somewhat of an art. The trick is to make the modification without damaging the circuit. Surface charging is a big problem, not only can it obscure the e-beam image but can also damage the device electrically and physically. Devices utilizing dummy metal structures for CMP flatness are particularly susceptible. The dummy metal structures charge up and, with no ground connection, can accumulate enough charge to actually tear themselves out of the surface of the device due to the repulsion of like charges. This phenomenon can generate a lot of underlying damage. Some device editors will coat the device with carbon to help bleed off the unwanted charge. The carbon can be ashed off in an oxygen plasma after the edit is done, removing the shorting layer. Exposure to a plasma can generate other problems in some devices, so other editors will do the modification at very low beam currents to prevent damage. The editor will need to have an understanding of the circuit to decide how best to proceed.

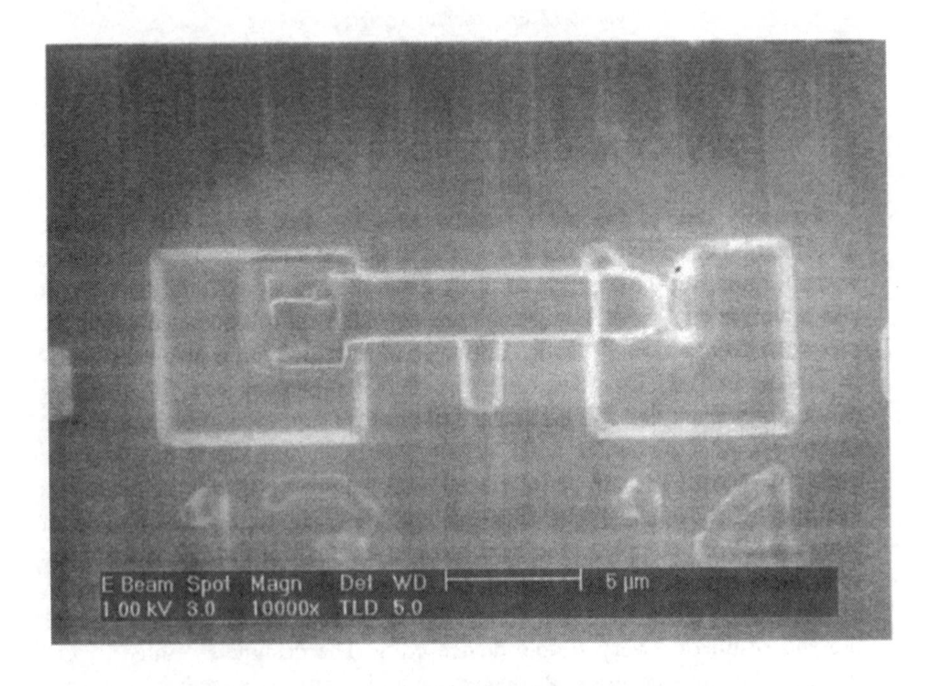

Figure 6.18. A simple device edit

6. MEMS DEVICES

Micro-elecromechanical systems (MEMS) devices and their sister devices, micro-optical-electromechanical systems (MOEMS) devices are classes of device that are extremely hard if not impossible to analyze without an FIB. The 3-dimensional, moving nature of MEMS devices precludes using other cross-sectioning techniques. Cleaving through unsupported or released structures will pop them right off the surface of the device, losing what you're interested in. Also, quite a few of the structures are fabricated using polycrystalline silicon, which doesn't cleave neatly like single-crystalline silicon and can shatter. Standard mechanical polishing can't be done on unsupported/released structures, either, without some type of encapsulation. And this encapsulation will generally ruin the device by either floating the structures off the surface or crushing them. FIB tools can cross-section MEMS class devices without damaging the unsupported/released structures. Figure 19 shows a tilted view of one pixel of a DLP™ device, which is a MOEMS device. (Davis, 2002)

These devices, manufactured by Texas Instruments, Inc., consist of roughly 1.2 million of these pixels on the surface of an integrated circuit. As the figure shows, this is a 3-D aluminum structure supported by two narrow metal hinges hanging in empty space. The pixel tilts +/-10° due to electrostatic attraction by the address electrodes underneath the pixel and reflects light either into or away from a projection lens. The only method of cross-sectioning these pixels that doesn't destroy them is by dual beam FIB. Since the metal in these structures is thin and easily milled, a single beam FIB (which uses only the ion beam for imaging and navigation) causes the pixel surface to warp and curl just trying to get to the location of interest. Figure 20 shows a cross-section through a pixel without disrupting the adjoining pixels or the hinges of the cross-sectioned pixel.

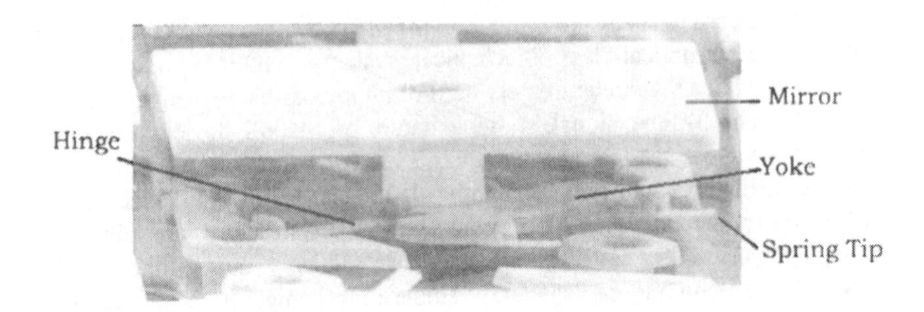

Figure 6.19. Tilted view of DLP™ device pixel. Reprinted with permission, ISTFA 2002 Proceedings, Fig. 2, p. 292, ASM International.

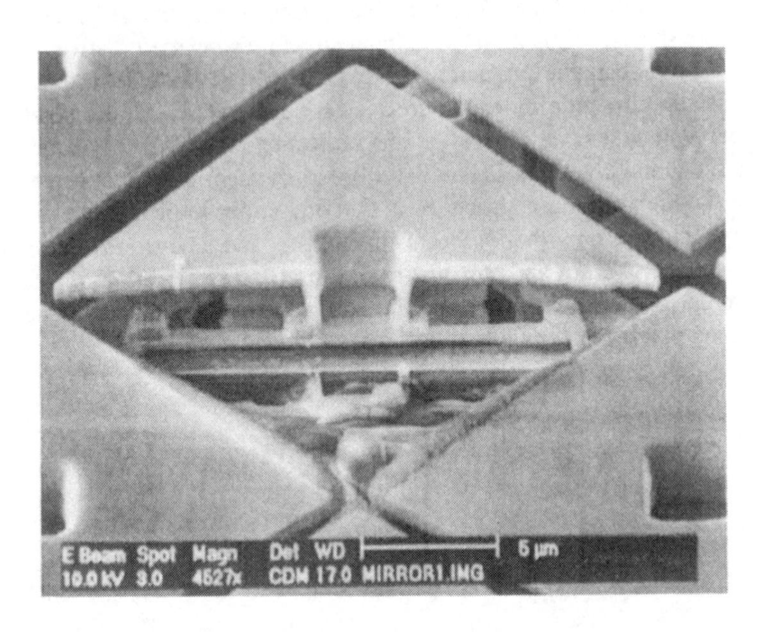

Figure 6.20. Cross-section through a DLP™ pixel

The next figure, Figure 21, shows another type of MEMS device, this one a polysilicon rotary gear from Sandia National Labs. (Walraven, 2002)

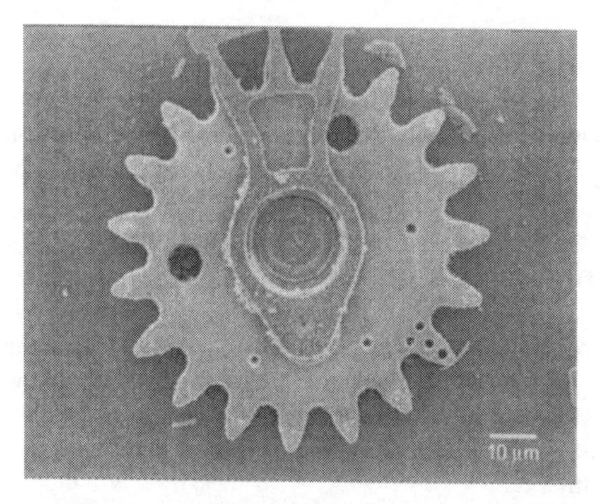

Figure 6.21. Polysilicon rotary gear. Reprinted with permission, Microelectronics Failure Analysis Desk Reference, 2002 supplement, Fig. 2, p. 75, ASM International.

Using the FIB, it is possible to cross-section these types of gears and assess the wear of the gear and detect any unusual damage, as in Figure 21. This type of cross-section would not be possible with any other method.

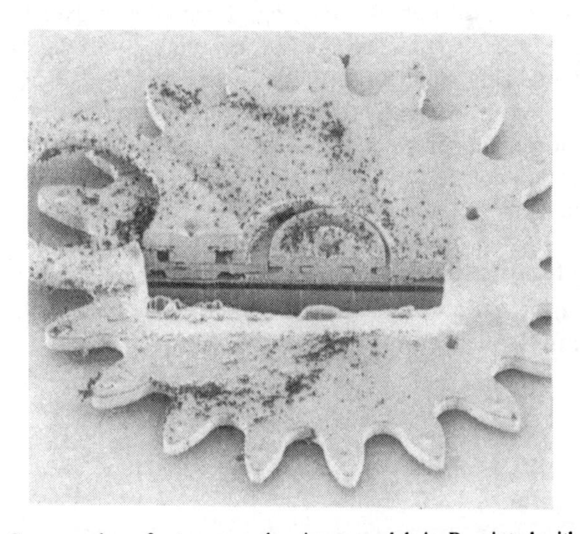

Figure 6.22. Cross-section of rotary gear showing wear debris. Reprinted with permission, Microelectronics Failure Analysis Desk Reference, 2002 supplement, Fig. 17, p. 85, ASM International.

For more information on MEMS/MOEMS devices, see De Wolf (2002) and Waterson (2002).

7. USER CAVEATS

So the dual beam FIB is the greatest thing since sliced silicon, right? Well, almost. There are things to be aware of, since nothing in FA is "one-size-fits-all".

7.1 Cross-section Dimensions

The biggest drawback to cross-sectioning with the FIB is the limited size of the cut. Cutting through the entire thickness of even a thinned silicon die is very time-consuming. The same is true for very large (more than 100μm) X and Y dimensions. If one needs these types of cuts, be prepared to use up a lot of FIB time.

7.2 Charging Damage

A 30kV ion beam induces quite a charge on the sample surface. Ungrounded structures will be damaged. The user must remember to ground the structures adequately if they have no ground path and use a carbon coating, if possible, to suppress charging. Carbon is an excellent choice for a coating because the sputter yield is low, meaning it will stay on the surface longer without being milled off by the beam, and it can be removed by ashing in an oxygen plasma without damaging the sample. Figures 23 and 24 show what can happen when the structure isn't grounded.

This structure consists of two levels of metal connected by a single via. The lower-level metal is at right angles to the top-level metal. This whole structure is isolated from the substrate of the die by several levels of dielectric. Figure 24 shows the damage that will occur if one tries to cross-section through the via without grounding one of the metal levels. This will generate confusing data during the course of FA.

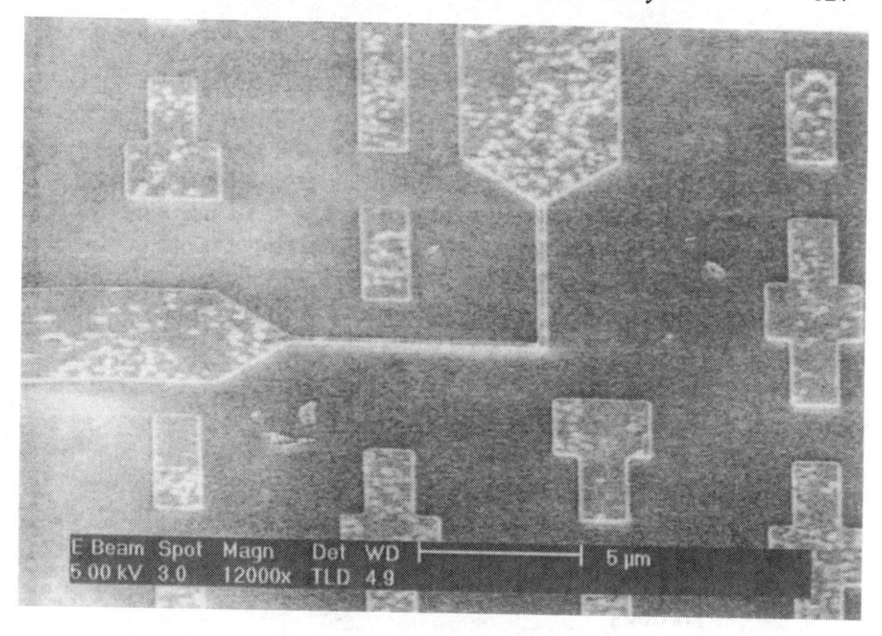

Figure 6.23. Top view of kelvin structure

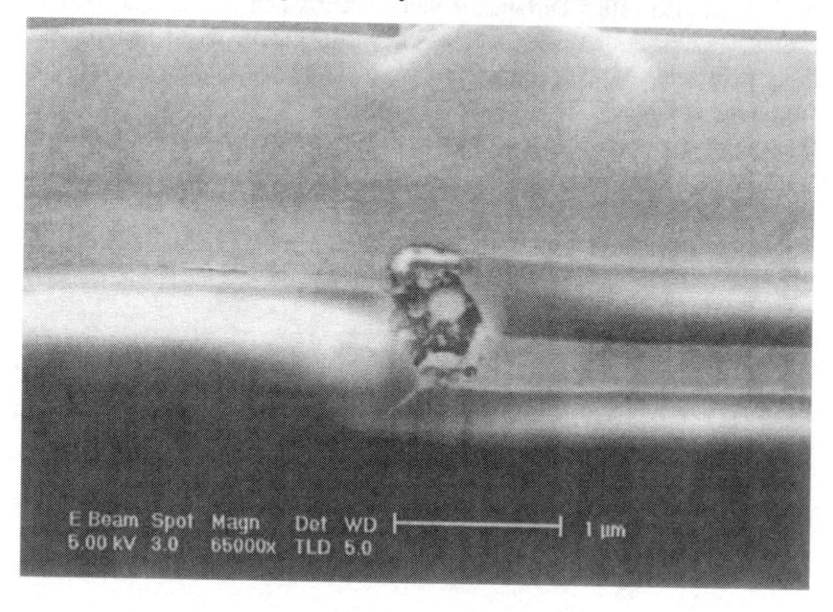

Figure 6.24. Damaged kelvin via

7.3 Ion Implantation

Imaging with the ion beam is sometimes a good idea. One can see the metal grain structure much easier in this mode and charging will be suppressed. It's also much easier to tell metal from oxide in this mode. If the cross-section face is imaged with the ion beam, be aware that gallium ions (or whatever the ion source is) will get implanted in the surface and change the surface characteristics. This will affect any wet etching or decorating of the cross-section face. If any wet chemistry might be done on the cross-section face, don't image it with the ion beam.

7.4 Cross-sectioning artifacts

Nothing in this life is perfect and that includes cross-sectioning with the FIB. FIB cross-sections, like other methods, can generate artifacts. The trick is to be aware that these occur and take steps to minimize them.

7.4.1 Striations

Striations, also called curtains or veils, occur when milling materials of differing hardnesses or thicknesses as shown in Figure 25. Changing the milling parameters and/or depositing a platinum planarizing layer can control these artifacts.

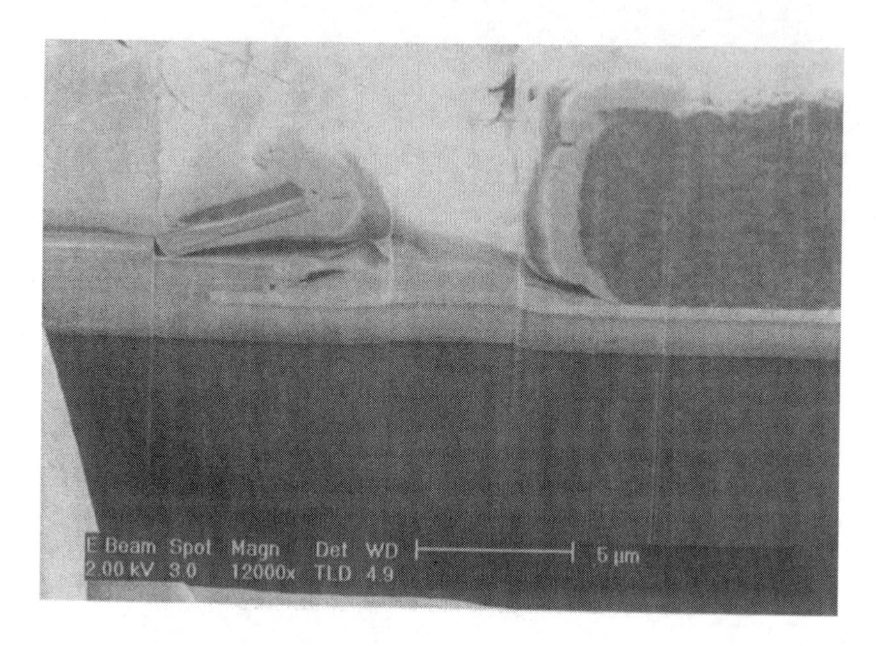

Figure 6.25. Example of striation artifacts. The vertical lines are an artifact of the section.

7.4.2 Redeposition

Redeposition of milled material can generate erroneous data. Figure 26 shows this artifact. The material highlighted by the arrow was not present in the blister prior to milling. This material was deposited into the void during a large spot size milling pattern. This can also occur in high-aspect-ratio holes. This type of artifact can be controlled by reducing the milling rate, making the hole bigger and/or employing some type of gas-assisted etching.

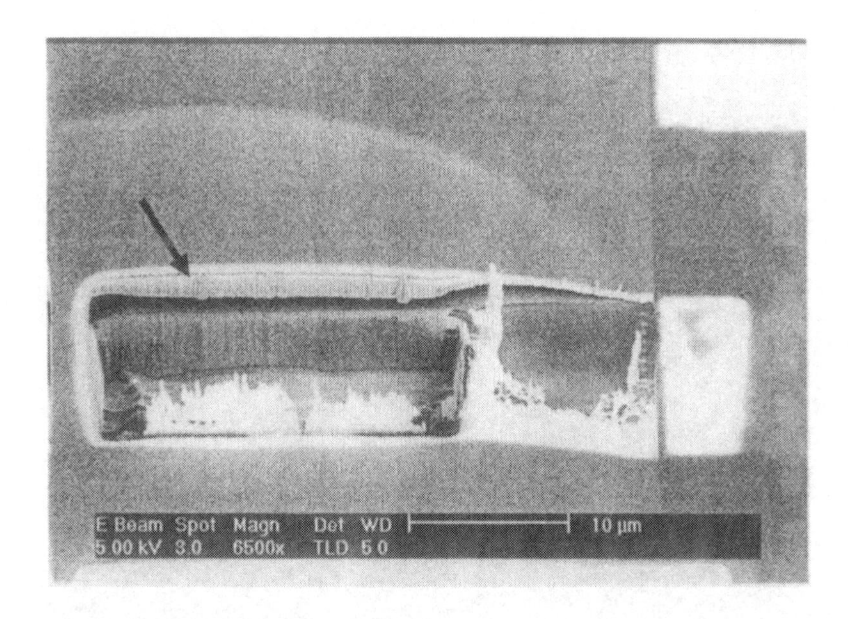

Figure 6.26. Redeposition artifact. Material at arrow was not present before milling.

7.5 A Word (or more) about Copper

Copper is a wonderful metal for IC metallization, but it can behave strangely during ion beam milling. [i]The grains in the metal are oriented at different angles to the milling beam and this leads to differing milling rates for different grains. This is due to a combination of ion channeling and gallium-Cu phase formations that resist the milling process. (Phaneuf, 2002; Yuan, 2002; Herschbein, 1998; Su, 2001) This can be a big headache at device edit, as some grains are milled away before others even start to show wear. This effect makes it difficult to cut precise holes with flat bottoms in copper runs. Some FIB vendors have proprietary methods for overcoming this problem. Copper also doesn't self-passivate as aluminum does and is extremely sensitive to the halogen gas chemistries used in some of the gas-assisted milling techniques. The iodine used in some metal etch chemistries will corrode the copper many millimeters from the mill site if used on a bare copper layer.

8. CONCLUSION

The dual beam FIB is an indispensable tool for state-of-the-art microelectronic and MEMS failure analysis. And it's extremely handy for not-so-state-of-the-art device FA. Its uses are not limited to the techniques discussed here. (Herschbein; Cole, 2002) With practice and experience, the user can produce artifact-free cross-sections of almost any material and come up with techniques this author hasn't even thought of yet.

ACKNOWLEDGEMENTS

I would like to thank my team, Silicon Technology Ramp and Advanced Characterization (Si-TRAC) for their support and encouragement. I would especially like to thank Fred Clark, manager of Kilby PFA lab, for his friendship, support, and the use of his DualBeam 830.

REFERENCES

Campbell A, "Voltage Contrast in the FIB as a Failure Analysis Tool", Microelectronics Failure Analysis Desk Reference, 4th edition, EDFAS/ASM International

Casey JD Jr., Gannon TJ, Krechmer A, Montforte D, Antoniou N, Bassam N, Huynh C, Silva B, Hill R, Gu G, Ray V, Saxonis A, Chandler C, Megorden M, Carleson P, Phaneuf M, Li J, "Advanced sub-0.13 μm Devices – Failure Analysis and Circuit Edit with Improved FIB Chemical Processes and Beam Characteristics", Proceedings of the International Symposium for Testing and Failure Analysis 2002, ASM International (2002)

Cole E, Campbell A, Henderson C, "Focused Ion Beam Technology and Applications to Microelectronics", Seminar Slides, The International Symposium for Testing and Failure Analysis 2002, ASM International (2002).

Davis C, Holdford B, Mahin W, "Failure Analysis of the Digital Micromirror Device", Proceedings, International Symposium for Testing and Failure Analysis (ISTFA) 2002, ASM International (2002).

De Wolf I, Jourdain A, Witvrow A, Fiorini P, Tilman HAC, van Spengen WM, Modlinski R, "Reliability and Failure Analysis of RF MEMS Switches", Proceedings of the International Symposium for Testing and Failure Analysis 2002, ASM International (2002).

Herschbein SB, Fischer LS, Shore AD, "Basic Technology and Practical Applications of Focused Ion Beam for the Laboratory Workplace", Microelectronics Failure Analysis Desk Reference, 4th edition, EDFAS/ASM International

Herschbein SB, Fischer LS, Kane TL, Tenney MP, Shore AD, "The Challenges of FIB Chip Repair and Debug Assistance in the 0.25 μm Copper Interconnect Millennium", Proceedings of the International Symposium for Testing and Failure Analysis 1998, ASM International (1998).

Li SX, "Performing Circuit Modifications and Debugging Using Focused Ion Beam on Multi-layered C4 Flip-Chip Devices", Proceedings of the International Symposium for Testing and Failure Analysis 1998, ASM International (1998).

Phaneuf MW, Li J, Casey JD Jr., "Gallium Phase Formation in Cu and Other FCC Metal During Near-Normal Incidence Ga-FIB Milling and Techniques to Avoid this Phenomenon", Proceedings, Microscopy and Microanalysis 2002, Microscopy Society of America (2002).

Su DH, Lai JB, H. W. Yang HW, "Deprocessing, Cross-sectioning and FIB Circuit Modification of Parts Having Copper Metallization", Microelectronics Desk Reference 2001 Supplement, EDFAS/ASM International (2001).

Walraven J, Waterson B, De Wolf I, "Failure Analysis of Microelectricalmechanical Systems (MEMS)", Microelectronic Failure Analysis Desk Reference 2002 Supplement, ASM International (2002).

Waterson B, "Failure Mechanisms in MEMS Devices", Seminar Slides, The International Symposium for Testing and Failure Analysis 2002, ASM International (2002).

Wills KS, Pabbisetty SV, "Microsurgery Technology for the Semiconductor Industry", Microelectronics Failure Analysis Desk Reference, 4th edition, EDFAS/ASM International

Wilson S, Nair M, Vicker M, Meader RB, Smoot G, Sanders P, Keifer S, Steels J, Sutton M, "Debug and Fault Isolation of an RF/IF Circuit for 3G Cellular Applications With High Leakage, A Case Study", Proceedings of the International Symposium for Testing and Failure Analysis 2002, ASM International (2002).

Yuan C, Mahanpour M, Lin H, Hill G, "Application of Focused Ion Beam in Debug and Characterization of 0.13 μm Copper Interconnect Technology", Proceedings of the International Symposium for Testing and Failure Analysis 2002, ASM International (2002)

Chapter 7

HIGH RESOLUTION LIVE IMAGING OF FIB MILLING PROCESSES FOR OPTIMUM ACCURACY

Peter Gnauck, Peter Hoffrogge, M.Schumann
Carl Zeiss SMT, Inc., Nano Technology Systems Division 56, D-73447 Oberkochen, Germany

Abstract: The combination of field emission scanning electron microscopy (FESEM) and focused ion beam (FIB) is a future key technology for semiconductor and materials science related applications. Through the combination of the Gemini ultrahigh resolution field emission SEM column and the Canion31+ high performance FIB column a wide field of applications can be accessed. This includes structural cross-sections for SEM and transmission electron microscopy (TEM) applications, device modification, failure analysis, sublayer measurement and examination, as well as SEM and FIB related analytical techniques such as energy dispersive x-ray spectroscopy (EDS), wavelength dispersive x-ray spectroscopy (WDS), secondary ion mass spectrometry (SIMS) etc. Real time high resolution SEM imaging of the cutting and deposition process enables the researcher to perform very accurate three dimensional structural examinations and device modifications.

Key words: SEM, FIB, TEM, semiconductor, failure analysis, device modification, analysis, inspection, metrology, deposition

1. INTRODUCTION

High resolution investigation of the microstructure of materials is very often restricted to the study of the very surface of the sample. This is because most high resolution analytical and imaging techniques like scanning electron microscopy (SEM), atomic force microscopy (AFM) or scanning tunnelling microscopy (STM) only provide information about the surface microstructure of the sample. To locally investigate the internal microstructure of the sample at high resolution, the sample has to be opened up. This can be done very precisely by the use of a focused ion beam (FIB) for cutting into the sample and the use of a field emission SEM for high resolution imaging of the internal structure. The combination of field emission scanning electron microscopy (FESEM) and focused ion beam (FIB) is a future key technology for semiconductor and material science related applications. A new CrossBeam tool is discussed in this presentation that allows one to use both beams (SEM and FIB) simultaneously and independently from each other. Due to this feature it is possible to use the FIB for cutting, polishing or patterning while one can observe the process live in high resolution SEM imaging. This feature is extremely useful if the sample contains unexpected structures. Especially in the case of cross sectioning or TEM sample preparation the detail of interest even if it is at a nanometer scale can not be missed or destroyed.

2. SYSTEM OVERVIEW

Through the combination of the LEO Gemini field emission SEM column and the Canion31+ high performance FIB column a new Crossbeam tool was designed. Both beams coincide at a crossover point 5mm below the objective lens of the SEM. The geometrical layout and arrangement of the two columns enables perpendicular tilt of the sample to the ion beam in the coincidence point. To provide full eucentric tilt at all operating conditions a 6-axis super eucentric specimen stage is used.

The system is fitted with a computer controlled gas injection system that can handle up to five different gases for metal and insulator deposition or enhanced and selective etching. To avoid contamination of the field emission gun (FEG) on the SEM and the liquid metal ion source (LMIS) on the FIB by the use of aggressive gases both columns are protected with a differential pumped vacuum system (fig. 1). Especially for the SEM a modified variable pressure column is used, that can be operated at chamber pressures up to 10^{-3}mbars without affecting the UHV in the gun area. (Gnauck, 2000, 2001) To complete the analytical capabilities of the system the specimen chamber can be equipped with several analytical attachments like x-ray detectors for EDS and WDS analysis as well as a mass spectrometer for SIMS analysis.

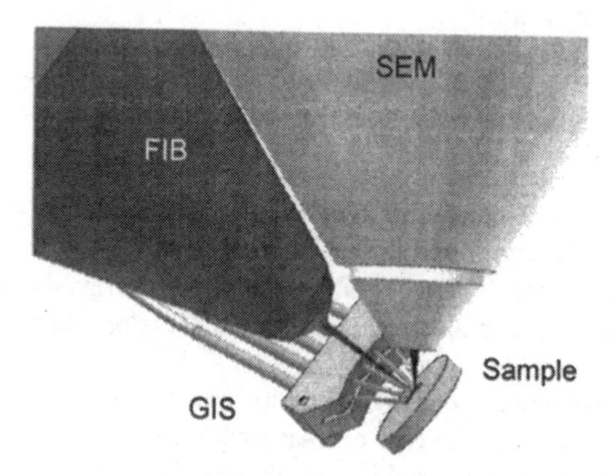

Figure 7.1. : Geometrical layout of the LEO 1500 CrossBeam system. The electron and the ion beam coincide at a crossover point 5mm below the objective lens of the SEM. The geometrical properties of the two columns allow ion milling on large samples (wafers) at angles up to 92° (in respect to the ion beam).

2.1 Electron Optics

To examine nonconductive samples like photo resist, oxide layers, ceramics etc. in their natural state (i.e. without significant sample preparation like coating etc.) in the SEM the technique of low voltage scanning electron microscopy (LVESEM) is used. The operation at low electron energies reduces the charging of the sample and gives very accurate information of the very surface of the specimen. However the resolution of a standard SEM is drastically reduced at low beam energies the special low voltage SEM column used in this system is capable of high resolution imaging even at very low beam energies. (Weimer)

The probe size d_p of an electron optical system is given by:

$$d_p = \sqrt{d_{So}^2 + d_C^{\ 2} + d_{Sp}^2 + d_D^2}\ , \tag{1}$$

with the virtual source size d_{so}, the chromatic aberration disc d_C, the spherical aberration disc d_{Sp} and the diffraction error d_D.

The resolution of such a system at low electron energies is mainly limited by the chromatic aberration.

$$d_C = C_C \cdot \frac{\Delta E}{E} \cdot \alpha, \quad (2)$$

with the chromatic aberration coefficient C_C, the energy spread of the gun ΔE, the beam Energy E and the aperture angle α.

This is caused by the increasing influence of the energy spread ΔE of the electron gun at decreasing electron energies (2). This problem is minimized by the special design of the GEMINI lens. Due to the design as a combination of a magnetic lens together with an electrostatic immersion lens (fig. 3), the aberration coefficients decrease with decreasing electron energy (fig. 2). This results in an ultrahigh resolution even at very low energies down to 100eV.

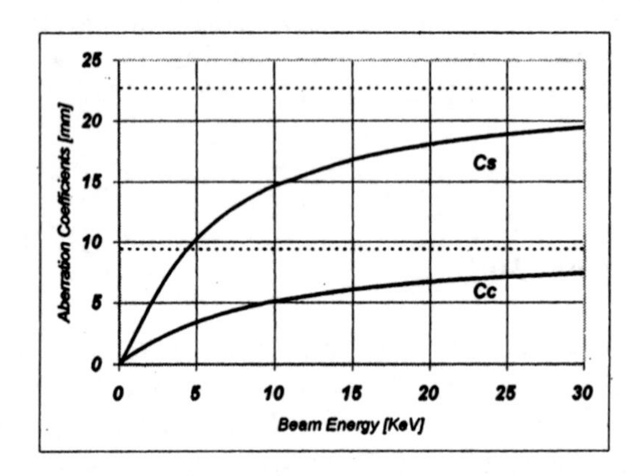

Figure 7.2. Aberration Coefficients of the GEMINI lens: Due to the design as a combination of a magnetic lens together with an electrostatic immersion lens, spherical and chromatic aberration coefficients CS and CC scale with the beam energy. The dotted lines represent the constant aberration coefficients CS and CC of the magnetic lens without the electrostatic immersion lens.

The schematic layout of the electron optics of the Gemini column is displayed in fig.3. The electron beam is generated by a Shottky filed emission gun. This kind of emission system is characterized by a high brightness, a low energy spread and a small virtual source diameter. The probe forming is done mainly by a magnetic electrostatic compound lens.

The condenser lens which is located directly below the gun area is mainly used for adjusting the optimum beam aperture in the low voltage operation. This results in a crossover free beam path with minimum Boersch effect and a minimized energy spread ΔE. The beam current is adjusted by the use of a software controlled electromagnetic aperture changer.

Another benefit of this design is the absence of a magnetic field at the sample. Therefore the electron optical and ion optical properties of the area between sample and final lens of the SEM never change regardless of focus sing or EHT change on the SEM.

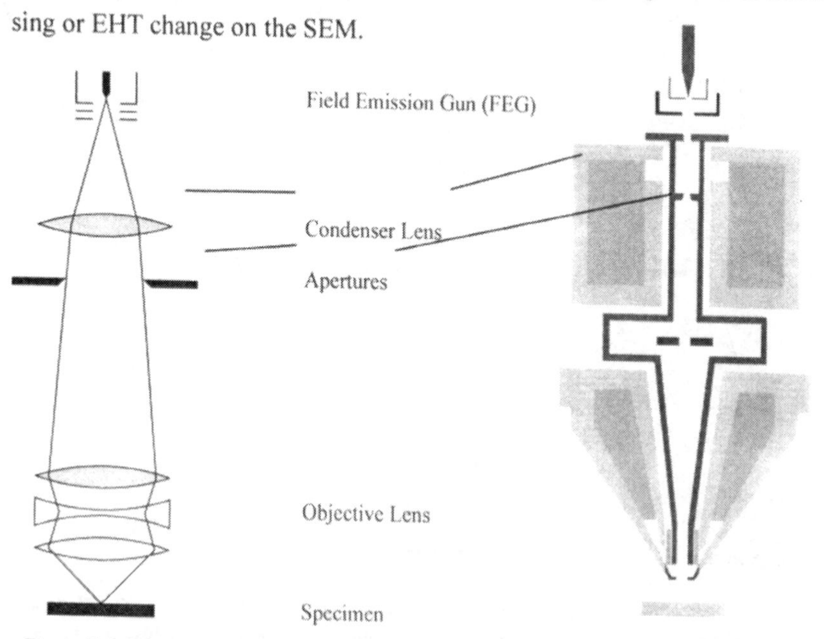

Figure 7.3. Schematic layout of the electron optics of the GEMINI column (right) and the light optical analogy of the column (left). The gray lenses represent the magnetic condenser lens and magnetic part of the objective lens. The white lenses represent the electrostatic immersion lens. Please note: This light optical analogy is only a rough approximation of the real electron optical properties of the system.

2.2 Ion Optics

The schematic layout of the ion optics is displayed in fig. 3. Gallium ions (Ga^+) are extracted from a high brightness liquid metal ion source and then accelerated to an energy of 5 - 30 keV. The ion emission is regulated by an extractor. The electrostatic condenser lens collimates and focuses the ion beam depending on the operating mode: Crossover beam path for highest beam current or non-crossover for best imaging performance. After passing the condenser lens the beam current is defined by a set of software controlled mechanical apertures. By using the different apertures in combination with different condenser settings the probe current can continuously be adjusted in the range between 1pA for and 50nA. The objective lens is designed as an electrostatic einzel-lens system. It focuses the beam onto the specimen surface. At the normal working distance of 12mm a resolution of 5 nm can be achieved.

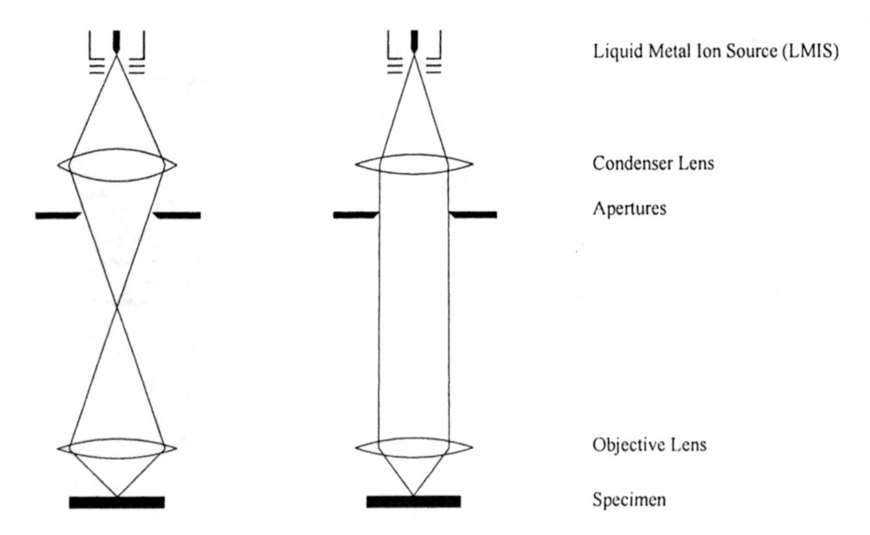

Figure 7.4. Schematic layout of the ion optics of the CANION 31 FIB column. The ion beam is generated by the LMIS and focused onto the specimen by an electrostatic two lens system. Crossover beam path for highest beam current on the left and the normal collimated beam path for optimum imaging performance on the right.

3. IMAGING MODES

The CrossBeam system can operate at three different imaging modes. By blanking the ion beam and only using the electron beam for imaging the system operates as high resolution field emission SEM. The second imaging mode uses the ion beam while the electron beam is blanked. The FIB imaging mode is used for grain analysis, voltage contrast imaging and defining of milling areas. The last imaging mode is the so called CrossBeam operation mode: Both beams are turned on and while the ion beam is milling a defined area, the SEM is used to image the milling process at high resolution in real time. This enables the operator to control the milling process on a nanometre scale and to perform extremely accurate cross sections and device modifications.

3.1 SEM imaging

If the system is used as a high resolution SEM only the ion beam is blanked and the SE - signal is synchronized to the SEM scan. In this operational mode the system can be used as a high resolution FE-SEM with no limitations. Figs. 4 and 5 show the principle of operation high resolution low kV SEM images in this operating mode.

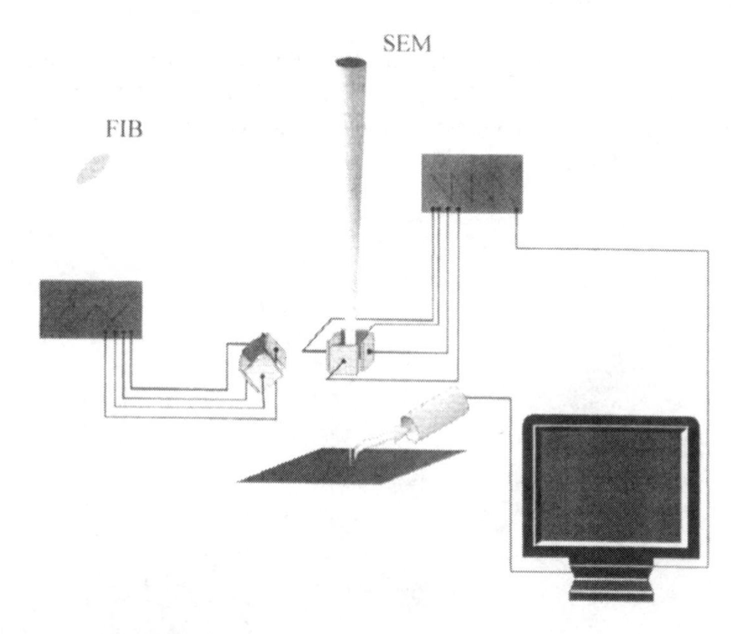

Figure 7.5. SEM imaging: The ion beam is blanked and the SE - signal is synchronized to the SEM scan.

Figure 7.6. High resolution low voltage (800V) SEM image of an uncoated photo resist structure on a silicon wafer (left). And a high resolution SEM image of a gold grain at 510V (right)

3.2 FIB Imaging

If the system is used as a FIB only the SEM beam is blanked and the signal is synchronized to the FIB scan. This mode is used for channelling contrast imaging, voltage contrast imaging and for defining milling patterns on the sample surface.

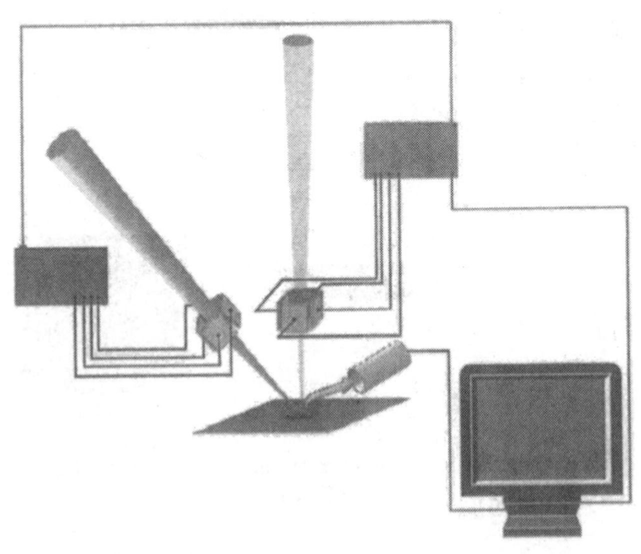

Figure 7.7. FIB imaging: The electron beam is blanked and the SE - signal is synchronized to the FIB scan.

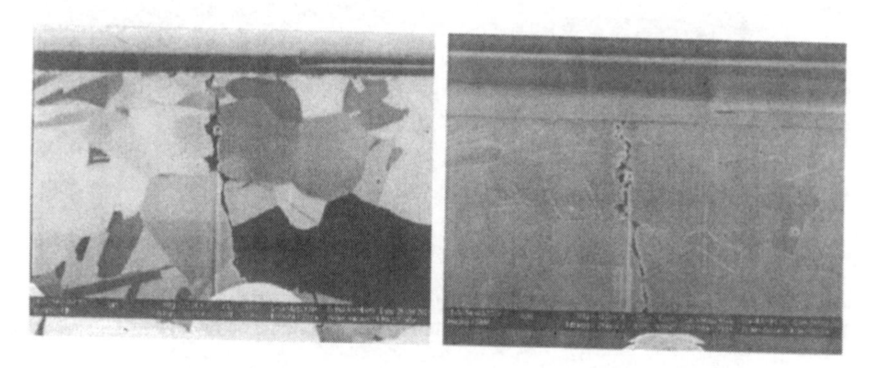

Figure 7.8: FIB imaging: Visualization of the grain structure in a copper sample by channeling contrast (left) same detail with SEM imaging on the right.

3.3 CrossBeam operation

To monitor the ion milling process in real time at high resolution in the SEM the CrossBeam operation is used. Both beams are turned on and while the ion beam is milling a defined area, the SEM is used to image the milling process at high resolution in real time. This enables the operator to control the milling process on a nanometre scale and to perform extremely accurate cross sections and device modifications. In this mode the SEM can also be used to compensate ion beam charging on non conductive samples.

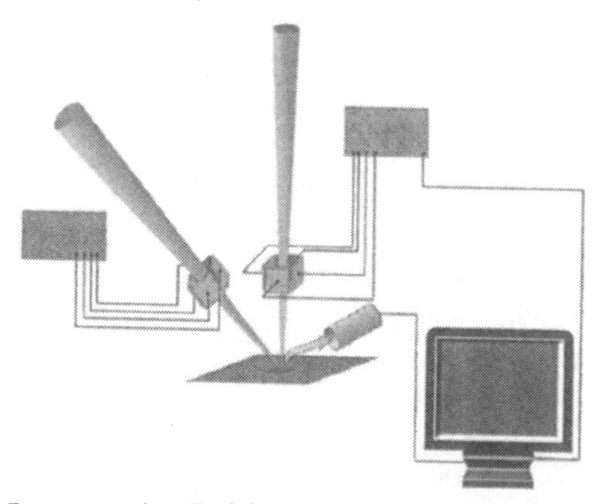

Figure 7.9. CrossBeam operation: Both beams are scanned completely independent form each other and the SED Signal is synchronized to the SEM scan. This results in the CrossBeam operation feature: The ion milling process can be imaged using the SEM in real-time.

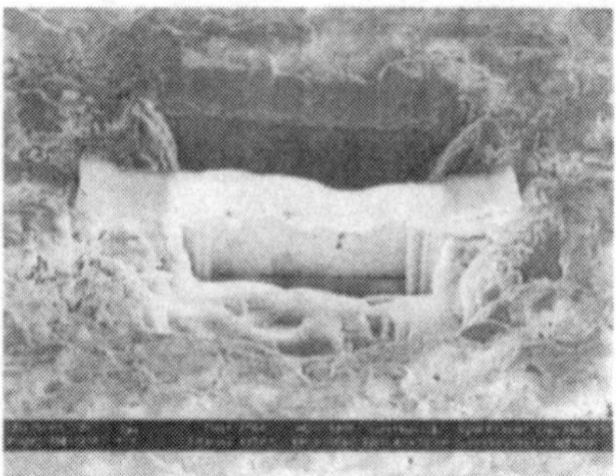

Figure 7.10. Live imaging of a Al_2O_3 2TEM sample during ion milling. Everhard Thornley detector signal on the left and the Inlens detector signal on the right.

REFERENCES

Gnauck P, Greiser J, "A new approach to materials characterization using low pressure and low voltage field emission scanning electron microscopy" DVM-Bericht 519, Mühlheim 2000, pp. 89-96 (2000).

Gnauck P, Drexel V, Greiser J, "A new high resolution field emission SEM with variable pressure capabilities" Microscopy and Microanalysis 2002, Long Beach, California, August 5 - 9 (2001).

Weimer E, Martin JP, "Development of a new ultra high performance scanning electron microscope", 13th Int. Congr. on Electron Microscopy ICEM, Vol.1, pp. 67-68.

Chapter 8

FIB FOR MATERIALS SCIENCE APPLICATIONS - A REVIEW

M. W. Phaneuf
Fibics Incorporated, 556 Booth St., Suite 200, Ottawa, Ontario, CANADA

Abstract: The application of focused ion beam techniques to a range of topics in materials science is reviewed. Recent examples in the literature are cited along with illustrations of numerous applications. Potential artifacts that can arise are discussed along with commentary on minimizing their impact.

Key words: coatings, contrast, corrosion, cracks, deformation, FIB, focused ion beam, fracture, friable, grain growth, materials, sectioning, tomography, wear

1. THE FOCUSED ION BEAM MICROSCOPE AS A MATERIALS SCIENCE TOOL

As noted in the introduction to this volume, the past decade and a half have seen tremendous growth in the number of focused ion beam systems, from about 35 in 1986 (Melngailis, 1987) to perhaps more than a thousand by the time of publication. While the vast majority of these systems are used for semiconductor applications, recent years have seen a marked increase in the use of these versatile instruments in the field of materials science.

The FIB microscope, in its most basic form, consists of a liquid metal ion source producing a beam of Ga^+ ions with a range of energies typically from 25 to 50 keV. These are directed to the sample as a fine focused probe using a process similar to that of the scanning electron microscope. Upon striking the samples, the incident beam ejects electrons, neutral atoms and ions of both charge species (along with molecular fragments). Typically either electrons or positive ions are detected in conjunction with a known beam position to form an image. In parallel with the incident ion beam, many FIB systems permit direction of low energy electrons at the sample for purposes of charge neutralization, and/or various gas chemistries for the purpose of ion beam gas assisted etching (GAE) or deposition. These processes are covered in detail elsewhere in this volume, as are applications of "dual-beam" instruments incorporating both electron and ion columns and capable of imaging with either beam.

One can, by varying lens and aperture settings, achieve beams with currents from approximately 1 pA to 30 nA or more. Beam diameters can range from about 5 nm to several micrometers, with modern instruments capable of delivering 30 nA in a sub-micrometer spot, achieving current densities in excess of 10 A/cm^2. When operating at lower beam currents, the FIB is a fine microscope in its own right, while at higher currents one may consider it a "nano-scale milling machine". The reader only recently acquainted with FIB is cautioned to note the term used is "nano-scale". FIB systems excel in sectioning regions that are a few tens of atoms to a few tens of microns in size. While it is certainly possible to section a region that is several hundred micrometers in length and over a hundred micrometers in depth, it becomes a time consuming process.

From a materials science standpoint, one great strength of FIB microscopy is the ability to prepare site-specific cross-sections in a virtually stress-free manner, essentially independent of the material. One can then observe these in-situ, reducing the beam current to a sufficient level to obtain the required resolution and tilting the cross-sectioned surface towards the incident ion beam (or electron beam in the case of dual-beam FIB systems). Imaging with the gallium ion beam can be a powerful technique in materials science. This chapter will illustrate some potential applications and provide references to many more available in the published literature.

Strengths of dual-beam systems, where the FIB prepared sample is subsequently imaged with the electron beam, are covered elsewhere in this volume, and this imaging mode will be readily understood by those with SEM experience. The physics of imaging with an incident focused ion beam is a less broadly understood topic, and was discussed earlier in this volume, to which the reader is referred as it covers in detail the subject of ion-sample interaction and provides an explanation of the resulting phenomena which lend FIB images a similar but distinctly different flavor from the more familiar secondary and backscattered electron images of the SEM.

From the standpoint of understanding applications of FIB imaging in materials science, we will briefly review the three primary FIB contrast mechanisms (topographic contrast, orientation (channeling) contrast and material contrast (Fig. 1)).

Topographic contrast is familiar to anyone who has seen an SEM image of an irregular object, and even those unfamiliar with images from beam instruments readily interpret topographic contrast images due to their similarity to optical images viewed with directed illumination. Topographic contrast produces image shadows and highlights readily interpreted by the casual observer.

Orientation or channeling contrast is perhaps the most striking feature of FIB microscopy of crystalline specimens (especially cubic metals), and produces an effect of grain contrast similar to that observed in a metallographically etched sample (Fig. 7). The relative intensity of each grain changes as a function of incidence angle of the gallium beam (Phaneuf 1999, among others). Although channeling contrast is exhibited in both the secondary electron and secondary ion images, secondary electron yield is significantly greater than that of secondary ions for most materials, and is less sensitive to changes in chemistry.

Material contrast arises from differences in the yield of secondary particles as a function of specimen chemistry. Although this effect can be significant in FIB secondary electron images, it is most readily observed in FIB secondary ion images where it is frequently the dominant contrast effect. Secondary ion yields can increase by up to three orders of magnitude for metallic species in the presence of oxygen (Benninghoven, 1976), making material contrast a valuable technique for studies involving corrosion or grain boundary oxidation of metals (Fig. 3). FIB secondary ion images tend to be of the "total detected positive ion" type, where the particle detector is biased to a sufficient negative voltage to repel electrons and attract all positive ions regardless of their mass. Mass (and charge) resolved ion imaging is the domain of FIB systems equipped with secondary ion mass spectrometer (SIMS) systems. Interested readers are referred to the Chapter on FIB-SIMS in detail. As FIB-SIMS requires additional equipment uncommon on FIB systems at the present time, this chapter will focus on applications of "total detected positive ion" imaging in materials science.

FIB imaging is an inherently "surface sensitive" technique, with the detected signal originating from only the top few nanometers in ion mode imaging or tens of nanometers in electron mode imaging. This effect can be used to a significant advantage in FIB imaging, but it must be offset against the fact that, unlike in SEM, when imaging with an ion beam, one is always removing material by sputtering as one images. This is an important fact to keep in mind, although the effect can be minimized. For example, at low beam currents it can require several image passes just to remove the native oxide that forms on the surface of a metal exposed to air and henceforth begin to observe the orientation contrast of the grains below. Another significant point is that with each image pass, as well as sputtering away material from the sample, one also implants gallium from the incident beam and ballistically mixes the near-surface monolayers. With care, this is not an issue in most applications, but one must always take it into consideration.

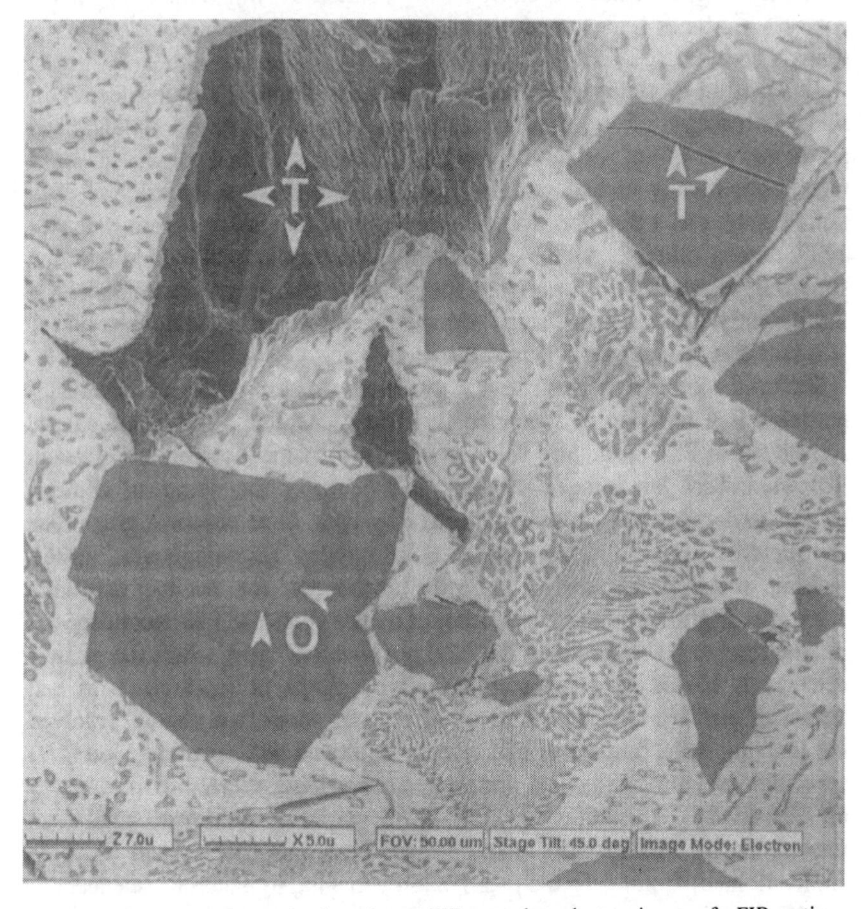

Figure 8.1. FIB contrast mechanisms from a FIB secondary electron image of a FIB section through a TiB_2 reinforced Zn-Al cast alloy after fracture toughness testing. Material contrast from the different phases dominates the bulk of the image, with the Zn η phase appearing light and the Al-rich α phase appearing in darker contrast Topographic contrast ("T") is visible down the void revealed in this section as well as in the crack running through a TiB_2 particle. Orientation contrast ("O") reveals multiple grains make up a larger TiB_2 particle. Orientation contrast is much more prominent in cubic metals than hexagonal, as Fig. 7, among others illustrates. Section 2.4 discusses methods of enhancing contrast in hexagonal metals using FIB gas assisted etching. A full discussion of the physics and mechanisms of contrast formation in FIB images is covered in another chapter in this volume.

1.1 Applications in Materials Science

Initial use of FIB systems in materials science involved scanning ion microscopy (Levi-Setti et al. 1983, Franklin et al., 1988) and high resolution chemical mapping using SIMS (Levi-Setti et al., 1985, 1986). At about the same time, the semiconductor industry was discovering the potential of the FIB (Cleaver and Ahmed, 1981; Cleaver et al., 1983; Kirk et al., 1987). As is documented in other chapters in this volume, rapid application development in the semiconductor industry along a number of fronts (lithography, photomask repair, circuit modification, diagnostics, failure analysis and specimen preparation) led to rapid evolution and commercialization of the FIB, which subsequently became a mainstay of the semiconductor industry.

FIB was slower in making its way into the field of materials science, hampered not by limitations in the technique but rather by the relative scarcity of these instruments outside semiconductor fabrication facilities. It should be noted that a great deal of characterization work that has been performed in the semiconductor industry should rightly fall into the category of "materials science", but for the purposes of this paper only non-semiconductor applications will be discussed.

The use of FIB in materials science was initially weighted heavily towards either SIMS or specimen preparation for analysis using other instruments (primarily TEM), but modern FIB instruments with high spatial resolution and the increasing availability of dual-beam systems have made FIB imaging of materials specimens a rapidly growing field. The ability of FIB to prepare TEM specimens (particularly site specific cross-sections with ~100 nm positional accuracy) from material systems that were previously extremely difficult to prepare for TEM has sparked the largest volume of publications to date.

FIB micromachining is covered elsewhere and has more recently been used to produce tools for materials science applications such as nano-indentor tips and micro-end milling tools (Adams et al., 2001), as well as for mechanical property testing (Davis, 1997).

FIB applications in materials science are still very much in their infancy. The following sections discuss FIB applications across a range of fields, with the goal of inspiring the reader as to the potential of the technique. Hopefully the future of FIB in materials science will see the introduction of useful instrument options that are already available to the SEM community such as hot and cold stages, orientation imaging microscopy in dual beam systems, variable pressure "environmental" systems capable of operating at nearer atmospheric pressure with different vacuum environments, etc.

1.1.1 Biology

Perhaps largely because FIB instruments tend to be located in semiconductor, electronics and material science groups, biological applications have been slow to appear. While many applications in biology will no doubt have to wait for the introduction of a "variable pressure, environmental FIB", the more mature science of preparation of biological samples for hard-vacuum SEM applications should make adoption of FIB in these fields relatively straightforward once the instruments become available to those in the biological sciences.

To date, the literature largely reflects a few attempts by FIB manufacturers and materials science groups to undertake brief forays into the field, often assisted by biologists curious about these "new" instruments. It is fair to say that these attempts have been successful as a proof of concept, but further investigation, particularly into beam damage and other potential artifacts of specimen preparation and imaging is required by those knowledgeable in the field.

Ishitani et al. (1995) illustrated the potential of the technique with some quick FIB cross-sections and images of biological specimens including the compound eye of a fly and portions of human hair. Hayashi et al. (1998) have investigated FIB cross-sections of human tooth enamel, as have Gianuzzi et al. (1999). Hoshi et al. (2001) followed this work with an investigation of human dentine by FIB sectioning for imaging and FIB XTEM (including samples sufficiently thin to permit energy filtered TEM). Hoshi states that dentine contains more organic substances than enamel, yet this did not stop successful specimen preparation. Young (2000) briefly shows results of FIB cross-sectioning samples of the inner ear for subsequent SEM imaging. Garnacho et al. (2000), using a FIB-SIMS system, demonstrated the detection of copper accumulation from environmental contamination at levels below SEM EDX/WDX detection limits in shrimp cuticle armor.

Our own limited efforts with organic specimens (mostly critical point dried plant samples and inked paper) have indicated simple sectioning to be relatively straightforward, requiring only relatively low doses to achieve large cross-sections, however the surface of the section did appear somewhat "melted" in each case, lacking fine, sharp detail when imaged with either a low current ion beam or in the SEM. Some of the results published in the literature appear to exhibit similar effects, which can no doubt be eliminated or minimized as effort is expended to optimize the FIB milling parameters for these types of specimens.

1.1.2 Polymers

As is the case with biology, polymer science applications of FIB are still poorly represented in the literature at the time of writing. Given the often difficult nature of specimen preparation in these soft or composite systems, optimization of FIB parameters to allow artifact-free sectioning should yield the same growth of applications as has happened in metallurgical systems. Initial evidence of this can be found in White et al. (2001), who prepared FIB XTEM specimens of two-component polymer coatings on silicon substrates. They state these samples were highly successful for interface studies, and a viable alternative to ultramicrotomy without mechanical damage frequently induced by this more common technique. In a similar result, Miyagawa et al. (2001) prepared FIB XTEM of clay reinforced polymer nanocomposites. Phaneuf (1999) illustrated FIB sectioning and imaging of polymer coatings on aluminum beverage cans and cross-sections of 200 ASA photographic film.

The semiconductor industry has produced a number of papers regarding FIB gas assisted etching (typically using xenon difluoride or water as the etch gas) to uniformly remove several micrometer thick polymer (typically polyimide) coatings prior to device modification. Although there has not yet, to the authors knowledge, been application of GAE to polymer applications published in the literature, preliminary investigations in our laboratory indicate GAE may allow differentiation (selected etching) of polymer phases or degrees of crystallinity which enhance subsequent imaging (by FIB or SEM) of FIB prepared polymer cross-sections.

As with biological specimens, the degree of beam damage during sectioning and other artifacts require further study to enhance the use of FIB in this field. Despite this, the relative ease of specimen preparation and potential for the technique are illustrated in Fig. 2 (Right). Here a portion of 200 ASA film has been sectioned and imaged in the FIB. While the section does appear to show some evidence of beam damage (and subsequent images continued to show further blurring of fine detail), total specimen preparation time was under 20 minutes. The retention of the relative morphology of the film was excellent, with no gross smearing or tearing, as is frequently observed in mechanically prepared specimens of this type.

1.1.3 Metals, Minerals and Ceramics

The vast majority of published applications of FIB in materials science are in the fields of metals, minerals and ceramics. The nature of many of these problems is frequently similar, and in many applications (metal matrix composites (MMCs) etc.) these materials are found in the same specimen. As a result, discussion of these materials will be dealt with together in this section in alphabetical order by sample application, rather than by material.

1.1.3.1 Coatings

Many thin coating specimens (less than approximately 50 μm in thickness) present ideal opportunities to showcase the potential of FIB, as they frequently are very difficult to prepare by other techniques while at the same time being small enough in scale that a cross-section of the entire coating can be prepared in a matter of a few hours and imaged directly or prepared as a TEM specimen. One early example of the success of FIB is zinc coated steel, either galvannealed (Saka et al., 1995; Phaneuf et al., 1997) or galvanized (Giannuzzi et al., 1998). A detailed study is documented in Malis et al. (2002)). Plasma sprayed metallic coatings have also been examined (Yaguchi, Kamino et al., 2000; Cairney et al., 2001). FIB specimen preparation of coatings for TEM analysis also comprises significant portions of other work relating to indentation, corrosion, wear, etc. (The reader is referred to references discussed in the subsections that follow). Fig. 2 illustrates the sectioning of a tin coating on steel and a photographic film composed of multiple layers polymers containing crystals of silver compounds, all on a polymer substrate.

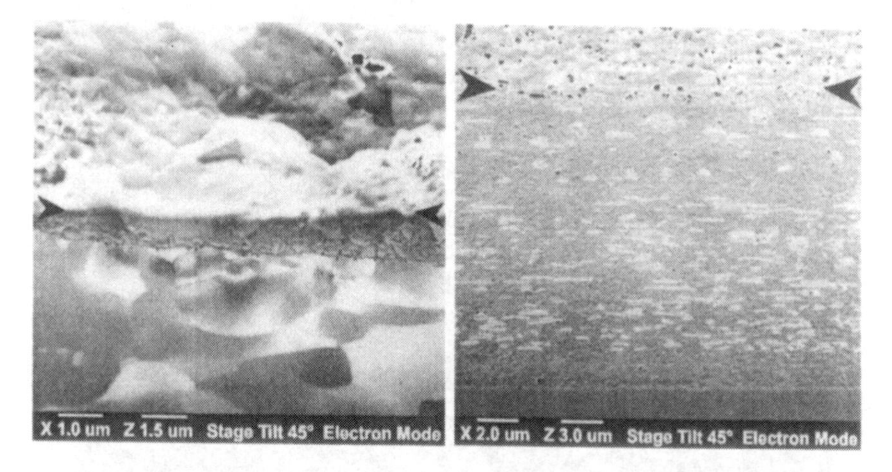

Figure 8.2. Left: FIB secondary electron image of a FIB section through food-grade tin plate coating on steel. Multiple image passes have ion-etched grain boundaries to reveal fine-grained intermetallics at the tin-steel interface. *Right*: FIB secondary electron image of a FIB section through 200 ASA photographic film. The polymer "carrier" is visible at the bottom., as are platelets of light sensitive chemicals located throughout distinct layers. Some evidence of beam damage (blurring and rounding of features) is visible in this image. In both images black arrows mark the position of the transition from surface to cross-section, a nearly 90° angle.

Protecting the top surface prior to FIB sectioning frequently improves the quality of the section, however if one is able to section without a protective coating, subsequent imaging at a tilt of 45° reveals both the top surface and the cross-sectioned face, Fig. 2. This can prove advantageous in interpreting the microstructure, but does often cause confusion to those unfamiliar with FIB. Note that at 45° both the top surface and the cross-sectioned face experience equal foreshortening due to tilt.

1.1.3.2 Corrosion and Oxidation

Despite the tremendous usefulness of oxygen enhanced yield effects in FIB secondary ion images to delineate the presence of corrosion (Phaneuf, 1999; Phaneuf and Botton, 2002), little has yet to appear in the literature exploiting this technique. More et al. (2000) have demonstrated FIB XTEM of thin alumina scales formed on an oxidized Ni-base alloy, and Wang et al. (1999) have examined stress-corrosion cracking profiles by FIB sectioning and imaging (see the following section). Botton and Phaneuf (1999) have demonstrated that FIB-prepared sections are well suited to EELS near-edge structure analysis of chemical states, a potentially powerful tool for analysis of corrosion. Combining FIB imaging to localize regions for site-specific FIB-TEM preparation with advanced TEM techniques holds promise for many grain boundary corrosion and oxidation studies.

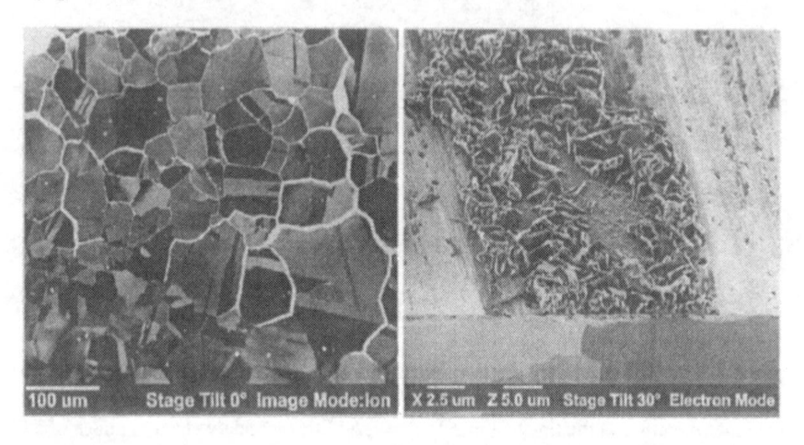

Figure 8.3. Left: FIB secondary ion image of a metallographically polished nickel alloy that has undergone intergranular corrosion. The presence of oxygen greatly enhances the secondary ion yield from the metal, resulting in corroded grain boundaries lighting up brightly. *Right*: FIB sectioning and secondary electron imaging of high-temperature corrosion resistant alloy after testing at 600°C in a sulphur-rich environment. Surface scale, visible on the original surface in this image, has formed where grain boundaries (revealed by FIB sectioning) intersect the surface. A faint "theater curtain effect" is visible on the section face.

1.1.3.3 Cracks and Fracture Surfaces

FIB sectioning avoids introduction of mechanical damage during sectioning, which can greatly assist the interpretation of crack and fracture specimens. The relatively small region disrupted during FIB sectioning allows access to multiple targets on the same sample in a manner not possible with conventional sectioning, and permits multiple sites to be examined over time during the course of a corrosion test. Redeposition (section **2.3**) must be taken into consideration when observing cracks. Site-specific specimen preparation for either TEM or FIB/SEM imaging permits exact analysis of crack profiles and structure. Saka and Abe (1997-1,2) have demonstrated FIB XTEM of crack tips in Si. Wang et al. (1999) have examined stress corrosion crack (SCC) profiles by FIB sectioning and imaging in SEM and FIB. Phaneuf (1999) discussed examination of SCC and cracking in oxide films on zirconium alloys. Cairney et al. (2000),

Giannuzzi et al. (2001) and Xu et al. (2002) have examined fracture surfaces in Fe-Al MMCs, aluminum and steel, respectively.

1.1.3.4 Deformation and Indentation

Site-specific and "stress-free" sectioning to study the profile and associated plastic deformation of micro- or nanoindentation has been the subject of several papers. Trtik et al. (2000) describe FIB sectioning and imaging of microindentation of cementious composites, noting the need for a variable pressure / environmental FIB to study specimen in a more ideal (hydrated) environment. Inkson et al. (2000) have performed FIB serial sectioning through a nanoindent in Cu-Al multilayers on an alumina substrate. Dravid (2001) discusses FIB sectioning nd FIB XTEM of nanoindents into W/TiN multilayers and directionally solidified eutectic oxides.

Muroga and Saka (1995) examined rolling contact fatigue microstructure by FIB XTEM. Evans et al. (1999) discuss FIB imaging of deformation and damage in tensile tested Al-based MMCs, also showing results of FIB TEM preparation. Here the authors discuss "mottled" contrast relating to degree of deformation in grains exhibiting an otherwise uniform orientation contrast. Burke et al. (2000) explored this effect in greater depth, imaging surface deformation in Ni-based superalloy tubing directly with FIB secondary electrons and verifying the presence of increased dislocation density with TEM at locations where the "mottled" contrast was observed using FIB. Further work is required to determine if this effect can be used in any more than a qualitative manner, but even without a means of directly quantifying the degree of deformation, the technique holds promise in terms of direct observation of plastic deformation in cubic metals. Fig. 4 illustrates this in a Ni-based superalloy. Here the outer diameter of a tube specimen is cross-sectioned by FIB and an intense mottled contrast dominates grains in the FIB secondary electron image. Subsequently, the outer 20 μm are removed by mechanical polishing and lapping with a fine diamond solution. FIB sectioning and imaging now shows a uniform grain orientation contrast devoid of the "mottling" observed in the outer grains .

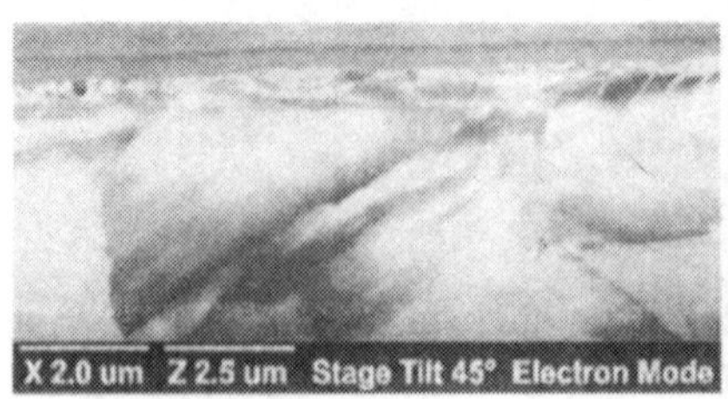

Figure 8.4. FIB section of a Ni-based alloy tube at the outer diameter of the tube. Note the streaks and mottled contrast due to deformation. *Right:* Same sample after the top ~ 20 μm are removed by mechanical lapping. Note the uniform orientation contrast of the "undeformed" material (after Burke et al., 2000). NB: Both surfaces were protected by FIB deposition of tungsten, visible in the images, prior to FIB sectioning. Note the thin band of mottled contrast immediately below the uniformly dark FIB tungsten at the top of the undeformed specimen (right), presumably from the mechanical polishing.

1.1.3.5 Fragile and / or Friable Materials

The "stress-free" nature of FIB sectioning makes possible the cross-sectional imaging of specimens whose mechanical strength or stability is not sufficient to permit conventional mechanical polishing. While embedding "weak" specimens in curable polymeric compounds does permit mechanical section of some types of specimen, many other specimens are difficult to infiltrate sufficiently to guarantee no deformation occurs during polishing. FIB sectioning is able to retain voids and fine structure and remove many questions regarding preparation-induced artifacts. When exploring the sectioning of an unfamiliar material, access to images from sections prepared by more conventional means are valuable in deciphering any potential FIB introduced artifacts.

Fig. 5 shows the FIB sectioning and imaging of a nickel hydroxide based battery electrode consisting of a nickel foam strut and spheroidal nickel hydroxide particles. Sectioning in this manner minimizes artifacts due to mechanical grinding and polishing and can be invaluable in the analysis of particle - strut delamination, internal cracking, or the "pulverization" of the hydroxide spheroids, which may result in battery failure after numerous charge / discharge cycles.

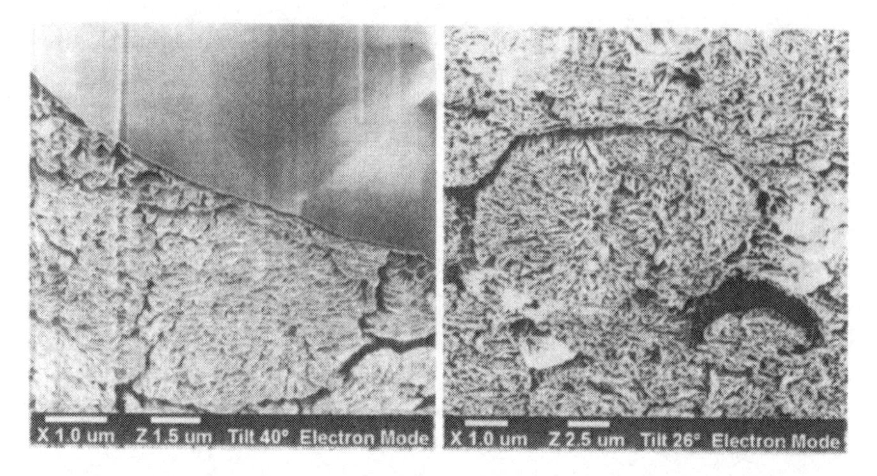

Figure 8.5. FIB secondary electron images of FIB sections through dried NiOH-Ni batteries after multiple charge/discharge cycles. *Left*: Section through the CVD Ni "strut" (see Fig. 7) and surrounding NiOH. *Right*: Further detail of NiOH particles. Note the extremely fine lamellae comprising each approximately spherical particle. Retaining this level of fine detail is extremely difficult using conventional polishing techniques.

1.1.3.6 Powders and Particulates

FIB specimen preparation allows direct selection, sectioning and imaging of powder specimens. No embedding media is required. Specimens can also be prepared for TEM analysis, either with or without embedding. Further discussion can be found in Fujikawa et al. (1995) who studied Mg-Zn mechanically alloyed powder particles, Kitano et al. (1995) who performed FIB sectioning and imaging as well as FIB XTEM of nickel powder, Prenitzer et al. (1998) (FIB TEM of Zn powder particles) and Lomness et al. (2001) (FIB TEM of Ti-Mg-Ni particles).

Fig. 6 illustrates direct FIB sectioning and imaging of nickel powders. Two powders with quite different external morphologies are sectioned, revealing that the "smooth" powder particles tend to consist of agglomerated, roughly equiaxed grains while the "spiky" powder consist of much finer angular grains or collections of heavily twinned grains. In each case specimen preparation was as straightforward as dispersing a small amount of powder onto some double sided conductive carbon tape and blowing off any excess, followed by a few minutes of sectioning in the FIB after suitable powder particles had been selected.

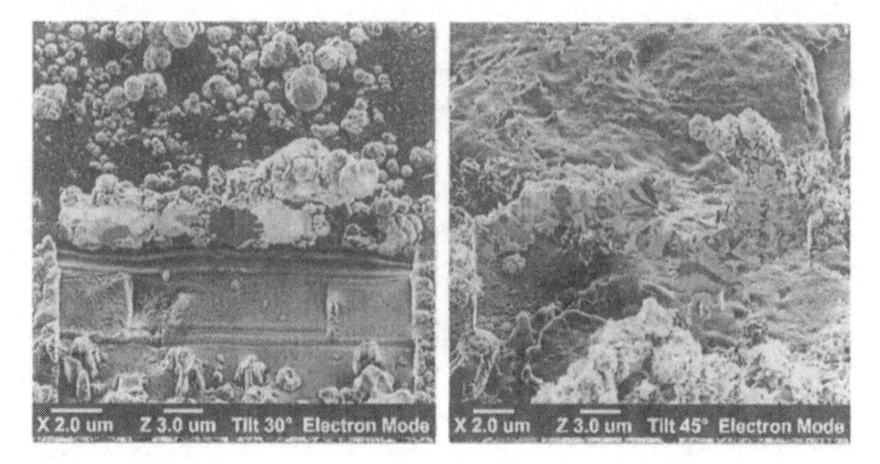

Figure 8.6. FIB cross-sectioning and imaging of Ni powders. In both cases the powders were sprinkled on conductive carbon tape and any excess was blown off. No coating, embedding, etc. of the powder is required. FIB sectioning is quick and FIB imaging in electron mode reveals the internal grain structure as it relates to the outer morphology.

1.1.3.7 Sectioning, Imaging and TEM Specimen Preparation of Diverse and / or Troublesome Materials

Mechanical polishing artifacts such as smearing or chemical reactions, broad ion beam artifacts such as differential thinning rates, rare or "ultaprecious" specimens, site specific locations: the majority of publications of material science applications of FIB originate with researchers who discover previously intractable problems of specimen preparation are solvable using FIB techniques. This field is the most rapidly growing segment of FIB publications. Several selected articles are referenced below; the reader is encouraged to perform their own up-to-date search as well as reviewing the following articles. Further details on the various methods for preparation of TEM specimens using FIB can be found in other chapters in this volume.

Gianuzzi et al. (1998) discuss applications of the FIB Lift-Out technique for TEM specimen preparation, illustrating the versatility of the technique with applications including galvanized steel and SiC fibres. Longo et al. (1999-1, 1999-2), discuss improving EDX quantification in Ni-Si system with novel approaches to FIB prepared TEM samples. Botton & Phaneuf (1999) illustrate FIB preparation for TEM is suitable for high spatial resolution chemical mapping and spectroscopy by EELS. Kim & Dravid (2000), discuss FIB preparation of continuous fiber reinforced ceramic composites for TEM analysis. Yaguchi et al. (2001) use FIB preparation and

TEM viewing in an iterative fashion using a specimen holder compatible with both instruments to produce site specific cross sections through the center of a precipitate in steel. Heaney et al. (2001) prepare FIB-TEM specimens of geological and planetary materials. Enquivist et al. (2001) document FIB sectioning, imaging and FIB XTEM of surface defects in tungsten carbide reinforced metal matrix composites (MMCs) after wear testing. Cairney et al. (2000) produce FIB-TEM specimens of aluminum based MMCs and tungsten carbide reinforced MMCs. Smith et al. (2001) have used FIB to produce porous copper catalysts for TEM analyses.

1.1.3.8 Tomography / Serial Sectioning / Three Dimensional Reconstruction

The concept of generating a three dimensional reconstruction from a stack of images acquired with a sputtering ion beam has been a standard technique in SIMS analysis for a number of years. Similarly, serial sectioning using mechanical or chemical means is also a well-known technique. FIB, however, brings a degree of lateral resolution to this that is frequently twenty times better than can be achieved using conventional approaches. Fig. 7 illustrates the "FIB serial sectioning" (slicing) approach: here a chemical vapor deposited (CVD) nickel foam has been marked off in 500 nm intervals ("slices") using the FIB, then in sequence each slice is imaged then precisely milled away down to the next marker. Tomography software can then be employed to generate a three-dimensional reconstruction of the relationship between each of the grains images in consecutive slices.

Utilizing the "slicing" approach, Sakamoto et al. (1998) report serial sectioning of particles in a dedicated dual beam instrument equipped with FIB SIMS. Tomiyasu et al. (1998), report similar serial sectioning results using FIB-SIMS. Phaneuf and Li (2000) show serial sectioning for three dimensional reconstruction of grain relationships in CVD nickel struts (Fig. 7). Inkson et al. (2000) describe FIB serial sectioning through a nanoindent in Cu-Al multilayers on an alumina substrate. Inkson et al. (2001) discuss three dimesional grain shape determination in Fe-Al nanocomposites using FIB serial sectioning.

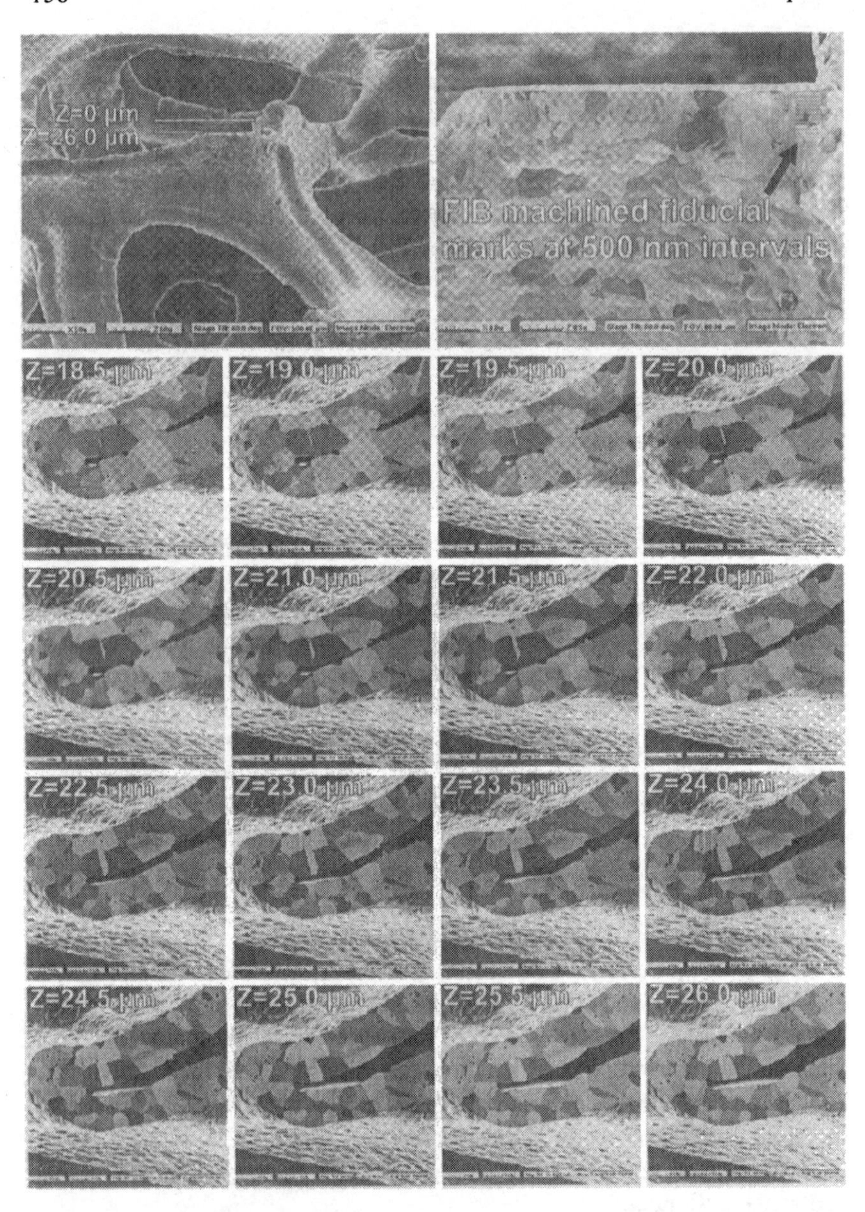

Figure 8.7. FIB serial sectioning for "FIB Tomography". *Top Left*: A portion of Ni foam "strut" with 26 μm of material iteratively sliced and imaged, as seen from the "slicing" orientation. *Top Right*: FIB machined fiducial marks at regular (500 nm) intervals, used to ensure each slice is a uniform thickness. *Bottom*: Sixteen images of consecutive slices. Tomographic reconstruction software can be applied to these images to deduce the three-dimensional shape of each grain in the strut.

Applying the "sputtering" approach, Hull et al. (2001), document spatially and chemically resolved depth profiling using both FIB SIMS and FIB secondary electron imaging for three-dimensional reconstruction. Interested readers are directed to research the SIMS literature under the topic of "three dimensional SIMS" for similar examples.

1.1.3.9 Wear

As with investigation of cracks and friable materials, site-specific, virtually stress free FIB sectioning is ideal to determine degree of damage induced during wear testing without introducing preparation related artifacts. Given the requirement to introduce extremely hard materials into many wear-resistant applications, the ability to uniformly FIB-section these composites is an additional advantage. Muroga and Saka (1995) discuss rolling contact fatigued microstructure examined by FIB TEM preparation. Chaiwan et al. (2000) report observation of cracks and sub-surface damage using site specific FIB sectioning and imaging of dry sliding wear on alumina. Enquivist et al. (2001) examine wear phenomena (cracking, subsurface damage, oxidation, etc.) of tungsten carbide based MMC face seal rings using FIB sectioning, imaging and TEM specimen preparation.

Fig. 8 illustrates this technique for a magnesium metal matrix composite reinforced with silicon carbide that had undergone wear testing. Surface material (likely oxide judging by its secondary ion yield (not shown)) has become entrained below the original surface and exhibits a number of cracks that could allow large regions to delaminate. Cracks in the SiC particles are also visible in several locations 20 μm or more below the surface.

2. POTENTIAL ARTIFACTS / PITFALLS TO BE AWARE OF AND POSSIBLE SOLUTIONS

As the previous section attempted to demonstrate, FIB has tremendous potential as a specimen preparation and imaging tool in materials science applications. However, as with all techniques one must become aware of artifacts that can arise, so as to initially recognize them for what they are, then work to eliminate or minimize them.

Many FIB artifacts arise from poorly tuned instruments. To achieve optimal performance it is necessary to begin with a well aligned beam, with minimal beam "tails", producing a tight round spot with no discernible halo. The various FIB manufacturers all suggest slightly different approaches to achieve this, but whatever the method suggested by the manufacturer of your instrument, it is a very important step to follow for each beam current (spot size) you require to complete your work.

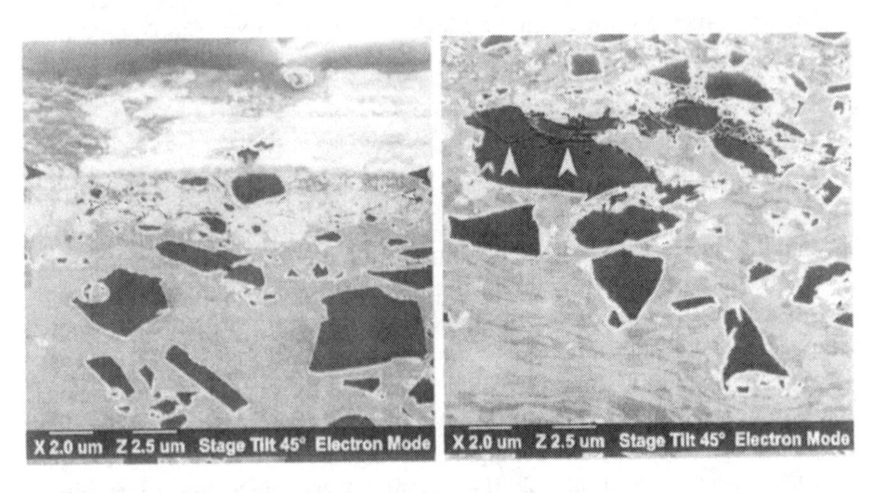

Figure 8.8. FIB sectioning and imaging of a silicon carbide reinforced magnesium alloy after undergoing wear testing. *Left*: Black arrows indicate the original surface while white arrows indicate a subsurface crack in material entrained below the sample surface during testing. *Right*: In an adjacent region, approximately 20 μm below the surface cracks have been formed in the SiC particulate (white arrows).

A second requirement for minimizing artifacts is selection of appropriate beam currents and milling parameters. Due to the variety (and age) of instruments presently available, let alone the range of specimens, no "universal" formula for optimizing these parameters can be stated, however experience, experimentation and innovation are necessary to achieve the best results.

While collectively we have yet to delve the depths of FIB artifacts in materials science, a number are common enough to be immediately recognized by the experienced user. These are listed below in most likely order of appearance. It should be noted that many problems arise as a result of a poorly aligned beam. Still others due to direct or near normal incidence implantation of gallium. While the latter cannot be helped when imaging, one should be aware of the potential for artifacts arising from FIB imaging and avoid imaging if it is not necessary. A poorly aligned beam should be avoided at all times.

The final two artifacts, FIB induced grain growth and anomalous gallium phase formation are relatively new observations that require a great deal of further study. They are included in this section not to induce a wave of panic regarding the suitability of FIB for specimen preparation (as this article illustrates a host of testimonials already exist in the literature as to the usefulness of the technique), but rather to reemphasize the point that FIB applications are still in their relative infancy and diligence must be exercised in the interpretation of any results, preferably drawing upon experience using other proven techniques as an aid to interpreting the FIB results.

2.1 Differential Sputter Rates and the "Theater Curtain" Effect

While FIB is often heralded as being able to mill cross-sections of disparate materials with ease, this is not strictly true. A truer comment would add "given sufficient time (beam dose) and assuming a well aligned beam". Early FIB publications concerning sectioning contain numerous examples of or references to the so-called "theater curtain" effect, where rather than obtaining the desired sharp, flat cross-section, the surface of the cross-section appears to be mapped onto the topography of an undulating curtain, as can be seen in the right hand image of Fig. 3. If the beam is well aligned, this type of effect is usually the result of insufficient milling dose. Applying a protective FIB-deposited metal layer will help alleviate this problem as it helps shield the sectioned face from beam tails, as will further milling (achieving a sharp, flat cross-section with FIB is really just an exercise in applying sufficient dose to "push" any undulations down below the field of view).

While a uniform degree of "curtaining" across the section surface is usually indicative of insufficient dose, the appearance of the theater curtain in one region of a nominally homogenous (by material) specimen is typically the result of orientation induced sputter rate variations due to ion channeling. Ion channeling was discussed elsewhere, and the specific case for copper is discussed by Kempshall et al. (2001) and Philips et al. (2000). Fig. 9 illustrates this for the case of a nickel-based polycrystalline specimen which has been protected with a layer of FIB-deposited tungsten. Image "A" shows a cross-section that has already received five doses of 8 nC/μm^2, for a total of 40 nC/μm^2 applied to the front face. Given the high degree of "curtaining" exhibited by the right hand grain of image "A", it is likely that a further dose of at least 80 nC/μm^2 will be required to achieve a smooth face using the standard technique. As is shown in image "B", only 8 nC/μm^2 further dose is required to polish this face flat if the incident angle is changed in such a manner that the offending grain is no longer oriented in a strongly channeling direction. This can be achieved while maintaining the original sample-beam orientation, wherein the cross-section face is perpendicular to the incident beam. Assuming the original section is aligned along the tilt axis of the FIB stage, by rotating the specimen 90° and tilting approximately 10° (see inset, Fig. 9) the cross-section face remains perpendicular to the ion beam (allowing further glancing angle polishing), yet this most likely moves the offending grain's relative orientation to the ion beam away from offending the channeling direction. Ten degrees is an arbitrary number that we have found works well, although anything larger than five will likely suffice.

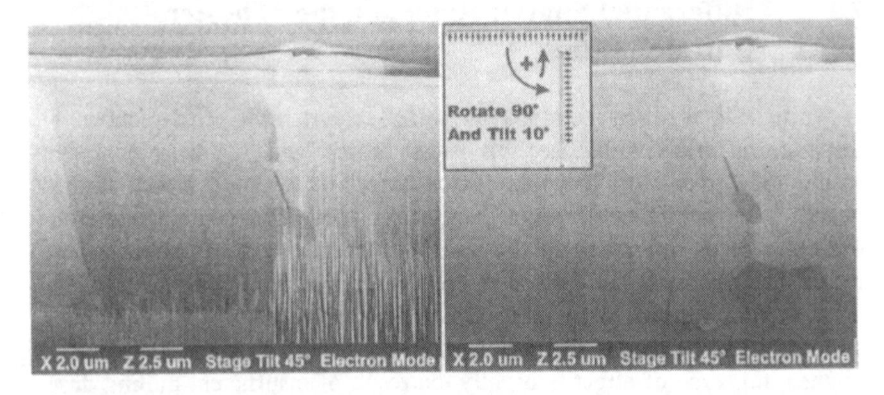

Figure 8.9. Left: FIB sectioning of a Ni-based alloy tilted 45° to view the section surface, after a dose of 40 nC/μm^2 has been delivered. Note the severe theater curtain effect visible in the rightmost grain. *Right:* The same sample after rotating 90° and tilting 10° (inset) so as to orient the offending grain so it is not aligned in a strongly channeling direction. After this combined rotation and tilt to a new geometry (inset), the original face is further polished as before (array of small arrows). This further milling of the original section with a dose of only 8 nC/μm^2 leaves the sectioned face smooth and free of artifacts. This approach delivers a smooth section free of the theater curtain effect at significantly lower dose than continuing to mill the section in the conventional manner.

2.2 Beam Induced Damage

The subject of FIB induced damage in TEM specimen preparation is well discussed in the literature, and is covered in another chapter. In the case of FIB sectioning of bulk samples for subsequent SEM (etc.) imaging, the same basic caveats apply. When possible, deposit a FIB protective coating, perform final thinning with as small a spot as is feasible to ensure minimal beam tails and do not observe the cross-sectioned face with the FIB beam once the section is within 200 nm of completion (larger distances in the case of beam sensitive or voided materials). When imaging occurs with the FIB beam, use as small a beam current and as short a dwell time as possible to still achieve desired results. Tune your beam elsewhere than on the sample to be imaged. After saving the "perfect" image, continue imaging for a number of cycles, noting any change in the image that may be indicative of ongoing damage. In general, recall that when sectioning, most of the energy of the FIB beam is directed down, across the face of interest in a glancing-angle manner. When imaging, that same energy is directed *into* the face being observed, potentially causing much greater issues.

2.3 Redeposition

Once again, this effect is well discussed in the FIB-TEM specimen preparation literature, however redeposition can still be an issue in bulk sectioning situations. Redeposited material will end up on the sectioned face when final thinning is performed with a beam with large tails, or when performed "timidly" when the users does not quite position the mill box against the sectioned face. Specimen drift can also result in final thinning taking place just before the sectioned face. Both drift and "timid" mill box positioning result in material being sputtered from the bottom and sides of the trench onto the sectioned face (redeposition). Redeposited material is always high in gallium concentration, which can be a convenient indicator of redeposition if the specimen is examined by on a system with chemical analysis capability. Typically, when viewed with a surface-sensitive technique such as FIB imaging, redeposited material shows little structure and tends to a uniform gray contrast. Unless the redeposition is severe, a few image passes will typically sputter it away, revealing the sample underneath. If possible, it is best to re-polish the face with the ion beam at glancing angle rather than imaging the surface revealed after "imaging away" the redeposition. A second form of redeposition occurs when voids are partially or totally filled with redeposited material. Once again, high gallium concentration can be a tell-tale marker. While void filling is a more difficult problem to solve, it can be greatly reduced by using as fine a FIB spot for milling as possible and reducing the milling dose to the minimum required while ensuring the milling box position is against (and slightly into) the face of the final section.

2.4 Lack / Loss of Orientation Contrast

Many cubic metals exhibit strong orientation contrast that persists through multiple imaging passes. Hexagonal metals such as zirconium, magnesium and zinc, however, frequently have little orientation contrast when imaged, or contrast that rapidly fades. This lack of orientation contrast is common when imaging hexagonal metals (note the relatively poor orientation contrast exhibited by hexagonal Zn in Fig. 1 and Mg in Fig. 8 compared to the cubic metals illustrated in several other figures). The exact mechanism as to why this happens has not been fully investigated, although the obvious culprit of beam induced disruption of the surface suppressing the channeling effect is frequently put forward. It has been our experience that orientation contrast can be returned and enhanced by using a continuous low pressure bleed of xenon difluoride gas during imaging, which presumably etches away any accumulating surface damage. Fig. 10 illustrates this effect; both images are acquired under the same conditions of beam current, detector gain, etc. after FIB sectioning of a Zr alloy. The right hand image was acquired after commencing a XeF_2 gas bleed that raised the chamber pressure from ~1x10^{-7} T to ~1x10^{-6} T. Similar results have been observed using Cl_2 gas on Zn alloys (Phaneuf and Li, 2000).

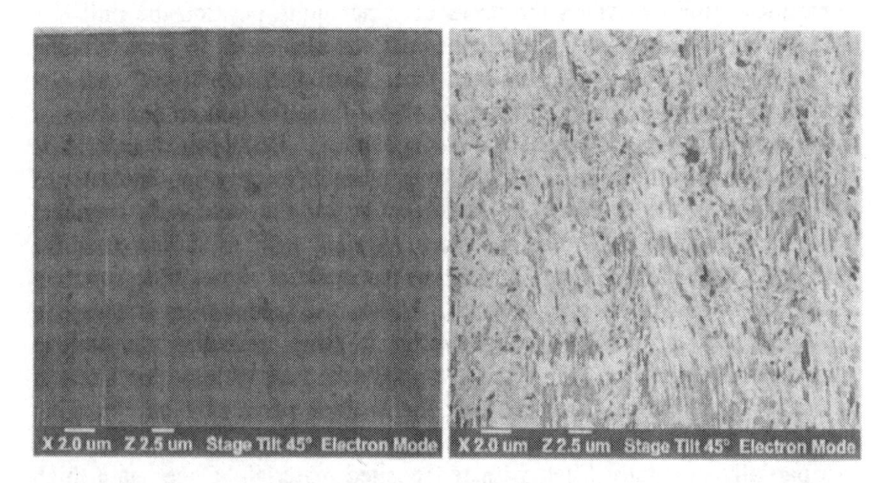

Figure 8.9. FIB secondary electron images of a FIB-prepared cross-section of a Zr alloy. *Left*: Imaged immediately after sectioning. *Right*: Imaged under otherwise identical conditions after commencing a low pressure bleed of XeF_2 gas shows significant improvement.

2.5 Beam Induced Grain Growth

FIB induced grain growth in fine-grained magnetic thin film specimens has recently been reported by Park and Bain (2002). We have also noticed this effect when sectioning nanocrystalline metals in our laboratory. In our experience, significant grain growth in these extremely fine grained systems can be observed after prolonged exposure to near-normal incidence imaging.

Fig. 11 illustrates this effect. Here a nanocrystalline nickel specimen was repeatedly imaged prior to sectioning in the FIB, and larger grains are visible at the surface in the cross-sectional image. A similar region, protected with a low dose FIB deposition of tungsten and never imaged before sectioning does not exhibit larger grains at the surface. Note that the growth of these larger grains are more likely due to the dose delivered by beam tails during sectioning and not from the original imaging, as the dose delivered during imaging was only a few times larger than that estimated to be delivered during the early stages of tungsten deposition, where no growth was observed.

To examine a "worst case" scenario, a shallow crater was milled into the specimen at normal incidence. Imaging this crater revealed the entire crater floor to be composed of large, twinned grains at a depth where only nanocrystalline grains were observed in a nearby cross-section. Park and Bain propose a momentum transfer argument rather than beam heating is responsible for this effect. We believe further study of this effect is required, but there is no doubt grain growth is observed, particularly in systems containing fine grain size with subsequently large driving forces for the reduction of grain boundary energy. The relationship between the appearance of "dark grains" due to grain growth (Fig. 11, top left) must also be rationalized versus potential anomalous phase formation.

2.6 Anomalous Gallium Phase Formation

Driven by the adoption of copper-based metallizations in the semiconductor inductry, a great deal of effort has been spent researching FIB sputtering of copper (Kempshall et al., 2001; Philips et al., 2000). These researchers examined channeling effects and differential sputter rates in copper, noting significant variation in sputter rate as a function of orientation. Kempshall et al. proposed using ion channeling contrast as a guide and tilt until the contrast is effectively equal across all grains. Phaneuf et al. (2000, 2002 -1,2) also studied this effect using single crystal copper of known orientations and discovered the formation of a copper-gallium phase when sputtering the relatively open lattice of (110) orientation grains. This effect can be observed as the formation and growth of dark grains during FIB imaging or sputtering, and appears to occur in a number of FCC metals (Phaneuf, 2002-1,2). Further investigation of this phenomena is required, however evidence points to this effect only occurring with near-normal incidence FIB milling, and not glancing-angle approaches such as those that occur during sectioning for FIB TEM preparation.

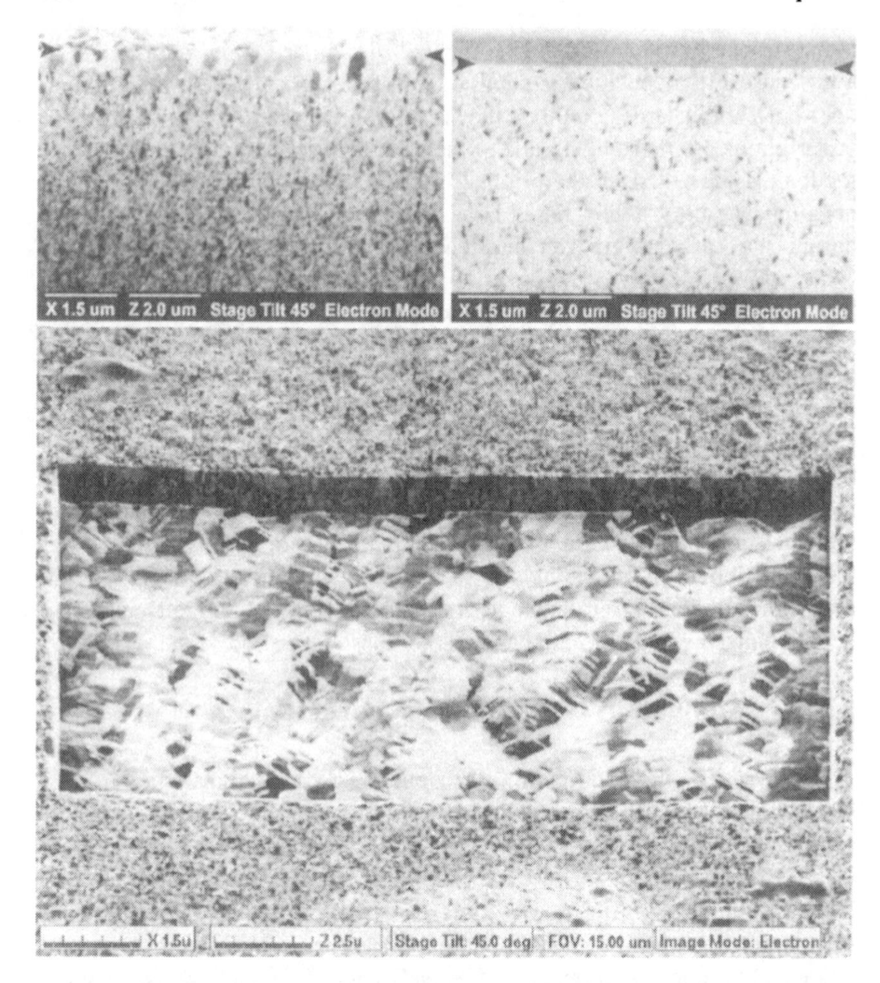

Figure 8.10. Top Left: FIB sectioning and imaging of an unprotected surface reveals larger grains at the surface, which had previously been imaged at normal incidence. *Top Right*: After low-dose deposition of FIB tungsten prior to sectioning, no large surface grains are observed. NB: The image in the top left was acquired after a second image pass, versus a single pass for the image in the top right, resulting in observation of a higher number of "dark" grains. Arrows denote the original surface. *Bottom*: FIB image after sputtering a crater into the same material at normal incidence. Note the large twinned grains on the crater floor.

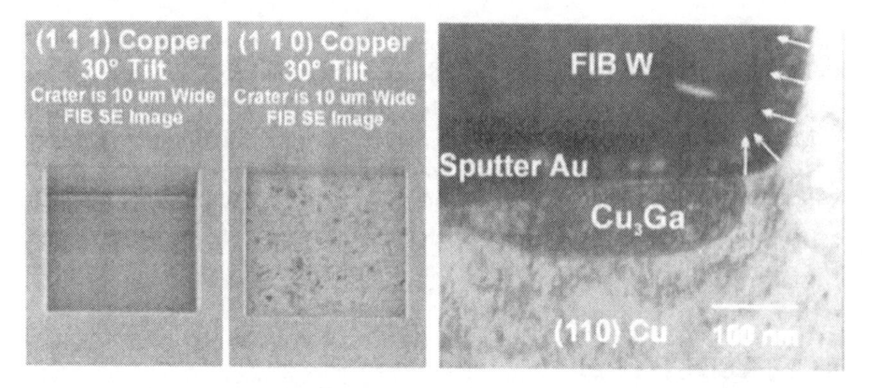

Figure 8.11. Left: Identical doses into single crystal (111) and (110) Cu result in trenches near 4 times deeper in (111) than (110). Note the dark speckled grains in the floor of the (110) crater. *Right*: FIB-prepared TEM cross-section of FIB-sputtered (110) single crystal after external deposition of a protective Au layer and FIB deposition of W. The dark speckled grains prove to be Cu_3Ga when analyzed by electron diffraction. Note that the Cu_3Ga ceases to form when the crater surface normal is no longer parallel to the incident beam (double headed arrows).

3. CONCLUSION

The application of FIB to materials science is a rapidly expanding field, full of opportunity for interesting research and development across a broad range of material systems. As with all specimen preparation techniques, FIB can introduce artifacts that the user must learn to recognize and account for. The relative immaturity of the FIB technique, particularly as applied to materials science, requires extra care and diligence, as well as a degree of experience and attention to detail, in order to derive the correct information.

ACKNOWLEDGEMENTS

The author gratefully acknowledges the efforts of the following individuals to contribute to research at Fibics Incorporated involving the use of FIB in materials science:

Pamela Dakers. Jian Li, Louise Weaver, Gregory McMahon, David Mayer and Robert Patterson (Fibics Incorporated), Graham Carpenter, Tom Malis, Gianluigi Botton, Marc Charest, Jason Lo, Zbig Wronski, Sylvie Dionne, Jim Gianetto, Elhachmi Es-Sadiqi, Mahi Sahoo, Glynis de

Silveira, Alf Crawley, Jennifer Jackman and Roy Sage (Materials Technology Laboratory, Natural Resources Canada), Conrad Zagwyn, Neil Bassom, David Casey, Ganesh Sundaram, Nicholas Antoniou, Bob McMenamin, Nick Economou and Steven Berger (FEI Company), Mary Grace Burke, Paul Duda, Barry Hyatt and Doug Symons (Bechtel Bettis Inc.), Neil Rowlands (Oxford Instruments), On-Ting Wu and Al Lockley (Atomic Energy Canada Chalk River Laboraotories), Shigeo Saimoto and Douglas Boyd (Queen's University at Kingston).

REFERENCES

Adams DP, Vasile MJ, Benavides G, and Campbell AN, " Micromilling of metal alloys with focused ion beam-fabricated tools". J. International Societies for Precision Engineering and Nanotechnology 25, 107-113 (2001).

Benninghoven A, Crit. Rev. Solid State Sci., 6, 291 (1976).

Botton GA, and Phaneuf MW, "Imaging, spectroscopy and spectroscopic imaging with an energy filtered field emission TEM", Micron 30, 109-119 (1999).

Burke MG, Duda PT, Botton G, and Phaneuf MW, "Assesment of deformation using the focused ion beam technique," Microsc. Microanal. 6 (Suppl 2:Proc.), 530-531 (2000).

Cairney JM, Munroe PR, and Schneibel JH, "Examination of fracture surfaces using focused ion beam milling." Scripta Materialia 42, 473-478 (2000).

Cairney JM, Munroe PR, and Sordelet DJ, "Microstructural analysis of a FeAl/quasicrystal-based composite prepared using a focused ion beam miller." J. Microscopy 201, 201-211. (2001).

Cairney JM, Smith RD, and Munroe PR, Transmission electron microscope specimen preparation of metal matrix composites using the focused ion beam miller. Microsc. Microanal. 6 (Suppl 2:Proc.), 452-462 (2000).

Chaiwan, S., Stiefel, U., Hoffman, M., and Munroe, P., "Investigation of sub-surface damage during dry wear of alumina using focused ion beam milling." J. Australasian Ceramic Society 36, 77-81 (2000).

Cleaver JRA and Ahmed H, "A 100kV ion probe microfabrication system with a tetrode gun." J. Vac. Sci. Technol., 19(4), 1145-1148 (1981).

Davis RB, "Preparàtion of samples for mechanical property testing using the FIB workstation". Microstr. Sci., 25, 511-515 (1997).

Dravid VP, "Focused ion beam (FIB): More than just a fancy ion bean thinner." Microsc. Microanal. 7 (Suppl 2:Proceedings), 926-927 (2001).

Evans RD, Phaneuf MW, Boyd JD, "Imaging damage evolution in a small particle metal matrix composite," J. Microsc. 196 (2), 146-154 (1999).

Engqvist H, Botton GA, Ederyd S, Phaneuf M, Fondelius J, Axen N, Wear phenomenat on WC-based face seal rings, Intl. J. Refract. Metals & Hard Matls 18, 39-46 (2000).

Franklin RE, Kirk ECG, Cleaver JRA, Ahmed H, "Channelling ion image contrast and sputtering in gold specimens observed in a high-resolution scanning ion microscope", J. Mat. Sci. Letters, 7, 39-41 (1988).

Garnacho E, Walker JF, Robinson K, and Pugh P, "Cuticular copper accumulation in Praunus flexuosus: location via a gallium SIMS on a FIB platform". Microscopy and Analysis 76, 76, 9-15 (2000).

Giannuzzi LA, Drown JL, Brown SR, Irwin RB, and Stevie FA, "Focused ion beam milling and micromanipulation lift-out for site specific cross-section TEM specimen preparation." Mat. Res. Soc. Symp. Proc. "Workshop on Specimen Preparation for TEM and Materials IV". R. Anderson (Ed.) (1997).

Giannuzzi LA, Drown JL, Brown SR, Irwin RB, and Stevie F, "Applications of the FIB Lift-Out Technique for TEM Specimen Preparation". Micr. Res.Tech. 41 285-290 (1998).

Gianuzzi LA, White HJ, Chen WC, "Application of the FIB lift-out technique for the TEM of cold worked fracture surfaces", Microsc. Microanal. 7 (Suppl 2:Proc.), 942-943 (2001).

Gianuzzi LA, Prenizer BI, Drown-MacDonald JL, Shofner TL, Brown SR, Irwin RB, and Stevie FA, "Electron microscopy sample preparation for the biological and physical sciences using focused ion beams", J. Proc. Analyt. Chem IV, No. 3,4, 162-167 (1999).

Heaney PJ, Vicenzi EP, Giannuzzi LA, and Livi KJT, "Focused ion beam milling: A method of site specific sample extraction for microanalysis of Earth and planetary materials". American Mineralogist 86, 1094-1099 (2001).

Hoshi K, Ejiri S, Probst W, Seybold V, Kamino T, Yaguchi T, Yamahira N, and Ozawa H, "Observation of human dentine by focused ion beam and energy filtering transmission electron microscopy", J. Microscopy 201, 44-49 (2001).

Hull R, Dunn D, and Kubis "A Nanoscale tomographic imaging using focused ion beam sputtering secondary electron imaging and secondary ion mass spectrometry" Microsc. Microanal. 7 (Suppl 2:Proceedings), 934-935 (2001).

Inkson BJ, Mulvihill M, and Mobus G. "3D determination of grain shape in a FeAL-based nanocomposite by 3D FIB tomography". Scripta Materialia 45, 753-758 (2001).

Inkson B.J., Steer, T., Mobus, G., and Wagner, T., 2001. Subsurface nanoindentation deformation of Cu-Al multilayers mapped in 3D by focused ion beam microscopy. J. Microscopy 201, 256-269.

Ishitani, T, Hirose H, and Tsuboi H, "Focused-ion-beam digging of biological specimens." J. Electron Microsc., 44, 110-114 (1995).

Ishitani T, Taniguchi Y, Isakozawa S, and Koike H, "Proposals for exact–point transmission-electron microscopy using focused ion beam specimen-preparation technique". J. Vac. Sci. Technol. B 16 (4), 2532-2537 (1998).

Kempshall BW, Schwarz SM, Prenitzer BI, Giannuzzi LA, Irwin RB, and Stevie FA, "Ion channeling effects on the focused ion beam milling of Cu". J. Vac. Sci. Technol. B 19 (3), 749-754 (2001).

Kim ST, and Dravid VP, "Focused ion beam sample preparation of continuous fibre-reinforced ceramic composite specimens for transmission electron microscopy". J. Microscopy 199, 124-133 (2000).

Kirk ECG, Cleaver JRA and Ahmed J, "In situ microsectioning and imaging of semiconductor devices using a scanning ion microscope". Inst. Phys. Conf. 87 (11), 691-693 (1987).

Kitano Y, Fujikawa Y, Kamino T, Yaguchi T and Saka H, "TEM observation of micrometer-sized Ni powder particles thinned by FIB cutting technique". J. Electron Microsc., 44, 410-413 (1995).

Kitano Y, Fujikawa Y, Takeshita H, Kamino T, Yaguchi T, Matsumoto H, and Koike H. "TEM Observation of Mechanically Alloyed Powder Particles (MAPP) of Mg-Zn Alloy Thinned by the FIB Cutting Technique", J. Electron Microsc. 44, 376-383 (1995).

Kitano Y, Fujikawa Y, Kamino T, Yaguchi T, and Saka H, "TEM Observation of Micrometer-Sized Ni PowderParticles Thinned by FIB Cutting Technique". J. Electron Microsc. 44, 410-413 (1995).

Kitano Y, Fujikawa Y, Takeshita H, Kamino T, Yaguchi T, Matsumoto H, and Koike H, "TEM observation of micrometer-sized Ni powder particles thinned by FIB cutting technique". J. Electron Microsc., 44, 376-383 (1995).

Kuroda K, Takahashi M, Kato T, Saka H, and Tsuji S, "Application of focused ion beam milling to cross-sectional TEM specimen preparation of industrial materials including heterointerfaces", Thin Solid Films, 319 92-96 (1998).

Levi-Setti R, Fox T, Lam K, "Ion channeling effects in scanning ion microscopy with a 60 keV Ga^+ probe", Nucl. Instrum. Methods., 205, 299-309 (1983).

Levi-Setti R, Crow G, and Wang YL, "Progress in high resolution scanning ion microscopy and secondary ion mass spectrometry imaging microanalysis". Sc. El. Micr. 535-551 (1985).

Levi-Setti R, Crow G, Wang YL, "Scanning ion microscopy: Elemental maps at high lateral resolution," Applied Surface Science. 26 249-264 (1986).

Lomness JK, Giannuzzi LA, and Hampton MD, "Site-specific transmission electron microscope characterization of micrometer-sized particles using the focused ion beam lift-out technique", Microsc. Microanal. 7, 418-423 (2001).

Longo DM, Howe JM, and Johnson WC, "Development of a focused ion beam (FIB) technique to minimize X-ray fluorescence during energy dispersive X-ray spectroscopy (EDS) of FIB specimens in the transmission electron microscope (TEM)", Ultramicroscopy 80, 69-84 (1999).

Longo DM, Howe JM, and Johnson WC, "Experimental method for determining Cliff-Lorimer Factors in transmission electron microscopy (TEM) utilizing stepped wedge-shaped specimens prepared by focused ion beam (FIB) thinning", Ultramicroscopy 80, 85-97 (1999).

Malis T, Carpenter GJC, Botton GA, Dionne S and Phaneuf MW, "Contributions of microscopy to advanced industrial materials and processing, In Industrial Electron Microscopy", Zhigang R. Li, (Ed.), Marcel Dekker Inc. (In press) (2002).

Melngailis J, "Critical Review: Focused ion beam technologies and applications", J. Vac. Sci. Technol., B. 5 (2), 469-495 (1987).

Miyagawa H, Chiou W-A, Daniel IM, "TEM sample preparation of polymer based nanocomposites using focused ion beam technique", Microsc. Microanal. 7 (Suppl 2: Proceedings), 946-947 (2001).

More KL, Coffey DW. Pint BA, Trent KS and Tororelli PF, "TEM specimen preparation of oxidized Ni-base alloy using the focused ion beam (FIB) technique", *Microsc. Microanal.* 6 (Suppl 2:Proceedings), 540-541 (2000).

Muroga A and Saka H, "Analysis of rolling contact fatigued microstructure using focused ion beam sputtering and transmission electron microscopy observation", Scripta Met. et Mater., 33, 151-156 (1995).

Park CM, and Bain JA, "Focused-ion-beam induced grain growth in magnetic materials for recording heads", J. Applied Physics 91, 6380-6832 (2002).

Phaneuf MW, Rowlands N, Carpenter GJC and Sundaram G, "Focused ion beam sample preparation of non-semiconductor materials", In: Mat. Res. Soc. Symp., 480, Anderson, R. (Ed.), MRS Pittsburgh, pp. 39-48 (1997).

Phaneuf MW, "Applications of focused ion beam microscopy to materials science specimens", Micron 30, 277-288 (1999).

Phaneuf MW, and Li J, "FIB Techniques for Analysis of Metallurgical Specimens", Microsc. Microanal. 6 (Suppl 2:Proceedings), 524-525 (2000).

Phaneuf, M.W. Li, J. Shuman, R.F., Noll, K., and Casey Jr., J.D. U.S. Patent Application 20010053605 (March 2000).

Phaneuf MW and Botton G, "Analysis of degradation in nickel-based alloys using focused ion beam imaging and specimen preparation combined with analytical electron microscopy." To be published in Proceedings of FONTEVRAUD 5: Contribution of Materials Investigation to the Resolution of Problems Encountered in Pressurized Water Reactors (2002).

Phaneuf MW, Li J and Casey Jr JD, "Gallium phase formation in Cu and other FCC metals during near-normal incidence Ga-FIB milling and techniques to avoid this phenomenon," Microsc. Microanal. 8 (Suppl 2:Proceedings), In Press (2002).

Phaneuf MW, Carpenter GJC, Saimoto S, McMahon GS, "A study of anomalous Ga phase formation during near normal incidence FIB milling of Cu". In Preparation (2002).

Phillips JR, Griffis DP, and Russell PE, "Channeling effects during focused-ion-beam micromachining of copper", J. Vac. Sci. Technol. 18 (4), 1061-1065 (2000).

Prenitzer BI, Giannuzzi LA, Newman K, Brown SR, Irwin RB, Shofner TL, and Stevie FA, "Transmission electron microscope specimen preparation of Zn powders using the focused ion beam lift-out technique", Metallurgy and Materials 29, 2399-2406 (1998).

Saka H, Kato T, Hong MK, Sasaki K and Kamino T, "Cross-sectional TEM observation of interfaces in a galvannealed steel", Galvatech '95: Proceedings of the Third International Conference on Zinc and Zinc Alloy Coated Steel Sheet., 809-814 (1995).

Saka H, and Abe S, "FIB/HVEM observation of the configuration of cracks and the defect structure near the cracks", Japanese Society of Electron Microscopy. 1 45-57 (1997).

Saka H and Abe S, "Plan-view observation of crack tips by focused ion beam/transmission electron microscopy", Materials Science Engineering. A, (234-236), 552-554 (1997).

Sakamoto T, Cheng Z, Takahashi M, Owari M, and Nihei Y, "Development of an ion and electron dual focused beam apparatus for three-dimensional microanalysis", Japanese J. Applied Physics 37, 2051-2056 (1998).

Smith AJ, Munroe PR, Tran T, and Wainwright MS, "FIB preparation of a sensitive porous catalyst for TEM elemental mapping at high magnifications", J. of Materials Science 36, 3519-3524 (2001).

Tomiyasu B, Fukuju I, Komatsubara H, Owari M, and Nihei Y, "High spatial resolution 3D analysis of materials using gallium focused ion beam secondary ion mass spectrometry (FIB SIMS)", Nucl. Instr.and Meth in Phys. Res. B 136-138, 1028-1033 (1998).

Trtik P, Reeves CM, and Bartos PJM, "Use of focused ion beam (FIB) for advanced interpretion of microindentation test results applied to cementitious composites", Materials and Structures, 33, 189-193 (2000).

Vasile MJ. Nassar R, Xie J, and Guo H, "Microfabrication techniques using focused ion beams and emergent applications", Micron 30, 235-244 (1999).

Wang YZ, Revie W, Phaneuf MW and Li J, "Application of focused ion beam (FIB) microscopy to the study of crack profiles", Fatig. Fract. Eng. Mater. Struct. 22, 251-256 (1999).

White H, Pu Y, Rafailovich M, Sokolov J, King AH, Giannuzzi LA, Urbanik-Shannon C, Kempshall BW, Eisenberg A, Schwarz SA, and Strzhemechny YM, "Focused ion beam/lift-out transmission electron microscopy cross sections of block copolymer films ordered on silicon substrates", Polymer 42, 1613-1619 (2000).

Xu S, Bouchard R, Li J, Phaneuf MW, Tyson WR, "Identification of cleavage origins using focused ion beam (FIB) sectioning", Microsc. Microanal. 8 (Suppl 2:Proceedings), In Press (2002).

Yaguchi T, Kamino T, Ishitani T, and Urao R, "Method for Cross-sectional Transmission Electron Microscopy Specimen Preparation of Composite Materials Using a Dedicated Focused Ion Beam System", Microsc. Microanal 5, 365-370 (1999).

Yaguchi T, Kamino T, Sasaki M, Barbezat G, and Urao R, "Cross-sectional specimen preparation and observation of a plasma sprayed coating using a focused ion beam / transmission electron microscopy system", Microscopy and Microanalysis 6, 218-223 (2000).

Yaguchi T. Matsumoto H, Kamino T, Ishitani T, and Urao R, "Method for cross-sectional thin specimen preparation from a specific site using a combination of a focused ion beam system and intermediate voltage electron microscope and its application to the characterization of a precipitate in a steel", Microscopy and Microanalysis 7, 287-291 (2000).

Young R J, "Application of the focused ion beam in materials characterization and failure analysis", Microstructural Science., 25, 491-496 (1997).

Young RJ, "Automation of Focused Ion Beam (FIB) Preparation", Microscopy and. Microanalysis. 6 (Suppl 2:Proceedings), 512-513 (2000).

Chapter 9

PRACTICAL ASPECTS OF FIB TEM SPECIMEN PREPARATION

With Emphasis On Semiconductor Applications

Ron Anderson and Stanley J. Klepeis
[1]Microscopy Today, PO Box 247, Largo, FL 33779. [2]IBM Microelectronics Division, East Fishkill, NY

Abstract: Polishing methods for TEM specimen preparation compete with FIB methods with regard to relative simplicity, success rate, and the ability to produce site-specific specimens in a timely manner—until the site-specific target resolution for the most advanced semiconductor specimens dropped below 0.5 μm. Beyond this point only FIB methods can yield specimens with the required degree of specificity. A procedure for preparing SEM and TEM specimens utilizing a combination of polishing and FIB techniques is presented that capitalizes on the best elements of both methods and offers the immense advantage of reducing or eliminating most of the artifacts associated with the FIB technique. Utilizing this hybrid technique offers a path forward when the incorporation of extremely fragile components into near-future semiconductor devices renders the lift-out technique marginally useful.

Key words: Semiconductor specimens, Rapid preparation, Precise preparation of preselected locations, Specimen preparation spatial resolution, Tripod polishing and FIB, FIB Artifacts

1. INTRODUCTION

The rastered-Ga-ion-beam manifestation of the FIB tool was developed to perform circuit modification and repair in semiconductor fabrication facilities. These early FIB tools were few in number and so expensive that their application was limited to this semiconductor repair function.

Following procedures first proposed by Hauffe (Hauffe, 1978, 1981, 1982, 1984)[1] for preparing specimens using argon ion beams, a small portion of available FIB time was used to prepare SEM and TEM specimens, most successfully of small, site-specific locations.

The advent of improved ion guns, stable mechanical stages, gas-assisted milling protocols, new methods (such as "lift-out"), tool automation, etc., and the introduction of less costly electron microscopy specimen preparation specific FIB models has made the FIB tool an important "player" in the field of electron microscopy specimen preparation. This chapter will focus on the preparation of TEM specimens. SEM sample preparation is included to the extent that it is the first step in making a TEM sample—where a desired plane of interest, usually a cross section of the specimen, is attained. A TEM specimen is created when this initial plane is preserved and a thin lamella is formed by cutting into the specimen from behind until it is thin enough for TEM analysis. This thin section may be self-supporting or, more likely, it is mounted on a support grid for insertion into the microscope.

2. PLACING FIB METHODS IN HISTORICAL CONTEXT

It is useful to place FIB methods in the context of other means of making TEM specimens. The method of Bravmen and Sinclair (Bravman and Sinclair, 1984) revolutionized TEM specimen preparation for semiconductor applications. In the Bravman and Sinclair (B&S) method, a stack of specimens was epoxied to each other, with the center two specimens in the stack face-to-face. A thin slice of the stack cross section was cut and first dimpled and then ion milled until a small center perforation occurred. The region around the perforation was suitable for TEM analysis. This method is still used in many laboratories. The main drawback of the B&S method is that it provides little control over where the thin region will occur. This is unfortunate because, in semiconductor laboratories, a high specimen preparation spatial resolution is called for. High specimen preparation spatial resolution means that, given a point (x, y, z) in a three dimensional sample, it is required to prepare a TEM specimen containing point (x, y, z) with a precision of a few tenths of a micrometers. Inasmuch as the desired

[1] Hauffe's 1981 and 1984 papers present the essential protocols for conventional FIB SEM and TEM preparation. These papers are rarely cited because he used an argon ion beam and called the technique "Ion Beam Microtomy."

target for preparation in the B&S method is invisible (being buried in a stack of similar specimens) new specimen preparation protocols were needed.

The tripod polisher method was invented in the authors' laboratory in the late 1980s (Klepeis, 1987, 1988; Anderson, 1989, 1992, 1995; Benedict, 1989, 1990a, 1990b, 1992). Tripod polishing became quite popular during the 1990s because it allowed for making TEM specimens with specimen preparation spatial resolution to 0.5 μm utilizing an inexpensive tool set. With tripod polishing, the specimen is mounted on a tool equipped with micrometers that allows for the tilting of the specimen so that the specimen is polished with the principal directions defined by the specimen parallel to the polishing wheel. After some introductory training, tripod polishing of pre-selected areas can be accomplished in a matter of two or three hours. The method yields extraordinarily large amounts of thin area, often the entire open 2 mm available in the typical TEM slotted grid used to mount the thin specimens is suitable for TEM examination. Beginning in the early 1990s, several similar tools were developed that performed essentially the same function as the original tripod polisher with about the same results. Hereafter, when "tripod polishing" is referred to, the term refers as well to any of these similar tool sets.[2]

Having a very large thin area available is an important contrast to specimens prepared with a FIB tool, where only about 10 – 30 μm of semiconductor cross section can be prepared in a routine specimen. In semiconductor failure analysis, the cause of failure is very frequently a defective interface in the complex stack of conductors and insulators built on a chip's substrate. Both tripod polishing and FIB methods can produce a TEM sample of the defective interface with high specimen preparation spatial resolution. With a FIB you see the defective interface and a few similar neighbor structures. In a tripod polished specimen you see hundreds or even thousands of similar neighbor structures stretching over millimeters of cross section edge. This is important because the defective interface, for example, might show a 3 nm thick unexpected amorphous layer, which probably accounts for the failure. What is missing with FIB'ed samples is the ability to analyze many neighboring structures for the same unexpected layer. With tripod polished samples, many neighboring structures can be analyzed and it is often found that the same unexpected layer is present in all of them—differing in thickness only. This is important information. The point of all this is to make the case that the synergy of complimentary specimen preparation methods is very important. Using a FIB or tripod

[2] The tripod polisher was invented and named by us in the authors' laboratory. The term "Tripod Polisher" has been copyrighted by South Bay Technology, Inc. in San Clemente, CA. There is no business relationship between SBT and the authors.

polishing or any other specimen preparation method exclusively is a substantial limitation in a TEM specimen preparation facility. At the very least, a complimentary technique, with its set of specimen preparation artifacts will aid in sorting-out the specimen preparation artifacts of your standard method.

3. WHERE THE FIB METHOD IS MOST USEFUL

It is instructive to consider those situations where a FIB tool is most useful. First on this list, of course, would be that using the FIB is the method of choice, albeit an expensive one, for preparing semiconductor device structures of pre-selected areas. For those specimens that must be prepared with specimen preparation spatial resolutions over 0.5 μm , FIB methods and tripod polishing compete in spatial precision and time required to make specimens. However, when the specimen preparation spatial resolution requirement is smaller than 0.5 μm for site-specific areas, which it most emphatically is for virtually all TEM operations supporting semiconductor manufacturers since the late 1990s, the direct preparation of TEM specimens at these spatial precision levels by tripod polishing is not possible and the FIB method is alone in meeting this requirement.

Figure 9.1. SEM of FIB prepared plated ceramic structure showing delamination.

The second area where the FIB is most useful is the preparation of specimens that are mechanically unstable due to delamination, or those specimens constructed with mechanically weak component layers. An example of the former is seen in figure 1 , where we see a SEM image of a plated ceramic substrate. It was suspected that there was delamination between two of the plating layers but SEM examination of fractured specimens or specimens prepared by mechanical polishing failed to show any delamination. In the first instance because several layers delaminated on fracturing and the specimen was destroyed and in the second case because polishing the soft metal plated layers smeared metal into the fracture and it was not visible. The FIB image clearly shows the delamination. Note the interesting FIB artifacts.

The third area, where FIB TEM preparation excels is the ability to perform "dry" specimen preparation, where the final specimen is untouched by water or any other liquid solvent. The fabrication of the stack of conductors and insulators on a semiconductor chip is largely a photographic process where photoresist is deposited, developed, and removed repeatedly as the stack is built-up. Many photoresist recipes contain chlorine. Should a small deposit of photoresist be left behind after a cleaning process and then buried by the construction of subsequent layers of the stack, the trapped chlorine will be free to etch whatever it comes into contact with, possibly causing a chip failure. TEM failure analysis of the failed site utilizing any technique where the trapped chlorine is exposed to water as a final step will yield an empty, etched-out cavity with no trace of what caused the cavity—the chlorine-containing reaction product is dissolved. With a FIB, it is a simple matter to use mechanical wet polishing up to *but not into* the fail location and then perform dry final thinning in the FIB tool, preserving the photoresist residue for analysis. These comments on the preparation of water or solvent sensitive specimens apply in several fields, geology for example.

4. FIB METHOD ARTIFACTS

No TEM specimen preparation method is without its share of artifacts. Artifacts detract from the science of a particular investigation and it is crucial that the researcher fully understands the nature and control of the artifacts associated with the method being used.

4.1 The FIB "Theater Curtain" or "Waterfall" Effect

The rate of specimen material removal by the grazing-angle-of-incidence FIB beam is sensitive to the hardness, atomic number, and topology of the surface of the specimen first encountered by the FIB beam. Grooves are cut into the specimen when the beam encounters soft or low atomic number specimen surfaces and the specimen surface is slow to thin when the beam first encounters hard or high-Z specimen portions. The resulting FIB'ed surface cut in the specimen will exhibit ridges and grooves and looks somewhat like a theater curtain or waterfall. The elimination of this artifact is straight-forward: the initially encountered specimen surface is made homogeneous with a coating of tungsten or platinum (or similar) high-Z deposition carried out in the FIB, in a line marking where the final specimen will be taken, as the first step in specimen preparation. The homogeneous, high-Z top surface will then control the rate of removal of what lies beneath—eliminating the artifact.

4.2 Electrostatic discharge in the FIB tool

The specimen undergoing FIB preparation is subjected to high energy ion bombardment and portions of the specimen, or the entire specimen, can accumulate several thousand volts of charge. High temperature superconducting specimens prepared by the FIB method in the authors' laboratory exhibited large craters and local melting due to electrostatic discharge. Creating a discharge path for the electrostatic buildup can essentially eliminate this artifact. This can be accomplished with the same heavy metal (W or Pt) FIB-deposited layer, deposited to control the theater curtain effect, and/or coating the entire specimen surface with carbon.

4.3 Limited Specimen Area and TEM Instrumental Tilt

This is not really a FIB "artifact" but more of a limitation of the FIB method. It has already been mentioned that the FIB method will yield only several tens of square micrometers of thin area and how this may limit the accumulation of specimen statistics if critical dimension measurements are required, etc. Limited TEM instrumental tilt capability refers to the fact that early TEM specimens prepared by FIB were in the form of a thin membrane or lamella centered in a roughly 100-micron thick specimen that could only be viewed when the specimen was tilted so that the electron beam traveled down the channel cut into the thick specimen. Little or no tilting of the specimen was possible, which limited the resulting TEM analysis. The instrumental tilt limitation can be eliminated completely by making FIB cuts into specimens that have been polished very thin initially or by using the FIB lift-out method.

4.4 Ga Contamination and sputtered specimen material contamination

The physics of particle and specimen interaction predicts that some FIB tool Ga ions will be implanted in the specimen despite the near grazing angle of incidence of the FIB ion beam. This is easily seen by taking x-ray energy dispersive spectra of the FIB prepared specimens and noting the Ga peaks. Likewise it is expected that specimen material sputtered off the specimen by the Ga beam in one portion of the specimen may land on another part of the specimen creating sputter-contamination artifacts that obstruct a clean view of the specimen. The amount of the implanted Ga and sputter deposited specimen material can be minimized by utilizing proper FIB protocols for the final thinning steps as discussed elsewhere in this book. These artifacts can be eliminated completely by utilizing a FIB protocol that allows the resulting sample to be argon ion milled prior to TEM observation.

4.5 Amorphous Surface Layers

The creation of amorphous layers on some classes of specimens, most notably silicon and germanium semiconductors, has received wide attention and is discussed thoroughly elsewhere in this compilation. This is probably the greatest problem confronting TEM examination of thin semiconductor TEM specimens. Fortunately, by choice of FIB technique, the thick amorphous layers created by FIB milling can be removed by argon ion milling. (To be more precise, low-keV, low-angle, argon ion milling will also create an amorphous layer. But this argon ion milled amorphous layer will be more than an order-of-magnitude thinner than the amorphous layer produced by Ga ion milling. Hence we are *replacing* a thick amorphous layer with an inconsequential thin amorphous layer.).

5. TABLE SUMMARY OF FIB METHODS

The pros and cons of the FIB methods used to make TEM specimens are summarized in Table 1. With regard to the "Speed" row in Table 1, The times cited are for experienced operators and include all pre- and post-FIB-tool specimen site selection, cutting or polishing as appropriate, and mounting the resulting specimen. The specimen preparation spatial resolution question, discussed earlier, is abbreviated in the table as "target specificity." All the methods are capable of high target specificity, it just takes longer to prepare very small pre-specified locations. Likewise, very hard materials, like jet engine turbine blades, take far longer to process in every phase compared to silicon specimens.

In the first column we see the original method proposed to make TEM specimens (Kirk et al., 1989), which some researchers call the "H-bar" method because a SEM view of the finished sample looks like the capitol letter "H." Briefly, the sample is located, rough cut from its matrix, and polished to a thickness of about 50 to 100 μm. The specimen is mounted on a half-grid (a large, single-hole, aperture grid with one side cut away to allow the FIB beam to strike the edge of the specimen). In the FIB, the location of the finished TEM specimen is coated with a W or Pt line and large trenches are cut on either side of the desired location using large apertures. Final cuts are made to thin and clean the resulting specimen. Because of the TEM tilt limitations and the probability of the specimen suffering sever FIB artifact contamination, this method has little to recommend it in the face of newer protocols.

The combined tripod and "H" pattern method is actually a trivial modification of the conventional "H" pattern method. The difference between the two protocols is that the specimen is initially polished to a thickness less than 10 μm instead of 50 to 100 μm. This was first suggested by Basile (Basile et al., 1992). The procedure for doing this will be explained in greater detail in the next section. The end result of making the specimen thinner before FIB'ing, besides shorter FIB times, is that the resulting specimen geometry allows the specimen to be ion milled after thinning from the substrate side. The benefits of ion milling the FIB'ed specimen are substantial. First, the resulting specimen can be tilted to the full range of the TEM instrument's tilt capability. Second, the specimen is made thinner, frequently making HRTEM imaging of atom columns in the specimen possible. Third, the thick (20 nm or so) amorphous layer created by the 30 to 50 keV FIB beam can be replaced by the negligible, order-of-magnitude thinner, amorphous layer created by argon ion milling at 2 or 3 keV at near grazing incidence. And lastly, any FIB back sputtered artifactual material on the specimen can be removed. Both "H" pattern methods have the disadvantage of requiring a mechanical polishing step but the times to make specimens compare well with the lift-out methods.

Table 1. Comparison of FIB Methods

	Conventional "H" Pattern	**Combined Tripod and "H" Pattern**	**Lift-Out Method** **External Grid Mounting**	**Lift-Out Method** **Internal Grid Mounting**
Advantages	Can put back into FIB for additional thinning Multiple specimen sites per initial prep	Can put back into FIB for additional thinning *Can be ion milled to make thinner and to remove FIB artifacts* *Full TEM tilt capability* Multiple specimen sites per initial prep	*Full TEM tilt capability* Serial sectioning, 3D reconstruction *Automated initial prep on multiple specimen sites unattended* Bulk specimen may be returned for further processing	*As for Lift-Out external mounting* *Ion milling possible for additional thinning and artifact reduction*
Disadvantages	Destructive to bulk specimen Limited tilting capability Preliminary prep needed FIB artifacts	Destructive to bulk specimen Preliminary prep needed	No additional FIB or ion milling possible FIB Artifacts Possibly inapplicable to fragile specimens	Additional ion milling may be one-sided only, pending invention of new procedures Possible ion mill artifacts
Speed	1 to 4 Hours depending on target size	1 to 4 Hours depending on target size	*0.5 to 2 Hours depending on target size*	1 to 4 Hours depending on target size

Italics denotes important advantages

The lift-out methods are covered in considerable detail elsewhere in this book. Ex-situ vs. in-situ grid mounting refers to where the specimen is placed on the TEM grid. Ex-situ removal is when the specimen is plucked from the substrate outside the FIB tool via electrostatic pick-up on a glass (or metal) filament and then placed on a carbon-film coated TEM grid. In-situ mounting is accomplished in the FIB tool via fastening the specimen to a transfer fixture, plucking it from the substrate, and then fastening it to a TEM grid all inside the FIB. The lift-out, ex-situ grid mounting method is very fast, produces specimens that do not limit the TEM tilt capability, and automation of the initial preparation steps is possible. The main disadvantage is that no further thinning or artifact removal is possible on the specimen sliver resting on the thin carbon film. Whatever you get is all there is. If you aren't happy (usually only a small fraction of the time) with the TEM specimen you must start over again. The in-situ method is newly developed and very promising. It combines all of the advantages of the ex-situ method but offers the possibility of subsequently ion milling the resulting specimen to further thin the specimen and to remove artifacts.

The simple lift-out methods work very well with Si/Si-oxide and similar semiconductors. The method is expected to run into trouble with the incorporation of extremely fragile multiple layers of low-κ dielectrics into near-future semiconductor devices. Some device manufacturers are beginning to experiment with a long list of polymeric and related materials to replace Si-oxide as a dielectric. As a group, materials with dielectric constants from 2.5 to 3 have at most one-tenth the strength and toughness of Si-dioxide. The next generation after these, with dielectric constants less than 2.5 are another order of magnitude weaker still. It is unlikely that these fragile devices can withstand the stresses associated with the lift-out method. While new lift-out protocols may be forthcoming, it would seem that small FIB cuts made into thin, tripod polished specimens might provide a way forward for preparing these fragile specimens thanks to the thicker portions of the specimen holding the ensemble together.

An interesting variation of the lift-out, ex-situ grid mounting method, which offers a means of low-angle, argon ion milling of the specimen, has been described by Langford and Petford-Long (2001). The thin specimen is prepared and cut away from the substrate as per the standard lift-out method. The specimen, somewhat larger than the standard 10 μm cut, is then placed on a Cu, very fine mesh (10 μm grid opening) non carbon-coated grid. The specimen is nudged about on the surface of the grid with the micromanipulator pluck-out needle so that it spans a grid hole with a bit of specimen overhanging the neighboring grid holes. The top surface is then Ar ion milled at 2 degrees and 2 keV for a short time to replace the 20 nm amorphous region and Ga implant damage with the <5 nm amorphous and damaged region corresponding to Ar milling as discussed above. The specimen area of interest must be centered in the covered grid hole and the specimen is not rotated during Ar ion milling to minimize the amount of support grid Cu material deposited on the specimen. The micromanipulator is then used to turn the specimen over on the Cu grid by going under the plane of the specimen in one of the neighboring grid holes, lifting the specimen up, inverting it, and re-nudging it back into the centered-on-a-grid-hole location. The ensemble is then returned to the ion mill where the second side of the specimen is Ar milled. If the operator loses track of which side has been milled during the inverting operation, the process is repeated with one minute milling operations until, statistically, both sides have been milled.

6. COMBINING MECHANICAL POLISHING AND FIB METHODS

As just discussed, producing thin specimens in the FIB tool that can be subsequently ion milled essentially eliminates concerns over too-thick specimens and amorphous or sputtered artifactual layer contamination. We shall see that the method is adaptable to both cross section and plan-view specimens. As initially practiced in the authors' facility, (Anderson et al., 1996), the specimens were thinned to something like 3 μm in thickness and then subjected to brief FIB cuts on either side to produce a specimen like that seen in figure 2. A low magnification TEM view is seen in figure 3, which shows trench structures 7 microns deep into the specimen. A remnant of the protective W FIB deposition can be seen in both the SEM and TEM images as well as in the higher magnification view of the trench tops in figure 4. The geometry's of the specimen are such that ion milling, from the substrate side is possible at angles less than 10 degrees.

The degree of skill necessary to consistently tripod polish samples to 3 μm in thickness, centered on a pre-selected target, can be applied consistently after training and a fair amount of practice. It was desired to simplify the procedure such that less skilled technical staff could routinely prepare this type of sample on a high volume basis. To that end, FIB tool protocols, leading to automated macros where possible, were created to provide assistance and simplify the procedure. These protocols provide FIB deposited W layers (or sometimes FIB cut trenches) that act as visual polish-stop indicators for the tripod polishing operation and two sets of FIB marks to indicate where the next-to-final and final FIB cuts are to be taken to yield a TEM specimen that includes the desired device area, as seen in figure 5. A protective W or Pt coating is applied to protect the specimen surface. The polish-stop FIB metal depositions, or trenches, are 2 μm wide and 8 μm long on either side and centered on the desired specimen plane. The sample is mounted using "super glue" on the tripod polisher, adjusted so that the plane defined by the final specimen location is parallel to the polishing wheel, and polished with, typically, 30, 15, and 3 μm diamond lapping film until the polish-stops on one side are encountered, as seen in figure 6, a light-optical photo. The specimen is soaked off the tool in solvent and remounted to polish from the opposite side until the same two polish-stops are encountered from the back. Figure 7 shows the appearance of the top of the specimen as it first appears in the FIB. The condition of the polished surface is of no consequence—a 3 μm finish is fine, and no great care need be taken to terminate tripod polishing exactly at the ends of the polish-stops—just as long as the specimen area is not polished into. Likewise, it is not important that the angular alignment of the specimen is as precise as seen here—angular misalignment can be compensated for in the FIB. The two polishing steps take 10 to 20 minutes each.

Figure 9.2. Specimen tripod polished to about 3 μm before finishing in the FIB

Figure 9.3. Low magnification TEM image of the specimen in figure 2.

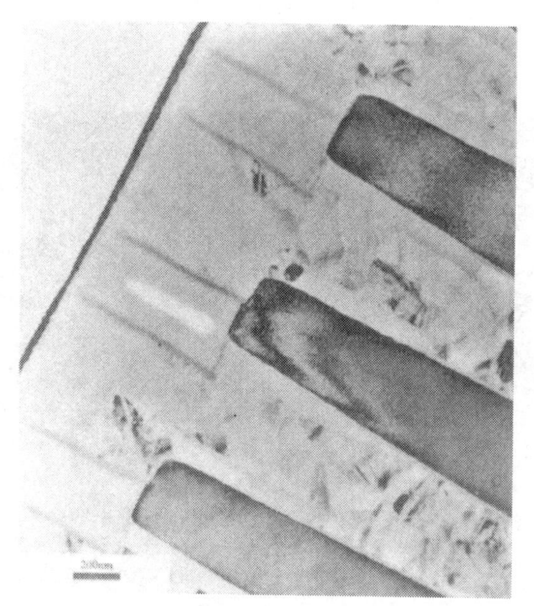

Figure 9.4. High magnification view of the trench tops in figure 3.

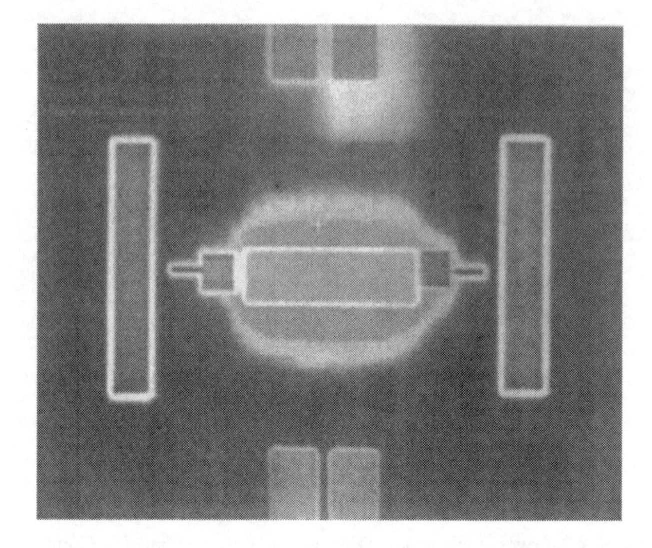

Figure 9.5. Specimen surface with tripod polishing macro in place showing W polish stops.

Figure 9.6. Light optical image showing W polish stops serving as a signal to terminate polishing.

Figure 7 shows a medium magnification view of all the FIB cuts in a finished sample with W polish-stops. From the geometry in the picture, this specimen could be ion milled from the substrate side at an angle of 6.7 degrees to accomplish near-low argon angle milling to thin the specimen. A short, approximately 30 second, argon ion milling at 20 degrees is most effective in removing any sputtered-material artifact. Figure 8 shows a finished sample with trench polish-stops. Cutting trenches into the original specimen surface is faster than depositing W rectangles for polish-stops and a small sacrifice in the ability to see the polish-stops while tripod polishing.

Essentially eliminating the FIB-deposited amorphous layer and sputter artifact from silicon semiconductors is the most visible benefit of post-FIB argon ion milling for these materials. Post-FIB argon ion milling is also effective for specimens much more sensitive to radiation and gallium implant artifacts from FIB processing than silicon. In figure 9, we see an as-FIB'ed GaN layer on a sapphire substrate direct from FIB processing. Admittedly, the specimen spent too much time being observed in cross section by ion imaging in the FIB, which exacerbates radiation and Ga implant damage. In figure 10 we see the same specimen area after argon ion milling from the substrate side, now free of FIB induced artifacts.

In order to facilitate the preparation of large numbers of FIB specimens and to reduce the amount of handling each specimen is subjected to, we have designed a FIB multi-specimen holder that can accommodate four (or more) specimens to be FIB'ed. The specimens are mounted in Ar ion mill holders so that they can be moved back and forth between the Ar ion mill and the FIB without risk of breakage. Details of the holder design may be found in the appendix to this chapter.

Figure 9.7. View in FIB after tripod polishing.

Figure 9.8. Medium magnification view of finished specimen with trench polish stops. Trench polish stops are faster to apply and contaminate the surface less.

Figure 9.9. GaN on sapphire substrate specimen after FIB milling with serious FIB artifacts in evidence.

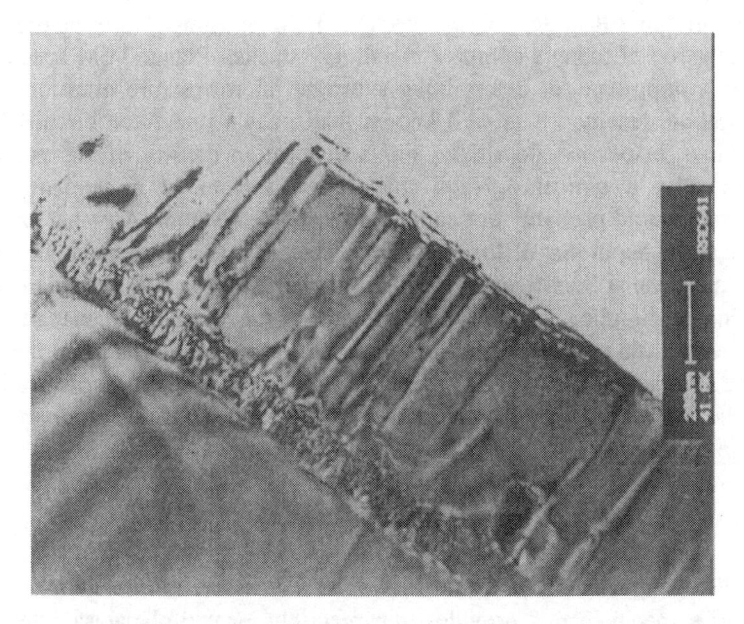

Figure 9.10. The same specimen area seen in figure 10 after 2 minutes of argon ion milling on each side. Arrtifacts have been removed.

7. PLAN VIEW SPECIMENS

There are a number of unique types of SEM and TEM specimens that can be prepared using a FIB tool that are not possible to prepare by any other means. Seeing cross sections in an SEM of a semiconductor specimen that has had two FIB cuts made at right angles to each other is an example. In such a specimen the structure of a device parallel and perpendicular to a given plane may be examined. Another FIB-unique specimen is the preparation of multiple plan view specimens in a single sample.

Plan view, or planar, TEM samples were the mainstay of semiconductor TEM analysis until the 1970s when the first cross section semiconductor specimens were made. By and large, these early plan view specimens were made by masking and acid-etching the silicon or germanium substrate from the back side. Etching was halted when a thin layer of substrate was left, if the aim of the experiment was to examine the substrate, or the substrate was etched away in a small circle allowing the device metallurgy/insulator stack to be examined. With the advent of integrated circuit technology, with multiple layers of conductors and insulators, planar TEM samples went out of favor because it was very difficult to unequivocally locate one particular layer in a conductor stack or a particular depth in the substrate. Cross section TEM specimens are favored for examination of today's complex metallurgy stacks. Planar TEM specimens are still important for determining substrate microstructure questions, like dislocation density. It is well known that today's integrated circuits have very low dislocation densities. For a dislocation density of 10^6 per cm^2, statistically, a typical 0.5 μm thick cross section of a semiconductor substrate would probably not contain a single dislocation. A planar sample, made at the depth that dislocations are expected, is therefore needed so as to have a chance at investigating the dislocations. There is still a problem: acid etching or dimpling and ion milling methods for preparing planar samples have very little specimen preparation spatial resolution. It is often the case that a satisfactory planar TEM specimen is prepared that shows dislocations, but the analyst has very little idea how deep into the substrate any particular image represents.

7.1 The 4 in 1 Specimen

FIB to the rescue! In figure 11 we see a cross section image taken in a FIB of a specimen that provides four precisely located planar samples in a silicon substrate. (Anderson and Klepeis, 1997) The specimen is the same trench structures we saw in figure 3. The specimen was made by tripod polishing into the cross section until the trench structures seen on top were revealed. The specimen was then tripod polished from the back side to a level approximately four microns below the seven-micron deep trenches, as seen in the figure. Four FIB "H" pattern samples were cut into the cross section face at depths below the Si surface where dislocations were expected by virtue of abrupt changes in the substrate doping or trench microstructure. When the resulting thin lamella were examined in the TEM (TEM beam perpendicular to the substrate top surface) four precisely located planar samples were revealed. There is, of course, no limit on the number of specimens one could cut into this single tripod polished sample. Figure 12 shows one of the resulting TEM images, which can be unequivocally stated to be at a precisely measured depth below the surface.

Figure 9.11. Planar semiconductor TEM specimen with four FIB TEM specimens cut at precise distances from the specimen top surface.

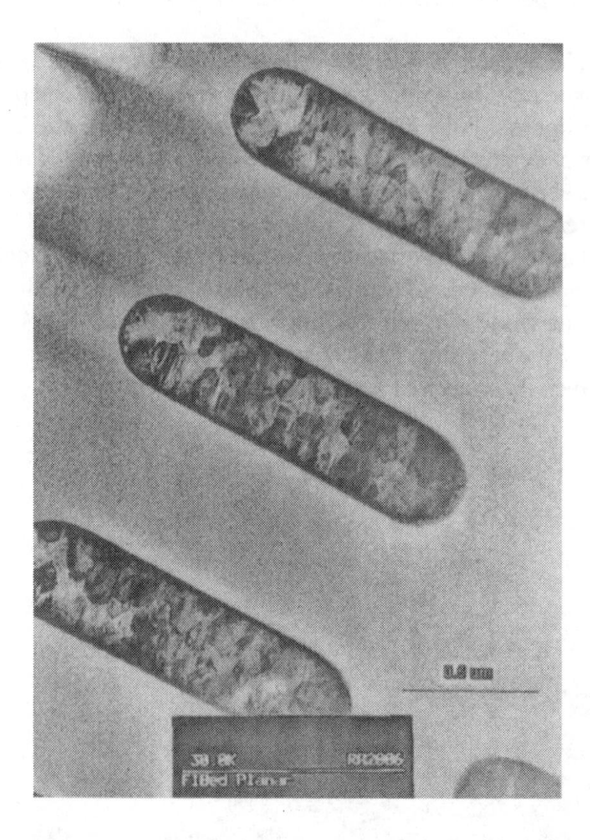

Figure 9.12. TEM image of one of the FIB planar specimens seen in figure 11.

7.2 FIB Dimpling

There is another challenge with regard to making planar specimens by the FIB method: the ability to prepare useful TEM planar specimens at very small pre-selected locations—the specimen preparation spatial resolution issue again. With the extreme density of modern integrated circuits it is becoming routine to require the preparation of a single failed component in an array of a billion or more similar components. This specimen preparation protocol was developed by Klepeis, et al. (2000).

Sophisticated electrical testers can find the faulty component and transfer its coordinates to a FIB tool where the failure can be marked by an array of FIB cuts. Cross section TEM specimens are appropriate when it is suspected that the failure is caused by a defect in the metallization stack on top of the device substrate. Planar specimens are appropriate when the electrical signature of the failure indicates that the fault lies in the substrate itself—dislocations or some other crystal defect. One approach for making a precisely located planar TEM sample, once the location of the failure is marked with the FIB tool, is to tripod polish the Si from the back of the substrate, leaving a few micrometers of Si plus the metallization stack (Or some fraction of the metallization stack if tripod polishing was employed to remove some top layers of metallization.) Low angle ion milling of the thin planar specimen can then be employed to finish the sample preparation. This tripod polishing and ion milling protocol will yield extremely large transparent regions when successful. However, all that thin area almost always cracked or peeled-up due to stress evolution in the specimen, destroying it. The FIB dimpling technique is a way to overcome these difficulties.

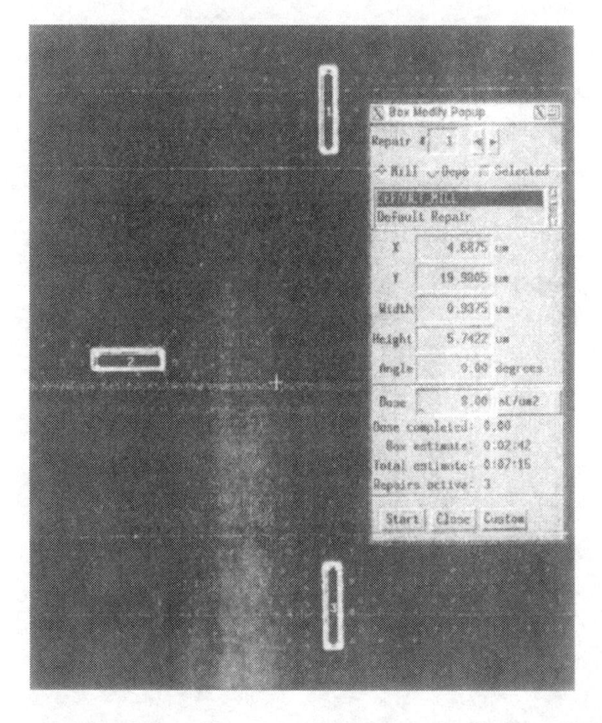

Figure 9.13. The FIB dimpling procedure begins by having the required defective cell marked by FIB cuts. This can often be accomplished automatically with failed cell coordinates provided by a circuit tester.

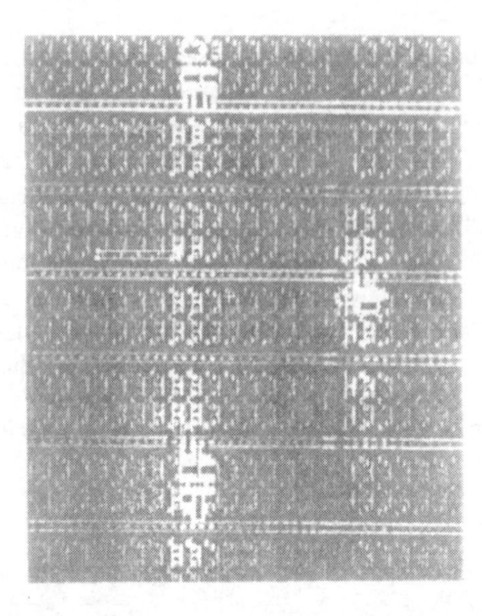

Figure 9.14. The specimen is turned, circuit side up, and parallel polished down to the tops of the gate structures.

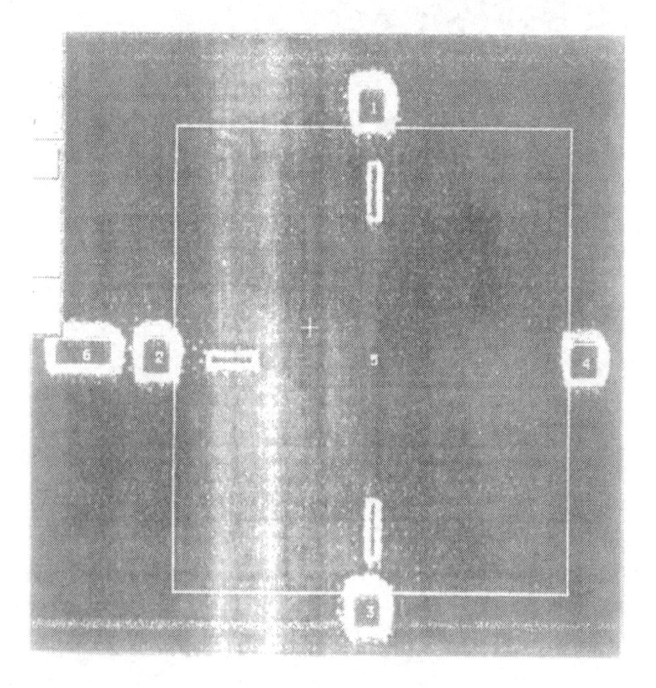

Figure 9.15. A second set of deep FIB marks are cut outside the first set.

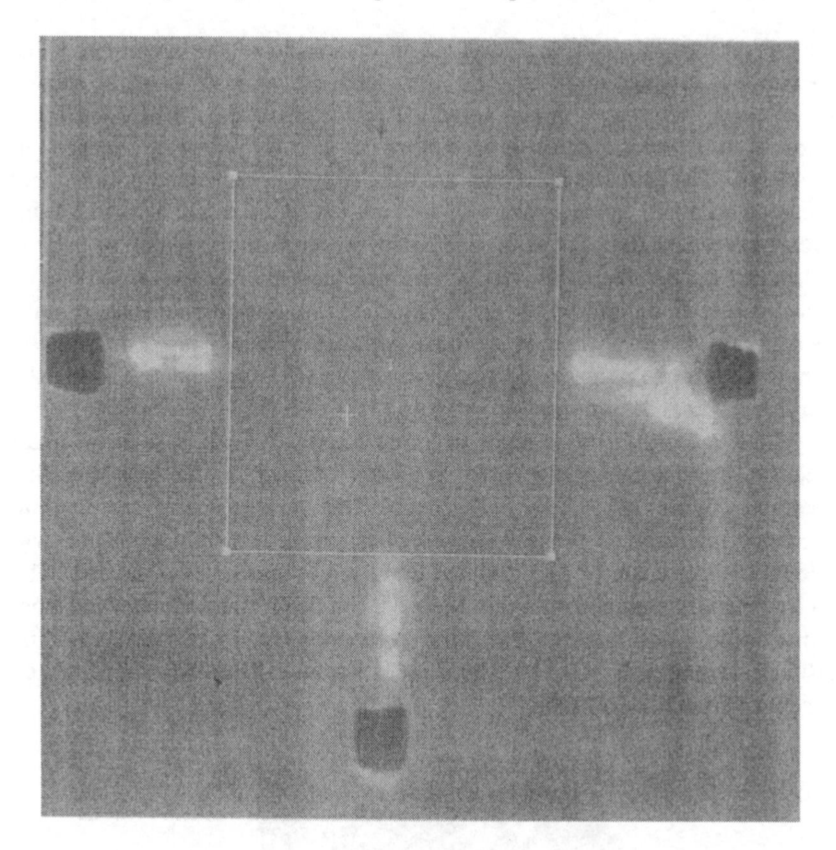

Figure 9.16. A square dimple raster is set up on the backside of the specimen, inside the second set of deep marks to locally thin the specimen at the pre determined location of the fail.

The FIB dimpling procedure is as follows: The failed cell is identified and marked with FIB cuts deep enough to reach the substrate Si (figure 13), the failure is at or near the intersection of lines through the FIB cuts. For this example the electrical signature of the fail suggested that it was at the lowest, gate stack level, of the specimen metallization stack or in the nearby substrate. Accordingly, the metallization stack on top was tripod polished down to just above the gate stack in a region centered on the FIB marks. (figure 14). The specimen is returned to the FIB and a second set of deep FIB marks are cut into the top surface, well into the Si substrate, just outside the first set (figure 15). A Mo TEM single aperture grid is epoxied to the top surface of the specimen with the grid opening centered on the region of interest. After the epoxy is cured, the open grid aperture is filled with glycol phthalate, or wax, in order to provide mechanical strength to the specimen during subsequent polishing steps on the backside of the specimen. The specimen is inverted and tripod polished from the backside until the second set of deep FIB cuts is encountered. The specimen should be 3 or 4 μm thick at this point, which is sufficient to prevent cracking or peeling problems. The grid and specimen are removed from the tripod polisher and mounted on a FIB holder. The FIB tool is then set up to cut a 20 to 50 μm square raster into the back of the specimen, centered on the region of interest delineated by the deep FIB marks cut from the top side. (figure 16). The square raster is monitored by the tool's endpoint detector, or visually until there is an abrupt change in contrast within the square raster (especially conspicuous when preparingsilicon-on-insulator (SOI) specimens) and/or the top side features become visible. (figure 17).

The end result, after eight or more hours of work, is a very small electron transparent planar area precisely located at the pre-specified location. (figure 18). The FIB thinned area has sufficient surrounding specimen thickness to prevent stress-induced specimen failure. More than one raster square can be FIB dimpled into a single specimen as needed. The entire structure mounted on a thin Mo grid, can be low angle ion milled from either side, to further thin the specimen and to remove FIB artifacts—principally Ga ion implant. Figure 19 shows the finished specimen from the grid side

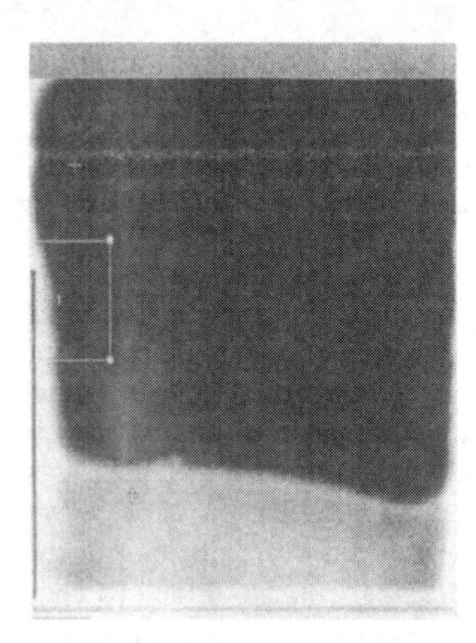

Figure 9.17. The dimple FIB raster cutting operation is terminated when the top side structures are visible. This is particularly easy to detect when preparing SOI samples as in this case.

Figure 9.18. The dimple opened centered on the fail site.

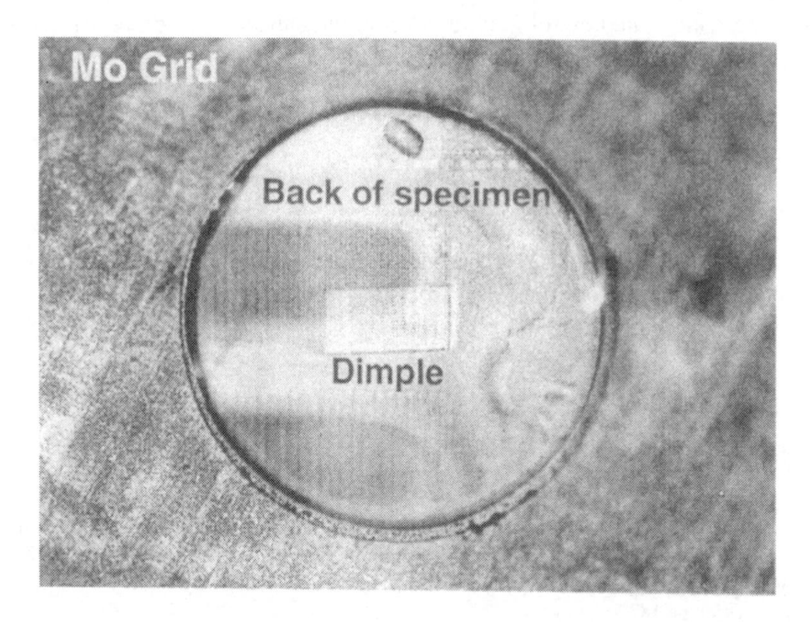

Figure 9.19. Low magnification view of FIB dimpled specimen from grid side.

8. CONCLUSION

From the perspective of a TEM specimen preparer in the late 1970s or early 80s, a modern FIB tool would seem like magic. High priced magic. But as stated earlier in this chapter, in the integrated circuit semiconductor world, as critical dimensions shrink under 0.5 μm and as esoteric materials are incorporated into chip design, there simply is no other way of preparing site-specific TEM specimens of the latest designs other than using a FIB. This may well prove to be true in other fields as well. We have seen an evolution in FIB TEM preparation protocols from the first conventional "H" pattern method through to the in-situ mounting lift-out technique—so new that the method's capabilities are still being explored as this is written. We are also seeing the development of unique specimen preparation methods, like the 4 in 1 planar method described above, that can only be executed with a FIB tool. Others are sure to appear.

It is always true that in any laboratory preparing TEM specimens the synergy of complimentary techniques is to be strived for. The key objective is the preparation of artifact-free specimens with, very frequently, the requirement of extraordinarily demanding specimen preparation spatial resolution. Whether one chooses a lift-out or a FIB-after-polishing method is immaterial. The key thing is to develop the capability to do either and to apply the best method for the task at hand.

ACKNOWLEDGEMENTS

The authors wish to thank our colleagues in the Microelectronic Division East Fishkill TEM department. Specifically, John Benedict, a co-inventor of the tripod polisher, Anthoney Domenicucci, Phil Flaitz, and Donald Hunt.

REFERENCES

Anderson RM and Klepeis SJ "Combined Tripod Polishing and FIB Methods for Preparing Semiconductor Plan View Specimens". Specimen Preparation for Transmission Electron Microscopy of Materials IV, (eds.) R. Anderson, et.al., Materials Research Symposium Series 480, pp 187-192 (1997).

Anderson RM, et al., Microscopy of Semiconducting Materials 1989, Proceedings of the Physics Conference held at Oxford University, 10-13 April 1989, ed. by Cullis, A. G. and Hutchison J. L., Institute of Physics Conference Series Number 100, Bristol and New York (1989).

Anderson RM, et al., Proceedings EUREM 1996, Steer M. Cottrell D (eds.) p 79-80 (1992).

Anderson, RM, "Precision Ion Milling of Layered, Multi-Element TEM Specimens with High Specimen Preparation Spatial Resolution" In Specimen Preparation for Transmission Electron Microscopy of Materials-III, ed. Anderson, et al., Mater. Res. Soc. Proc. 254, Pittsburgh, PA USA pp. 141-148 (1992).

Anderson, RM, et al., Microbeam Analysis--1995, Proceedings of the 29th Annual Meeting of the Microbeam Analysis Society, ed. E. Etz, p. 135 (1995).

Basile D, et al. "FIBXTEM – Focussed Ion Beam Milling for TEM Sample preparation." Specimen Preparation for Transmission Electron Microscopy of Materials III, (ed.) R. Anderson, Materials Research Symposium Series 254, pp 23-41 (1992).

Benedict JP, et al., "A Method for Precision Specimen Preparation for Both SEM and TEM Analysis." EMSA Bulletin, 19, 2, pp. 74-80 (ovember,1989)

Benedict JP, et al., "Recent Advances in E.M. Specimen Preparation: Cross-Section samples." in International Symposium on Electron Microscopy, ed. by Kuo K and Yao J, World Scientific, p. 450-460 (1990a).

Benedict JP, et al., " Procedure for Cross Sectioning Specific Semiconductor Devices for both SEM and TEM Analysis." In Specimen Preparation for Transmission Electron Microscopy of Materials-II, ed. Anderson, R., Mater. Res. Soc. Proc. 199, Pittsburgh, PA USA pp. 189-204 (1990b).

Benedict, JP, et al. "Recent Developments in the use of the Tripod Polisher for TEM Specimen Preparation." In Specimen Preparation for Transmission Electron Microscopy of Materials-III, ed. Anderson et al., Mater. Res. Soc. Proc. 254, Pittsburgh, PA USA pp. 121-140 (1992).

Bravman JC, Sinclair R "The Preparation of Cross-section Specimens for Transmission Electron Microscopy." J Electron Microsc Tech 1:53-61 (1984)

Hauffe W Thesis B, Technical University of Dresden, Dresden (1978)

Hauffe W Proc. 10th National Conference on Electron Microscopy, Phys. Sci. GDR, Leipzig, p. 307-308 (1981)

Hauffe W, DDR Patentschrift DD218954 B1, 29 (September 982)

Hauffe W, "Ion Beam Microtome for Preparation of TEM Samples." Proc EUREM 1984, **1**, pp 105-106 (1984).

Kirk EC et al. "Cross-sectional transmission electron microscopy of precisely selected regions from semiconductor devices." Inst. Phys. Conf. Series, 100, p 501-506 (1989) and numerous other references by many authors in the following years.

Klepeis SJ, et al., "A Grinding/Polishing Tool for TEM Sample preparation." In Specimen Preparation for Transmission Electron Microscopy of Materials, ed. Bravman, et al., Mater. Res. Soc. Proc. 115, Pittsburgh, PA USA pp. 179-184 (1987).

Klepeis SJ, et al., "A Technique for Preparing Semiconductor Cross Sections for both TEM and SEM Analysis," EMSA Proceedings, ed. by Bailey, G.W., San Francisco Press, p. 712-713 (1988).

Klepeis SJ, Proceedings of the 14th International Congress on Electron Microscopy, Vol III, Electron Microscopy 1998, p 541-542 (1998).

Klepeis SJ, et al., "FIB Dimpling: A Method for Preparing Plan-View TEM Specimens," EMSA Proceedings, ed. by Bailey, G.W., Springer-Verlag Presspp. 506-507 (2000).

Langford RM and Petford-Long AK "Broad ion beam milling of focused ion beam prepared transmission electron microscopy cross sections for high resolution electron microscopy," J Vac Sci Technol A 19(3), p 982-985 (2001).

APPENDIX: A MULTI-SPECIMEN FIB HOLDER

In order to increase FIB thoughput, a multi-specimen holder (see figure 20) was designed that can accommodate four or more FIB specimens simultaneously (Klepeis, 1998). Inasmuch as our objective is fabricating FIB specimens that can be subsequently Ar ion milled to remove FIB artifact layers, the multi-specimen holder utilizes GATAN[3] PIPS (Precision Ion Polishing System) spring-style DuoPost specimen holders to mount the specimens to be FIB'ed and subsequently Ar ion milled. Once mounted in the DuoPost holders, the specimens need not be touched until they are ready for the TEM.

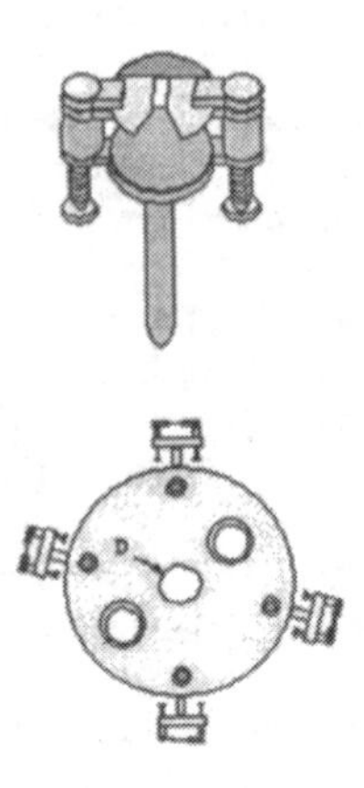

Figure 9.20. GATAN DuoPost with specimen held with spring clips for low-angle Ar ion milling and FIB cutting (top) and the Multi-specimen holder with four specimens

The holder can be mounted in the Micrion and FEI FIB tools in the authors' laboratory and should be easily accommodated in other machines. The holder is machined such that it is at a +15 degree angle relative to the stage so that the specimen, in the DuoPost holder, can be tilted from –15 to +45 degrees to facilitate FIB milling. Detailed plans and specifications can be obtained from author S.K.

[3] The GATAN PIPS and the DuoPost holder are trademarks of GATAN Inc., Pleasanton, CA.

Chapter 10

FIB LIFT-OUT SPECIMEN PREPARATION TECHNIQUES

Ex-Situ and In-Situ Methods

L.A. Giannuzzi[1] B.W. Kempshall[2], S.M. Schwarz[2,3], J.K. Lomness[3], B.I. Prenitzer[2], and F.A. Stevie[4]

[1]FEI Company, Hillsboro, OR 97124, [2]NanoSpective Inc., Orlando, FL 32826, [3]University of Central Florida, Department of Mechanical Materials and Aerospace Engineering, Orlando, FL 32816, [4]North Carolina State Universiy, Analytical Instrumentation Facility, Raleigh, NC 27695

Abstract: In this chapter, we review methods and applications of the FIB lift-out specimen preparation technique. A historical overview of the development of the technique is given. The ex-situ and in-situ lift-out techniques are described. Examples, advantages, and disadvantages of each of the techniques are presented.

Key words: Focused ion beam, FIB, lift-out, LO, TEM, in-situ LO, INLO, ex-situ LO, EXLO, micromanipulation

1. INTRODUCTION

In another chapter in this volume Anderson and Klepeis provide a thorough review of FIB usage for electron microscopic specimen preparation. Thus, this chapter will be dedicated solely to the discussion of the lift-out technique. It should be noted that there are advantages and disadvantages to any specimen preparation technique. The lift-out technique, and the use of FIB in general, should not be assumed to be a panacea for all specimen preparation applications (but it certainly has made specimen preparation easier!).

The use of focused ion beam (FIB) instruments has become a mainstay for the preparation of specimens for subsequent analysis by a number of analytical tools. The versatility of the FIB makes it advantageous for numerous applications. In particular, FIB milling techniques are generally quite rapid (on the order of ~ 0.5 to 1 hour for many applications). FIB preparation for TEM analysis can take longer (e.g., from 0.5 to 4 hours). In addition, because of the small probe size (e.g., < 10 nm) and a broad range of beam currents (e.g., ~ 1 pA to 30,000 pA) achievable by new FIB instruments, FIB specimen preparation techniques presently offer the best spatial resolution where site specificity is required. Since FIB milling procedures for specimen preparation are often performed at glancing angles to the surface of interest (an incident angle of ~ 89° from the sample surface normal), high quality specimens may be milled from samples that may include multiple layers of materials oriented either parallel or perpendicular to the ion beam. Although small deviations in specimen thickness from one phase or grain to another due to preferential milling may occur (which can cause curtaining effects), this usually does not limit analyses by TEM or other methods.

A primary advantage to the "lift-out" (LO) technique over other specimen preparation methods mentioned in this book is that specimens may be prepared from the starting bulk sample with little or no initial specimen preparation. Assuming that the material in question is compatible with FIB milling, the only initial sample requirement for the lift-out technique is that the bulk sample must be small enough to fit comfortably within the FIB vacuum chamber. If an insulating material is to be FIB milled, a conductive coating may be applied to the sample to prevent charging as a result of ion bombardment (or electron bombardment in the case of a dual beam instrument). Alternatively, all FIB milling of an insulator can be performed while collecting ion induced secondary ion images. In addition, further usage of a charge neutralizer in conjunction with secondary ion imaging may also enhance the ability to mill non-conducting samples without prior application of a conductive coating. Using the LO technique requires the addition of either a micromanipulator station for ex-situ LO, or a vacuum compatible probe assembly for in-situ LO as described below.

The "lift-out" or removal of an FIB prepared electron transparent membrane from a larger bulk sample was first discussed by Overwijk et al. (Overwijk et al., 1993). They used an FIB to sputter away trenches of material, leaving behind a thin electron transparent membrane. They reported using a sharp tungsten (W) needle with x, y, and z micrometer controls to remove the thin membrane and place it on a specimen holder consisting of a Si_3N_4 sheet on a Si frame. Leslie et al. (Leslie et al., 1995) used a similar removal procedure as Overwijk et al., and also noted some practical problems and artifacts of the technique. They also discussed the successful preparation of planar TEM specimens. However, little details were provided on how to use the W probes to remove the thinned specimen. Stevie et al. also described an FIB method for obtaining an isolated electron transparent membrane, but suggested no method for the removal of the TEM specimen (Stevie et al., 1995). Herlinger et al. also used a W probe to remove an electron transparent specimen from a bulk sample, but provided few details on how to actually perform the specimen removal (Herlinger et al., 1996).

In 1997, Giannuzzi and co-workers documented a routine FIB lift-out technique for the preparation of site-specific TEM specimens (Giannuzzi et al., 1997). This technique was first developed specifically for applications in the microelectronics industry, however, the FIB LO technique has been routinely used by the authors with slight variations for numerous material systems and geometrical configurations as summarized in Table 1. In addition, FIB LO has been used for analytical techniques other than TEM (e.g., SEM, Auger, SIMS, XPS) (see Stevie et al., 2001, Ferryman et al. 2002). As evident by the other Chapters in this book, Table 1 is by no means an exhaustive list of FIB usage. A current literature search on FIB applications will yield hundreds of papers.

In the *ex-situ* LO (EXLO) technique, the thin membrane is prepared using the FIB. Then the sample containing the electron transparent membrane is removed from the FIB and the membrane is micromanipulated (or "lifted-out") from the bulk sample to a carbon (or formvar) coated, 400 mesh 3 mm TEM grid. The micromanipulator system consists of a light optical microscope and a glass rod attached to a hydraulic micromanipulator (Giannuzzi et al., 1997). We shall refer to this type of lift-out as *"ex-situ"* since the membrane is removed from the milled trenches after the sample is removed from the FIB vacuum chamber. The final thickness of the LO specimen may be tailored to suit requirements of the analytical technique that will be performed to characterize the specimen.

The lift-out method also has significant implications for the overall analytical approach to a site-specific problem. As discussed above, the removed specimen can be analyzed by other techniques. The Ar sputter gun present in most AES systems can remove the Ga implantation remaining from the ion milling process as discussed in the Chapter by Principe. A specimen holder that can accept the 3 mm diameter TEM grids has been designed to transport samples easily from one analytical instrument to another. Note that the LO process for many of the analytical methods is much easier than for TEM, because the specimen does not have to be thinned to electron transparency and is therefore typically removed at a thickness of 1 μm or greater. A discussion on the FIB/AES method is presented in detail by Principe elsewhere in this volume.

The use of computer automated milling of EXLO specimens as described by Young and Moore elsewhere in this volume for microelectronics applications have been employed in our laboratory with as many as 12 specimens prepared overnight. Automated milling is beneficial for preparing multiple specimens of similar type. However, it is sometimes faster and easier to make specimens manually in the case of one-of-a-kind and/or novel specimens because material specific milling parameters must be optimized prior to automation setup. The automation procedure can be extended to rough surfaces by milling fiduciary FIB marks onto a previously deposited layer of Pt or W. The Pt (or W) layers deposited on either side of the region of interest fills the gaps and holes in the rough surface, thereby creating a smooth surface for the fiduciary marks (as per FEI technical notes).

Table 10-1. Example uses of TEM LO by the Authors

Material System	Reference(s)
Semiconductor materials	Giannuzzi et al., 1997, 1998, 1999a, 1999c, 2002; Kempshall et al., 2002; Merchant et al., 2000; Readinger et al., 1999; Rossie et al., 2000; Stevie et al., 1995; Stevie et al., 1998; Stevie et al., 2001
Monolithic ceramics	Giannuzzi et al., 1998, Lomness et al., 2001b
Glasses	Francois-Saint-Cyr et al., 2001, Stevie et al., 2001
Composite materials	Giannuzzi et al., 1999d, 2000, 2001a
High Temperature Coatings	Carim et al., 2001
Polymeric materials	White et al., 2000
Biological materials	Giannuzzi et al., 1999b; Clayton et al., 2002
Geological materials	Drown-MacDonald et al., 1999; Heaney et al., 2001
Powder particles	Drown-MacDonald et al., 1999; Lomness et al., 2001b; ; Prenitzer et al., 1998
Fibers	Giannuzzi et al., 1998
Thin film cross sections	Giannuzzi et al., 1997, 1998; Readinger et al., 1999; Rossie et al., 2000; Stevie et al., 1995; Stevie et al., 2001
Metals, alloys	Giannuzzi et al., 1998, 2001b; Hampton et al., 2002; Kempshall et al., 2001; Lomness et al., 2001a,2001b; Merchant et al., 2000; Prenitzer et al., 1998; Schwarz et al., 2003
Plan view specimens	Stevie et al., 1998
Fracture surfaces	Giannuzzi et al., 1999d, 2000, 2001a, 2001b

Yaguchi et al. pioneered the lift-out of samples within the FIB chamber (Yaguchi et al., 1999, 2000, 2001). We shall refer to this type of lift-out as the *"in- situ"* method since the sample is removed from the bulk material entirely within the FIB instrument. Kamino and co-workers report on the *in-*

situ FIB LO in detail elsewhere in this volume on their discussion on micro-sampling techniques. As elegantly shown by Kamino and co-workers, innovation in micro-sampling (or in-situ LO) has enabled multiple TEM specimens to be prepared in both cross-section and plan view from the same location.

Procedures of both the ex-situ and in-situ LO are described below using an FEI 200TEM FIB workstation. This instrument is a single beam FIB outfitted with automated computer scripting algorithms, Pt deposition, gas assisted etching, and selective carbon mill FEI accessories. The UCF FIB is also outfitted with a probe assembly (courtesy of Omniprobe) to perform in-situ LOs. The steps described below may be applied to other types of FIB instruments and probes in their entirety or with slight deviations as deemed necessary. For example, these or similar techniques have also been employed by our group using the FEI 611, FEI 800, FEI DB235, and an Hitachi FB2000A FIB instruments.

1.1 *Ex-situ* FIB Lift-Out Procedures

Figure 1 shows a schematic diagram of the basic steps used in the manual production of a FIB *ex-situ* LO (EXLO) specimen. Figure 2 shows actual FIB images of some of the steps used in EXLO. The first step is to decide on the final dimensions of the specimen. The final dimensions of the FIB LO will depend on (i) the dimensions of the area of interest, (ii) the characterization technique that will be used to analyze the LO specimen (e.g., for TEM. the thickness is generally < 200 nm, however, specimens may be left thicker for SEM, SIMS, Auger, XPS, etc.), and (iii) what size mesh hole openings are used in the carbon (or formvar) coated grids to support the LO specimen, and (iv) the FIB milling time available. Generally, the dimensions of a FIB EXLO specimen will be 10 - 35 μm in length, 3 – 10 μm in height (deep into the sample) and 50 – 100 nm thick.

Most FIB cross-sections begin by using the ion beam assisted chemical vapor deposition (CVD) process to deposit a 0.5 – 1 μm thick metal line as shown in figure 1a (e.g., usually Pt or W). The metal line serves two purposes. Firstly, the metal line may be used to mark the region of interest, and secondly, the metal line may be used to protect the underlying region from being sputtered away during subsequent milling steps. If the outer most sample surface is of interest, it is necessary to protect the surface from FIB damage that may be incurred during the Pt deposition operation (Kempshall et al., 2002). This may be accomplished by sputter coating a protective metal layer on the surface prior to inserting the sample into the FIB. It has been determined that a sputter coated layer of continuous Cr protects the surface more effectively than the characteristic island-like coverage of an Au-Pd layer, since FIB damage will penetrate the side-walls of the islands (Kempshall et al., 2002). As an alternative, a Pt layer deposited at lower FIB beam energies induces less damage than at higher ion energies (Kempshall et al., 2002). In addition, electron beam induced chemical vapor deposition via a dual beam instrument induces less surface damage than a Ga^+ beam assisted process at low energy (Kempshall et al., 2002).

It should be noted that FEI's FIB instruments use "material files" during any milling operation. This file instructs the system as to how long to mill for a given volume of material to be removed, and the material file is a function of the sputtering yield of the material. Any material file may be used for any material to be milled. For example, if one knows the relative sputtering yield of a material relative to Si, then either (i) the times (or depths) of milling relative to Si may be adjusted accordingly or (ii) a specific material file may be added to the system by using the appropriate sputtering yield factor in the material file set up.

Next, high beam currents with correspondingly large beam sizes are used to mill large amounts of material away from the front and back portion of the region of interest. The front trench is usually positioned ~ 2 μm from the Pt layer and is usually created using either the "stair step" (SS) or "clean up cut" (CUC) milling algorithm using the highest beam current available (e.g., 20,000 pA), as described elsewhere. For a given trench dimension, the CUC usually mills faster than the SS algorithm and also yields a "cleaner" cut at the surface of interest. (see Prenitzer et al, 2002). However, the CUC fills the milled trench behind the beam with redeposited material making specimen preparation more difficult for materials that have a large sputter yield. Thus, it is recommended that the SS cut be used for materials that mill very quickly relative to Si.

The trench milled on the back side of the region of interest is smaller than the front trench. The smaller back trench is primarily to save time, but the smaller trench also prevents the finished LO specimen from falling over flat into larger milled trenches which may make it difficult to remove the specimen during the micromanipulation operation. The back trench may be milled using any number of algorithms. A rectangular cut is preferred to a CUC trench so that sputtered material can escape the trench thereby reducing redeposition artifacts. Next, finer probe sizes are used to progressively thin the specimen to ~ 1 μm in thickness (see figure 1a).

When using a dual platform FIB + SEM (e.g., a DualBeam instrument), it may be useful FIB mill the back side trench with the same dimensions used for the front side trench. The larger trench opening will allow SEM viewing of the specimen from both sides.

It is generally recommended to observe every specimen surface that is prepared for TEM by tilting the stage to ~ 45° and collecting a FIB image and/or an SEM image in the case of a DualBeam instrument. This allows for a one-to-one comparison between the FIB image and a TEM or other image. This may require preparing the surface using finer beam sizes down to 100 pA to collect a high quality FIB or SEM image. Any FIB damage induced during this image acquisition via the ion beam would be removed during subsequent milling steps. To prevent FIB damage, it is generally recommended that the surface of the specimen not be imaged with the ion beam, even with a single beam scan, at a thickness less than ~ 0.5 μm. The exact specimen thickness that one should limit imaging the specimen surface will depend of course on the ion beam/material interactions for that specimen and the final thickness desired for the analytical technique of interest. Note that since the ion beam damage is virtually instantaneous (e.g., ~ 10's of ps - see chapter on ion beam material interactions), even a single scan of the beam may induce beam damage to the surface of the sample. Observation of the specimen surface with an electron beam (e.g., via a DualBeam instrument) prevents ion beam damage to the specimen and allows for thinner specimens to be prepared, since the region of interest may be monitored during FIB milling.

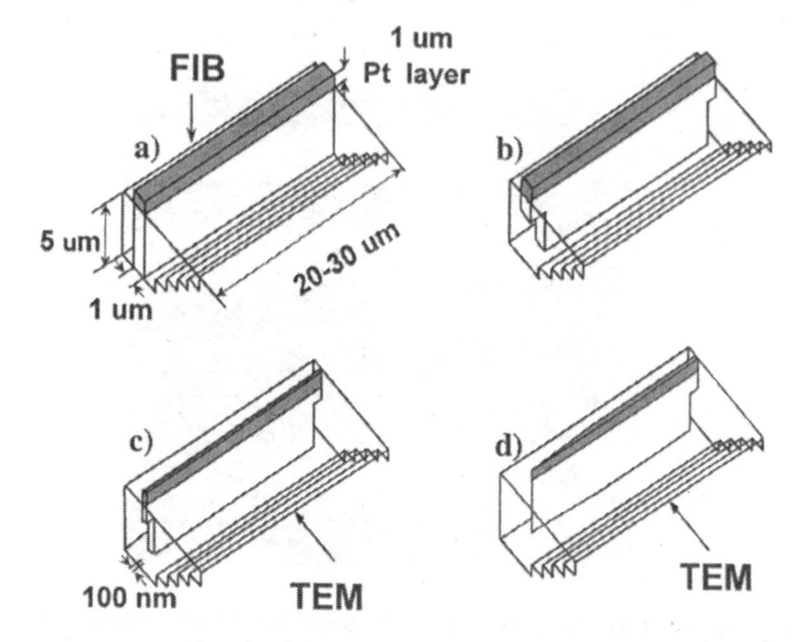

Figure 10.1. A schematic diagram of the general procedures for the EXLO technique. (a) cutting the initial trenches (b) undercutting the specimen (c) final thinning (d) preparation for HREM observation.

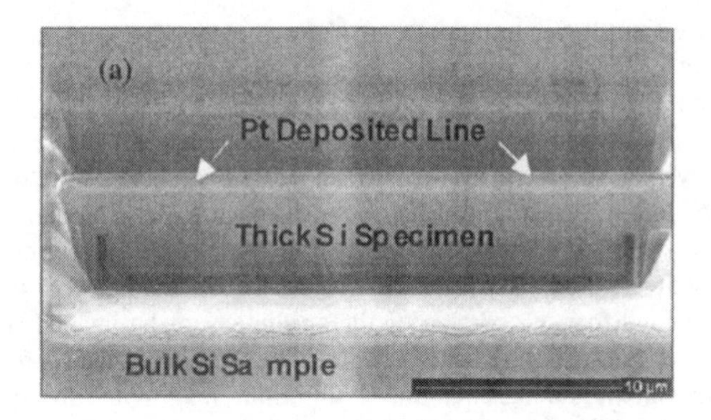

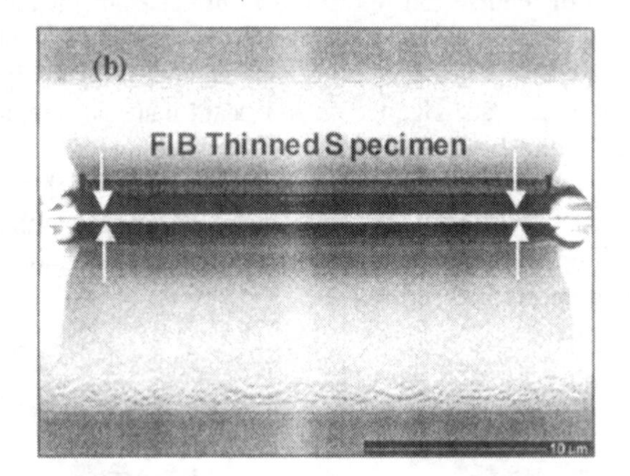

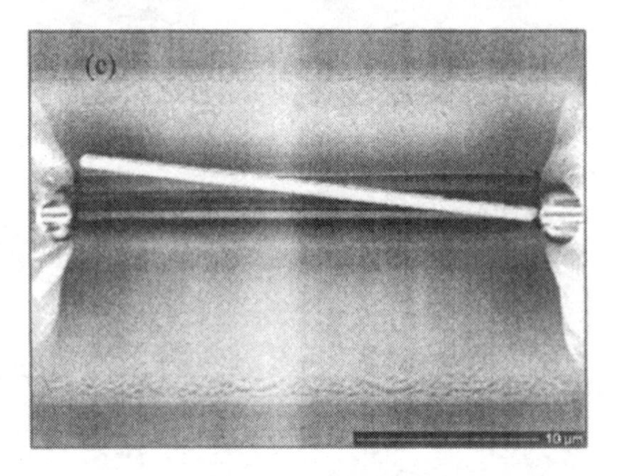

Figure 10.2. EXLO FIB steps for TEM specimen preparation. (a) sample tilted to 45° showing undercut and side cuts of the specimen, (b) final thinning of specimen, (c) freed specimen ready for EXLO (courtesy S. Rajsiri).

Once the specimen reaches a thickness of ~ 1 μm, the stage is tilted 45° (or greater) and the bottom and ~ 3/4 of the right and left edges of the specimen are cut free as shown in figures 1b and 2a, leaving just tabs of material holding the LO specimen. The small tabs allow the least amount of material to be milled free after the specimen is completely FIB polished, reducing the possibility of redeposition artifacts accumulating on the thin LO specimen. These cuts are usually performed with a beam current of 1000 pA. These cuts will usually be 0.5 μm – 1 μm in width depending on the sputter yield of the material. A wider cut may be necessary for materials with a larger sputter yield to avoid redeposited material from filling in the cut during subsequent milling steps as described later. Stresses in the specimen may cause it to warp as it gets thinner.

After the undercutting operations, the specimen is tilted back to normal incidence with respect to the ion beam. For some materials, special types of cuts may be performed when the specimen is ~ 1 μm thick to prevent warping by allowing for expansion of the specimen to reduce internal stresses (Walker 1997, 1998; Rossie et al. 2000). The cutting of a notch at each end of the specimen appears to provide the best results to date (Rossie et al. 2000). The specimen is then further thinned using progressively finer beam sizes (see figures 1c and 2b). Final FIB cuts are usually performed using a beam current of 100 pA or less. Any region(s) of the specimen may be thinned preferentially (as in figure 1d) if high resolution lattice imaging, energy filtered or electron energy loss information is required for the TEM analysis (Giannuzzi et al. 1999a). Once the final tabs of material are cut free the membrane may be observed to move or fall over slightly in the trench. If the specimen is observed to fall over, one can be sure that the specimen has been completely FIB milled leaving the micromanipulation operation to be relatively straight-forward, taking only ~ 3 minutes to perform. A completed EXLO specimen observed to be FIB milled free is shown in figure 2c.

Most LO failures occur because the specimen has not been completely FIB milled free from the trench. It has been observed that sputtered material may redeposit in the undercut portions of the LO, thereby, re-connecting the specimen back to the trench. This may be avoided by using appropriate beam currents for your specimen to limit redeposition effects. In addition, larger dimension cuts can be used when FIB milling a material with a large sputter yield. This problem may also be avoided by undercutting your specimen when it is as thin as possible, but not too thin as to impart beam damage into your specimen due to viewing at 45°.

The sample is then removed from the FIB and placed on a light optical microscope stage equipped with a Narishige hydraulic micromanipulator system. Any microscope that has a long objective working distance e.g., ~ 10 mm will do. Currently available micromanipulator and microscope systems have been developed specifically for EXLO (Micro Optics, FL). Figure 3 shows a typical lift-out system. A glass rod pulled to a fine point e.g., ~ 1 μm in diameter is used to just touch the LO membrane that has been milled free. Electrostatic forces cause the specimen to cling to the glass rod tip. The LO specimen is then positioned on a carbon, holey carbon, or formvar coated 400 mesh TEM grid. An example of a TEM specimen micromanipulated to a carbon coated Cu TEM grid is shown in the light optical micrograph in figure 4. The images in figure 4 show a low magnification light optical image of the 400 mesh TEM grid followed images of the grid containing the specimen at increasing magnification.

Figure 10.3. A micromanipulator system for EXLO.

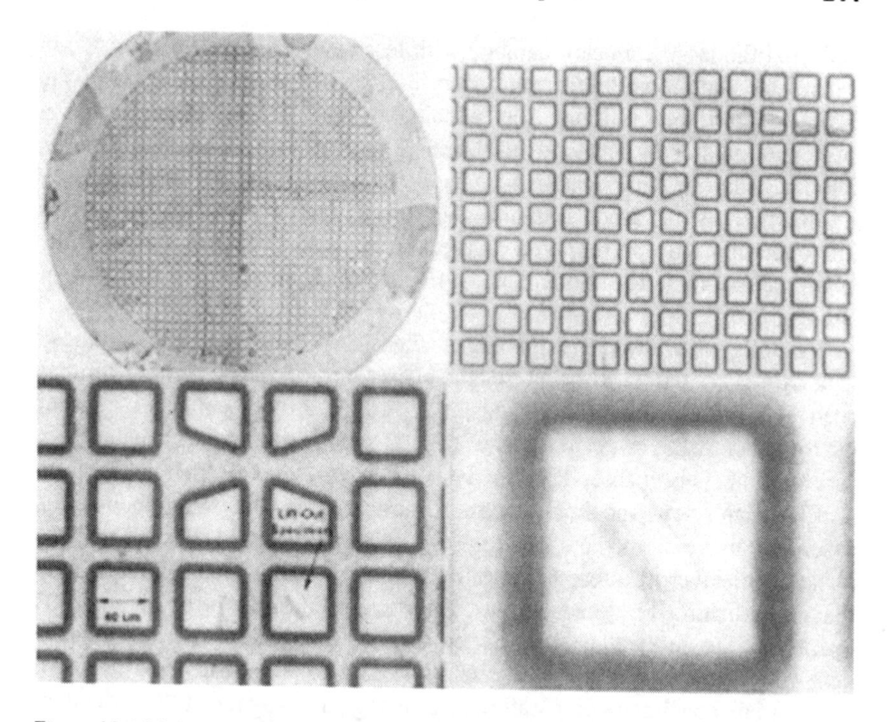

Figure 10.4. Light optical micrographs of increasing magnification of an EXLO specimen on a carbon coated TEM grid

If the specimen has been correctly FIB milled, the LO portion of the preparation may take only a couple of minutes. The success rate generally depends on one's experience with different types of sample systems. The success rates for the EXLO procedure can be greater than 90 %. Multiple specimens may be positioned on the same TEM grid to save time in TEM specimen exchange. These EXLO specimens tend to be very robust. For example, EXLO specimens have been successfully transported on airplane trips and through the mail. Specimens have also survived inadvertent falls to the floor during handling.

Generally, the LO specimen is positioned on a carbon coated grid, or holey carbon coated grid, such that it lies within a single grid opening. However, Langford and Petford-Long (2001a) recently reported the use of a large mesh grid (i.e., small grid hole openings) that does not have a carbon or formvar support film. In this approach, the EXLO specimen was placed directly over 1 or more grid openings. The EXLO specimen remains on the grid due to surface tension. The lack of a carbon or formvar coating enables the EXLO specimen to be ion milled on both sides of the specimen, thereby replacing the large Ga beam damage with less beam damage from an Ar ion mill. The sample assembly was ion milled from the specimen-side of the grid. After ion milling, the specimen was flipped over using a micromanipulator system and put back in the ion mill to remove the FIB damage from the other specimen surface. The advantages to this technique are that there is no carbon film to image through and beam damage from the FIB operation may be removed. Care must be taken when performing the micromanipulation to ensure that a grid bar does not obscure the area of interest.

Recently we showed that the edges of an EXLO specimen can be placed on a carbon coated TEM grid after the center portion of the carbon film has been FIB milled away (Giannuzzi et al. 2002). In this manner, tens of micrometers of the center portion of the LO specimen are unobstructed by a carbon film. This enables TEM analysis using, e.g., energy filtered TEM.

FIB plan view specimens may be prepared in the same manner as described above simply by mounting the specimen in the FIB such that the plane of the layer of interest is parallel to the ion beam. This technique may require polishing the specimen to a particular location prior to any FIB operation (Stevie et al, 1998). Alternatively, Langford et al. developed a plan view specimen preparation technique that has eliminated the need for any initial specimen preparation by using a two-step LO procedure (Langford et al., 2001b). This technique is actually quite similar to the micro-sampling or in-situ LO techniques described below. A two-step process is needed since each micromanipulator operation is performed outside of the FIB. The sample is positioned inside of the FIB in a conventional "cross-sectional" geometry. The plane of interest need not be the outer most surface layer of the sample. First, a wedge of material is FIB milled free such that it contains the plane of interest. Then it is removed from the FIB and micromanipulated onto any flat surface such that the plane of interest is positioned approximately normal to the flat surface. The assembly is then placed back into the FIB and positioned such that the plane of interest is parallel to the ion beam. The specimen position and orientation may be fine tuned using the sample stage. Then an EXLO specimen is FIB milled in the usual fashion. The assembly is removed from the FIB and the electron transparent membrane is micromanipulated to a TEM grid for analysis. The technique described by Langford et al. (2001b) takes full advantage of the EXLO technique by allowing multiple specimens to be prepared from any surface without initial specimen preparation.

1.2 *In-Situ* FIB Lift-Out Procedures

Kamino details micro-sampling techniques, which we will refer to as in-situ LO, elsewhere. The *in-situ* FIB LO (INLO) technique may be summarized as follows. The INLO method can be very similar to the EXLO method described above, however, the specimen size can be larger. The probe is positioned to touch a FIB milled piece of sample. The FIB is then used to attach the probe to the sample using the FIB CVD capabilities. Then the sample is lifted out of the bulk material. An example of a ceramic joint prepared using the INLO technique with an Omniprobe probe assembly mounted in an FEI 200TEM FIB is shown in figure 5. The probe/sample assembly is then positioned onto a 3 mm TEM slotted grid or washer that has been cut in half. The CVD operation is used to attach the sample to the grid. Once the sample is secured to the grid, the probe is FIB milled free from the sample. The sample may then be FIB milled to electron transparency using conventional FIB milling practices.

Figure 10.5. FIB image of an INLO across a ceramic SiC/SiC joint (specimen courtesy of C. Henager, PNNL.)

The INLO technique is the most flexible of the FIB specimen preparation techniques since it offers all of the advantages of the EXLO technique plus the advantages of conventional FIB specimen preparation. In general, an INLO specimen may be FIB milled thinner than its EXLO counterpart since one would tend to leave an EXLO specimen a bit thicker to ensure that a specimen would be retained. In addition, FIB redeposition artifacts tend to be less problematic in an INLO specimen compared to an EXLO specimen since the area surrounding the surface of interest is less confined. For more information see the discussion on redeposition in the chapter on ion-solid interactions.

The mechanism for EXLO is believed to be via electrostatic attraction between the glass tip and the thin membrane. Since the electrostatic attraction is a relatively weak force, it is difficult to extract TEM specimens much larger than ~ 50 μm in length. However, we have performed EXLO on a 150μm long Cu sample that was 1μm thick for other purposes (Contino 2001). Alternatively, the INLO technique relies on attaching the probe to the sample via a CVD process. Thus, much larger samples can be routinely lifted out using INLO. We have successfully prepared INLO specimens on the order of 150 μm in length (Schwarz et al. 2003). For reference, specimens of this size took approximately 4 hours to prepare.

We have sometimes observed that INLO specimens are not as robust as EXLO samples. Great care must be made with INLO samples to assure that nothing comes in contact with the inner portion of the TEM grid in the vicinity of the specimen either during handling or during storage. For example, we have observed that fibers from e.g., a "kimwipe" tissue-like piece of paper, can grab the delicate INLO sample and pull it off the grid. When necessary, greater success in handling INLO grids have been made by placing the samples on fiber-less papers such as cover slip paper. In addition, since INLO specimens are in contact with e.g., a carbon film, the thin region generally stays in tact. Conversely, thin regions in EXLO specimens can break off. This problem is analogous to the loss or "fall out" of thin regions that may be observed in conventionally prepared electropolished, or dimpled and ion milled, TEM specimens.

2. LIFT-OUT EXAMPLES

2.1 EXLO

FIB specimen preparation has its roots in the semiconductor industry. The LO approach also makes possible the removal of this specimen from an unbroken wafer, which may be as large as 300 mm in diameter. FIB modification of devices that are subsequently put back into the process line has already been performed without detrimental Ga contamination (Weiland et al., 2001). Figure 6a shows a semiconductor cross-section TEM image of a 5 layer metal integrated circuit prepared using the FIB EXLO technique. It was recently shown that sufficiently thin EXLO specimens may be used for

EFTEM analysis (Giannuzzi et al. 2002). Figure 6b shows s cross-section of a semiconductor device containing a Ti/TiN diffusion barrier layer. The location and thickness of the Ti and TiN layers is evident by comparing the Ti EFTEM map vs. the N EFTEM map.

Plan view applications of LO additionally demonstrate the versatility of this method (Stevie et al. 1998). Figure 7 shows a TEM image of over 200 W vias prepared by plan view EXLO. The barrier layers surrounding the W may easily be examined at higher magnification.

While the FIB LO technique has been exploited principally in the microelectronics arena to produce site-specific cross-section and plan view TEM specimens, the FIB LO method has also shown tremendous versatility by useful extraction of TEM specimens from a wide range of different materials as shown below.

One of the areas where either LO technique excels is in the TEM specimen preparation of the fracture surface of materials. The advantage here is that a small TEM specimen may be removed from the bulk material without disturbing the remaining surface. One possibility is serial sectioned TEM specimens of surfaces. Another possibility is TEM specimens of multiple regions of interest from a single sample that has been processed (e.g., heat-treated) in a different manner. Figures 8a and 8b show an FIB image and corresponding low magnification bright field TEM image from the fracture surface of a SiC fiber/SiC matrix composite. The composite that was exposed to subcritical crack growth conditions at an oxygen concentration of 2kppm 0_2, and a temperature of 1100 °C (Giannuzzi and Lewinsohn, 2000). Note that the dark contrast in the FIB image indicates charging of an insulating region and corresponds to the amorphous oxide phases located where the carbon interphase region has been oxidized as shown in the TEM image. One advantage of INLO over EXLO for the preparation of fracture surfaces is related to the depth of field of the FIB vs. the light optical microscope. Locating the EXLO specimen on a rough surface using a light optical microscope can be very difficult, but achievable, as shown by figures 8a and 8b.

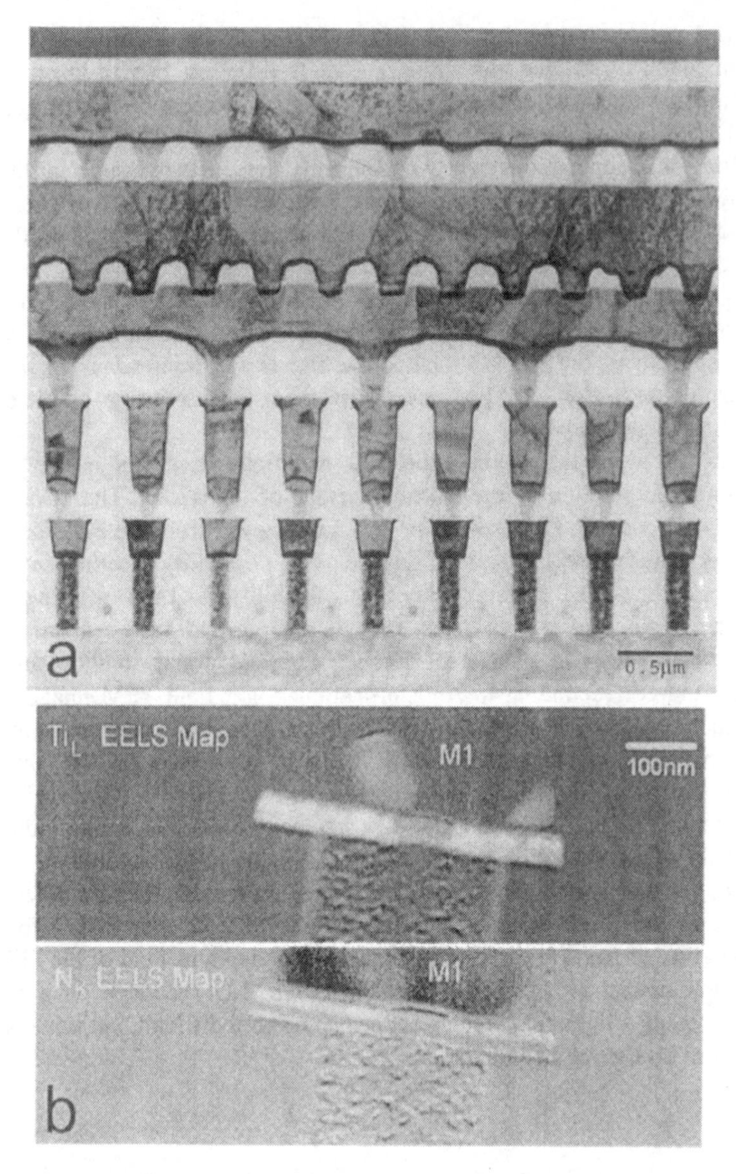

Figure 10.6. TEM analysis of integrated circuits showing (a) a bright field TEM image of a 5 metal layer structure and (b) Ti and N EFTEM maps showing the delineation between Ti and TiN layers.

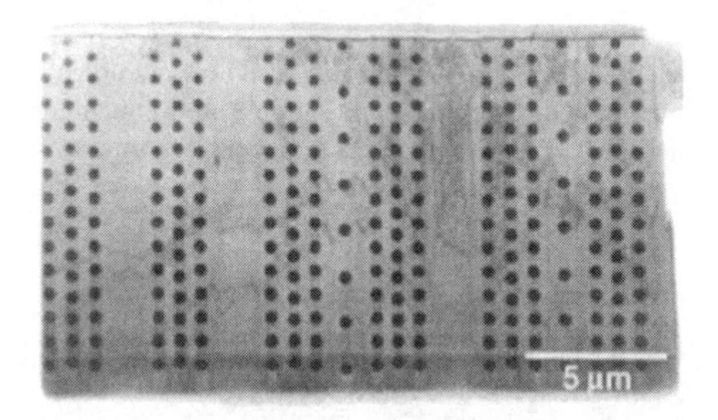

Figure 10.7. Low magnification TEM image showing plan view of over 200 W plugs prepared by EXLO.

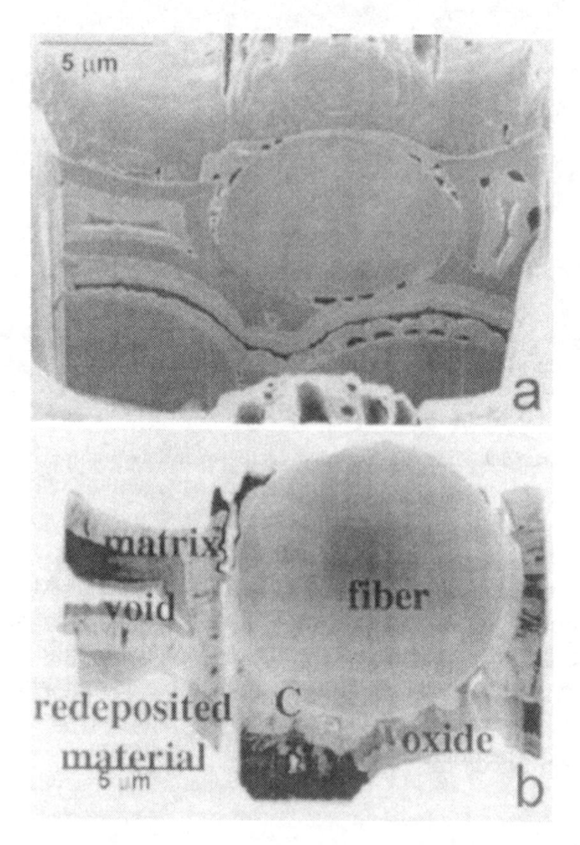

Figure 10.8. (left) FIB image obtained at a 45° tilt and (right) corresponding TEM image from the fracture surface of a SiC/SiC composite (sample courtesy of C.A. Lewinsohn).

An example of the EXLO technique used in a biological specimen is shown in figure 9. The TEM image in figure 9 shows the microstructure of abalone consisting of layers of $CaCO_3$ (aragonite) separated by organic material (Clayton et al., 2002).

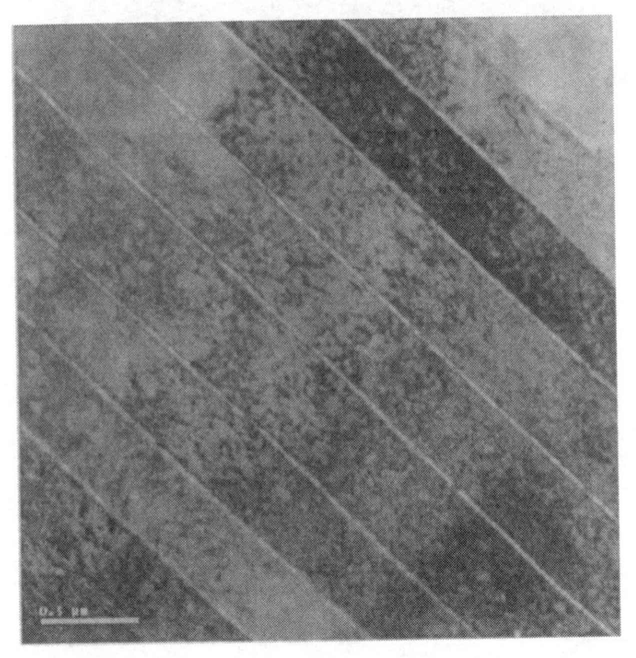

Figure 10.9. TEM image of nacre coating of abalone prepared by EXLO.

One other example of FIB EXLO is shown in the TEM image in figure 10. Figure 10 is an example of EXLO for polymeric materials. The alternating layers of polystyrene and poly-2-vinyl pyridine containing Au nano-particles deposited on a (100) Si wafer are evident (White et al., 2000). In this case, W was used as the protective CVD layer. The Si appears very dark due to diffraction contrast.

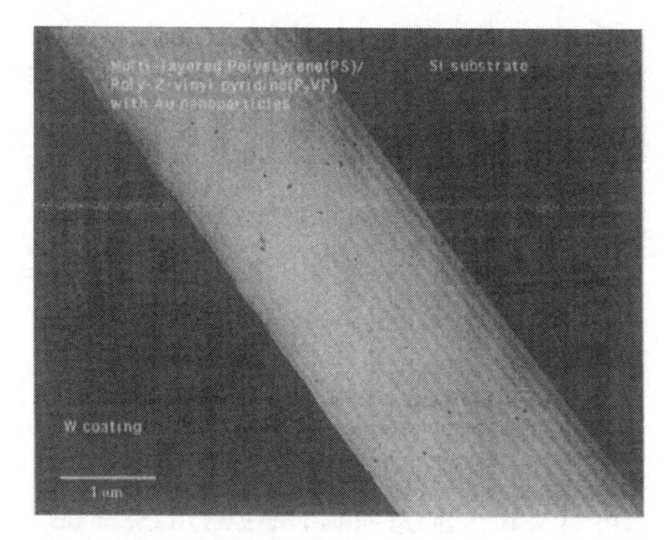

Figure 10.10. TEM image of alternating polymeric layers deposited on a Si substrate prepared by EXLO (sample courtesy of H. White).

2.2 INLO

2.2.1 INLO Example of a Ceramic Joint

The in-situ LO (INLO) technique may be used for any site-specific application where the EXLO technique may be used. In addition, the INLO yields the extra benefit of not having a carbon or formvar film to obstruct the TEM analysis. Furthermore, an INLO specimen can easily be Ar ion milled to remove Ga FIB damage. Another advantage to INLO is in the preparation of irregularly shaped surfaces or fracture surfaces. The poor depth of field of a light optical microscope can make EXLO of these TEM specimen quite challenging. However, since INLO relies on the excellent depth of field of the FIB and/or SEM imaging, INLO from these types of specimens are as easy as extracting a TEM specimen from a flat surface.

An example of a site-specific application of the INLO for the characterization of ceramic interfaces is shown in the high angle annular dark field (HAADF) STEM image in figure 11. The SiC/SiC joint is formed during a displacement phase transformation involving the reaction of TiC + Si to form $TiSiC_2$ + SiC. The phase regions labeled in figure 11 were corroborated using XEDS and electron diffraction. Note that the joint also shows some unreacted TiC. The $TiSi_2$ phase is also expected to form under these conditions as dictated by the Ti-Si-C ternary phase diagram (Radhakrishnan et al., 1996).

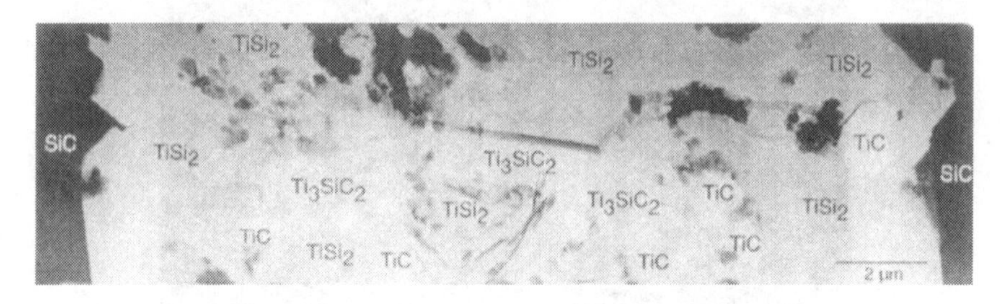

Figure 10.11. HAADF STEM image of a SiC/SiC joint region prepared by INLO as per figure 5 (sample courtesy of C. Henager, PNNL).

2.2.2 HRSTEM With INLO and a Low kV FIB Clean-Up

It has been previously mentioned that Ar ion milling a FIB prepared specimen can remove amorphization FIB damage and therefore improve high resolution TEM imaging. It has also been shown that similar improvements in conventional high resolution TEM imaging, as well as high resolution STEM imaging, may be achieved by polishing the specimen using the FIB at lower operating accelerating voltages (Sudurov, Moore et al., Engelmann et al., 2003). This last step in FIB milling has been referred to as a "low kV clean-up."

2.2.3 Backside INLO FIB Milling

One of the artifacts associated with FIB specimen preparation is what is commonly known as "curtaining." The curtaining effect has been discussed by Anderson and Klepeis, by Phaneuf, and by Holdford, in other chapters in this volume. Curtaining artifacts are most often observed in semiconductor materials where multiple patterned layers of materials having a low sputtering yield blocks a faster sputtering yield material. Curtaining may also be observed in materials exhibiting different topographic regions where changes in sputtering yields vary with the milling incident angle. This artifact can be especially problematic in electron holography of semiconductor gate structures where the phase image is dependent on specimen thickness as well as the desired dopant distribution (Gribelyuk et al., 2002; McCartney et al., 2002). In a bright field TEM image, curtaining appears as mass/thickness contrast where, e.g., the Si substrate appears darker directly under a gate as compared to a region far from the gate. In another chapter, Phaneuf showed that by FIB milling at different angles, the curtaining effect was eliminated. An alternative method for avoiding the curtaining effect is presented below. This method is specifically useful for determining 2D dopant profiles using electron holography. The technique is based on in-situ lift-out (INLO) of a semiconductor device that is FIB milled to electron transparency from the Si-side of the device. Backside milling by FIB INLO on an FEI single beam 200TEM FIB is described below. The electron holographic examination of these FIB prepared specimens are very encouraging and will be presented elsewhere (Schwarz et al., 2003, McCartney et al., 2003).

INLO of a Si-based gate structure was prepared as described previously. The in-situ probe was adhered to the top circuitry portion of the device. However, in order to FIB mill from the Si-side of the sample, a cut TEM grid was mounted "upside down" in the FIB holder. That is, the cut side of the grid was positioned such that it faced downward as shown in figure 13a. In figure 13a the probe is observed to just pass underneath the slotted Cu grid. The probe carrying the specimen passes partially through the grid opening and the specimen is mounted on the side of the Cu grid using ion beam assisted Pt deposition. In this geometry, the specimen is mounted as a cantilever. The specimen mount was then removed from the FIB instrument and the Cu grid was flipped 180° such that the grid cut opening, and hence the Si substrate, faced upwards. The holder was placed back into the FIB so that FIB milling could now be performed from the Si side of the specimen as shown in figure 13b. Gate structures in the specimen were identified and material below the gate, as viewed from figure 13b, was removed by FIB milling. The specimen was FIB milled from the Si side to a thickness of ~ 300 nm for electron holography. In another alternate procedure, specimens were FIB milled to ~ 800 nm and then further thinned in an ion mill in an attempt to remove FIB damage (McCartney et al., 2003). The ion milling operation was observed to remove the so called "dead layer" that is often associated with electron holography of FIB prepared specimens. An example of a backside FIB milled INLO specimen is shown in the FIB image of figure 13b. Note that there is no evidence of curtaining in the Si substrate in this FIB image.

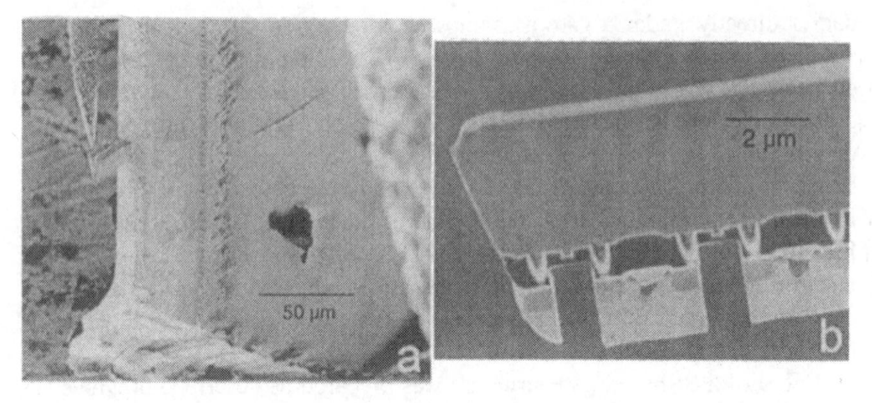

Figure 10.12. Backside INLO for reduction of curtaining. (a) FIB image of INLO probe attaching specimen to upside down grid. (b) FIB image of backside milling of Si substrate for electron holography of gate structure.

3. VARIATIONS OF THE FIB LO TECHNIQUE

The general steps of the LO technique may be altered slightly to accommodate specific types of specimens. For example, particles smaller than the conventional LO specimen size have been prepared by the EXLO technique by using the Pt CVD deposition to attach multiple particles together (Lomness et al., 2001a, 2001b). In this manner, site specific specimens have been successfully prepared from micrometer and sub-micrometer particles for TEM analysis. In addition, the ex-situ micromanipulator may be used to adhere specific particles onto a TEM half-grid for conventional FIB specimen preparation.

Figure 13a shows a light optical micrograph of the ex-situ probe dipped into M-Bond adhesive. The probe itself is not observed due to the poor depth of field of the light optical microscope, but the tip of the probe is seen as the dark spot dipping into the center of the adhesive in figure 14a. The tip of the probe was then used to smear some of the M-Bond on a TEM grid as shown in figure 14b. Then the probe was used to pick up a specific particle as shown in figure 14c and position it on the TEM grid where the M-Bond was located as shown in figure 14d. Multiple particles may be mounted along the TEM grid in this fashion. The heat from the microscope lamp is sufficient to cure the M-Bond used to affix the particle to the grid. The grid containing the particles was then mounted in the FIB for conventional specimen preparation. Figure 15 shows a FIB image of the particle mounted as shown in figure 14. This particle was mechanically alloyed using a mixture of Ti, Mg, and Ni. The convoluted microstructure due the mechanically alloying is evident in figure 15.

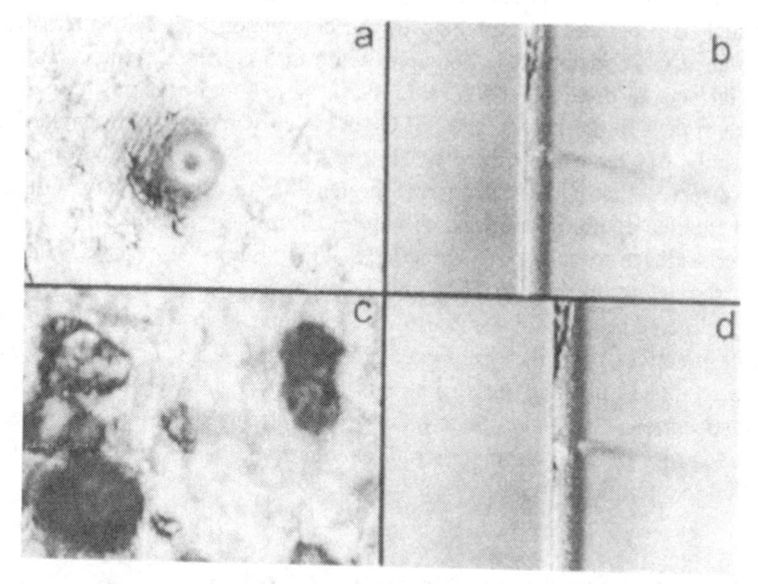

Figure 10.13. Light optical micrographs showing the use of the ex-situ micromanipulator probe for site-specific FIB/TEM specimen preparation of individual particles. (a) the probe touches M-Bond adhesive. (b) M-Bond is smeared onto a Cu TEM grid using the probe. (c) the probe is used to pick up an individual particle (d) the probe is used to place a particle onto the grid containing the M-Bond.

Figure 10.14. FIB image of a mechanically alloyed particle consisting of a mechanically alloyed mixture of Ti, Mg, and Ni. This single specific particle was placed on a TEM grid as shown in figure 14.

4. EXLO VS. INLO BEAM DAMAGE

Recent results have shown that redeposition artifacts can be worse for EXLO than for INLO specimens. These differences may be controlled using careful FIB practices as explained in the chapter on ion-solid interactions. Since an EXLO specimen is prepared while in a confined trench, sputtered material has a greater tendency to be retained and redeposited on the specimen than in the case of the INLO specimen where sputtered material is essentially free to escape directly via the vacuum system. Therefore, we have observed that EXLO specimens, in general, may contain large amounts of redeposited sputtered material containing significant amounts of Ga if one does not adhere to sound FIB practices. On the contrary, Ga is generally below the detection limit of TEM and X-ray energy dispersive spectrometry on specimens prepared by the INLO technique. However, careful FIB milling practices may also yield excellent quality EXLO specimens as well. These results imply that large amounts of Ga previously observed within TEM specimens are solely due to redeposition effects and are not an inherent characteristic of ion beam/material interactions. These results will be detailed in a publication in the near future.

5. SUMMARY AND CONCLUSIONS

The EXLO and INLO FIB methods have been summarized and examples of the techniques have been demonstrated. The LO techniques offer the advantage of rapid site-specific specimen preparation with little or no initial preparation needed. The INLO technique offers all the advantages of the EXLO technique plus the benefits of conventional FIB milling methods.

ACKNOWLEDGEMENTS

The authors would like to thank FEI, Hitachi, Micro-Optics, and Omniprobe for equipment support. Thanks to Dr. Kamino for helpful discussions. We would like to thank the I4/UCF/Cirent/Agere Partnership, NSF DMR #9703281, AMPAC, Florida Space Grant Consortium, Florida Solar Energy Center, and the DOE OIT program for financial support. We would especially like to thank the Cirent/Agere staff, high school students, and UCF students and staff who have performed FIB-related research over the years. The names include: Shawn Anderson, Mike Antonell, A. Pat Arauz, Christian Bechtold, Steve Brown, Bo Clayton, Catherine Contino,

Jennifer Drown, Erik Hogue, Rich Irwin, Christal Jolly, Tammy Matteson, Bob Mills, Preston Patterson, Brenda Purcell, Supphachan Rajsiri, Ben Rossi, Alex Schwitter, Terri Shofner, Brian Surdyck, Carrie Urbanik-Shannon, Zia Ur Rahman, Cathy Vartuli and Hanlin Zhang. Special thanks to Jeff Bindell for his undying support of the UCF/Cirent/Agere partnership.

REFERENCES

Carim AH, Dobbins TA, Mayo MJ, and Giannuzzi LA, "Interfacial Microstructure For As-Deposited And Cycled-To-Failure Thermal Barrier Coatings," in Elevated Temperature Coatings: Science and Technology IV, N. B. Dahotre, J. M. Hampikian, and J. E. Morral, eds., TMS, Warrendale, PA, pp. 45-59 (2001).

Clayton FB, Kempshall BW, Schwarz SM, and Giannuzzi LA, "Automated Crystallography and Grain Mapping in the TEM," Microsc. and Microanal. 8 (Suppl. 2), 656-657CD (2002).

Contino C, "Electromigration Behavior in Single Crystal Copper," Undergraduate Honors Thesis, University of Central Florida, (2001)

Drown-MacDonald JL, Prenitzer BI, Shofner TL, and Giannuzzi LA, "TEM FIB Lift-Out Of Mount Saint Helens Volcanic Ash," Microscopy and Microanalysis, vol.5, supplement 2, Proceedings: Microscopy & Microanalysis 99, 908-909 (1999).

Engelmann HJ, Volkmann B, Ritz Y, Saage H, Stegmann H, de Robillard Q, Zschech E, "TEM Sample Preparation Using Focused Ion Beam - Capabilities and Limits," in Microscopy Today, vol 11 No 2, p 22-24 (March/April 2003).

Ferryman AC, Fulghum JE, Giannuzzi LA, and Stevie FA, "XPS Analysis FIB Milled Si," Surf. Interface Analysis, 33, 907-913 (2002).

Francois-Saint-Cyr H, Elshot K, Le Coustumer P, Bourrat X, Richardson K, Giannuzzi L, Glebov L, Glebova L, Stevie F, "TEM Specimen Preparation Techniques and Analysis of Photo-Thermo-Refractive Glasses (PTRG)," Microsc. Microanal., 7 (Suppl 2: Proceedings), Microscopy Society of America, 432-433 (2001).

Giannuzzi LA, Drown JL, Brown SR, Irwin RB, Stevie FA, "Focused Ion Beam Milling and Micromanipulation Lift-Out for Site Specific Cross-Section TEM Specimen Preparation," Mat. Res. Soc. Symp. Proc., 480, 19-27 (1997).

Giannuzzi LA, Drown JL, Brown SR, Irwin RB, Stevie FA, "Applications of the FIB Lift-Out Technique," Microscopy Research and Technique, 41, 285-290 (1998).

Giannuzzi LA, Prenitzer BI, Drown-MacDonald JL, Brown SR, Irwin RB, and Stevie FA, Shofner TL, "Advances in the FIB Lift-Out Technique for TEM Specimen Preparation: HREM Lattice Imaging," Microstructural Science, Volume 26, The 31st Annual Technical Meeting of the International Metallographical Society, 249-253 (1999a).

Giannuzzi LA, Prenitzer BI, Drown-MacDonald JL, Shofner TL, Brown SR, Irwin RB, Stevie FA, "Electron Microscopy Sample Preparation For the Biological and Physical Sciences Using Focused Ion Beams," J. Process Analytical Chemistry, vol. IV, No. 3,4p. 162-167 (1999b).

Giannuzzi LA.and Stevie FA, "Focused Ion Beam Milling for TEM Specimen Preparation," Micron, vol, 30, No. 3, p. 197-204 , (1999c).

Giannuzzi LA and Lewinsohn CA, "Interphase Oxidation In SIC/SIC Composites At Varying Partial Pressures Of Oxygen," 23rd Annual Conference on Composites, Advanced

Ceramics, Materials, and Structures: B, Ceramic Engineering and Science Proceedings, eds. Ersan Ustundag and Gary Fischman, Vol. 20, Issue 4, p. 115-122 (1999d).

Giannuzzi LA and Lewinsohn CA, "Interphase Oxidation in SiC/SiC Composites," 24rd Annual Conference on Composites, Advanced Ceramics, Materials, and Structures: A, Ceramic Engineering and Science Proceedings, eds. Todd Jansen, Ersan Ustundag, Vol. 21, Issue 3, p 469-477 (2000).

Giannuzzi LA and Lewinsohn CA, "Fiber-Reinforced Composites: C and SiC Fibers - Oxidation Effects - Crack Growth Behavior and TEM Analysis of Interphase Oxidation in Boron-Enhanced SiC/SiC Composites," Ceramic engineering and science proceedings. 22, no. 3, p 617-624 American Ceramic Society (2001a).

Giannuzzi LA, White HJ, Chen WC, "Applications of the FIB Lift-Out Technique for the TEM of Cold Worked Fracture Surfaces," Microsc. Microanal., 7 (Suppl 2: Proceedings), Microscopy Society of America, 942-943 (2001b).

Giannuzzi LA, Kempshall, BW, Anderson SD, Prenitzer BI, Moore TM, "FIB Lift-Out for Defect Analysis," in Analysis Techniques of Submicron Defects, 2002 Supplement to the EDFAS Failure Analysis Desktop Reference, ASM International, Materials Park, Ohio 29-35 (2002).

Hampton MD, Lomness JK, and Giannuzzi LA, "Surface Study of Liquid Water Treated and Water Vapor Treated $Mg_{2.35}Ni$ Alloy," accepted and in press, International Journal of Hydrogen Energy, (2002).

Heaney PJ, Vicenzi EP, Giannuzzi LA, and Livi Kenneth JT, "Focused Ion Beam Milling: A Method of Site-Specific Sample Extraction for Microanalysis of Earth Materials," American Mineralogist, 86, 1094-1099 (2001).

Herlinger LR, Chevacharoenkul S, Erwin DC, "TEM Sample Preparation Using a Focused Ion Beam and a Probe Manipulator," Proceedings of the 22nd International Symposium for Testing and Failure Analysis, 18-22, Los Angeles, California, 199-205 (November 1996).

Kempshall BW, Schwarz SM, Prenitzer BI, Giannuzzi LA, and Stevie FA, "Ion Channeling Effects on the FIB Milling of Copper," Journal of Vacuum Science & Technology B, 19(3), 749-754 (2001).

Kempshall BW, Prenitzer BI, Giannuzzi LA, Stevie FA, and Da SX, "A Comparative Evaluation of FIB CVD Processes," Journal of Vacuum Science & Technology Journal of Vacuum Science & Technology B, 20(1), p 286-290 (2002).

Langford RM and Petford-Long AK, "Broad Ion Beam Milling Of Focused Ion Beam Prepared Transmission Electron Microscopy Cross Sections For High Resolution Electron Microscopy," J Vac Sci Technol A 19(3), p 982-985 (2001a).

Langford RM, Huang YZ, Lozano-Perez S, Titchmarsh JM, Petford-Long AK, "Preparation Of Site Specific Transmission Electron Microscopy Plan-View Specimens Using A Focused Ion Beam System," J. Vac. Sci. Technol. B 19 (3), 755-758 (2001b).

Leslie AJ, Pey KI, Sim KS, Beh MTF, Goh FP, "TEM Sample Preparation Using FIB: Practical Problems and Artifacts," ISTFA '95, International Symposium for Testing and Failure Analysis," 6-10 November 1995, Santa Clara, California, ASM Int., 353 (1995).

Lomness JK, Giannuzzi LA, and Hampton MD, "TEM Characterization of Sub-Micrometer Particles Using the FIB Lift-Out Technique," Microscopy and Microanalysis, 7, 418-423 (2001a).

Lomness JK, Kempshall BW, Giannuzzi LA, Watson MB, "TEM of Sub-Micrometer Particles Using the FIB Lift-Out Technique," Microsc. Microanal., 7 (Suppl 2: Proceedings), Microscopy Society of America, 950-951 (2001b).

McCartney MR, Li J, Chakraborty P, Giannuzzi LA, Schwarz SM, to be published, Proc. Microscopy and Microanalysis (2003).

Merchant HD, Wang JT, Giannuzzi LA, Liu YL, "Metallurgy and performance of electrodeposited copper for flexible circuits," Circuit World, vol. 26, No. 4, September p. 7-14 (2000).

Overwijk MHF, van den Heuvel FC, and Bull-Lieuwma CWT, "Novel Scheme for the Preparation of Transmission Electron Microscopy Specimens with a Focused Ion Beam," J. Vac. Sci. Technol., B 11(6), 2021-2024 (1993).

Prenitzer BI, Giannuzzi LA, Newman K, Brown SR, Shofner TL, Irwin RB, Stevie FA, "Transmission Electron Microscope Specimen Preparation of Zn Powders Using the Focused Ion Beam Lift-Out Technique," Metallurgical and Materials Transactions A., v 29 n 9, P. 2399-2406 SEP 01 (1998).

Prenitzer BI, Urbanik-Shannon CA, Giannuzzi LA, Brown SR, Irwin RB, Shofner TL, Stevie FA, "The Correlation Between Ion Beam/Material Interactions and Practical FIB Specimen Preparation," accepted and in press, Microscopy and Microanalysis (2002).

Radhakrishnan R, Henager CH Jr, Brimhall JL, Bhaduri SB, "Synthesis of Ti//3SiC//2/SiC and TiSi//2/SiC composites using displacement reactions in the Ti-Si-C system," Scripta Materialia, 34 1809-1814 (1996).

Readinger ED, Wolter SD, Waltemyer DL, Delucca JM, Mohney SE, Prenitzer BI, Giannuzzi LA, and Molnar RJ, "Wet Thermal Oxidation of GaN," J. Electronic Materials, Vol. 28, No. 3, p. 257-260 (1999).

Rossie BB, Stevie FA, Shofner TL, Brown SR, Irwin RB, Microsc. Microanaly., 6 (Suppl 2) 532-533 (2000).

Schwarz SM, Kempshall BW, Giannuzzi LA, and McCartney MR, to be published, Proc. Microscopy and Microanalysis, (2003).

Schwarz SM, Kempshall BW, Giannuzzi LA, "Effects of Diffusion Induced Recrystallization on Volume Diffusion in the Copper-Nickel System," accepted and in press, available on-line(http://www.sciencedirect.com/science/article/B6TW8-485P4SJ-3/2/570494c5708da0eb0ccf67600d0b3cd4), Acta Materialia, (2003).

Stevie FA, Shane TC, Kahora PM, Hull R, Bahnck D, Kannan VC, and David E, "Applications of Focused Ion Beams in Microelectronics Production, Design And Development," Surface and Interface Analysis, 23, 61-68 (1995).

Stevie FA, Irwin RB, Shofner TL, Brown SR, Drown JL, and Giannuzzi LA, "Plan View Sample Preparation Using the Focused Ion Beam Lift Out Technique," Proceedings of the 1998 International Conference on Characterization and Metrology of ULSI Technology, vol. 449, Gaithersburg, MD, March 23-27, 1998, eds. DG Seiler, AC Diebold, WM Bullis, TG Shaffner, R. McDonald, EJ Walters, p 868-871 (1998).

Stevie FA, Vartuli CB, Giannuzzi LA, Shofner TL, Brown SR, Rossie B, Hillion F, Mills R H, Antonell M, Irwin RB, and Purcell BM, "Application of Focused Ion Beam Lift-Out Specimen Preparation To TEM, SEM, STEM, AES, and SIMS Analysis," *Surface* Interface Analysis, 31, 345-351 (2001).

Walker JF, "Preparing TEM Sections By FIB: Stress Relief To Straighten Warping Membranes," Microscopy of Semiconducting Materials 1997. Proceedings of the Royal Microscopical Society Conference, Institute of Physics Publishing; Bristol, UK, Microscopy of Semiconducting Materials 469-72, (1997).

Walker JF, Electron Microscopy, 555 (1998).

Weiland R, Boit C, Dawes N, Dziesiaty A, Demm E, Ebersberger B, Frey L, Geyer S, Hirsch A, Lehrer C, Meis P, Kamolz M, Lezec H, Tittes W, Treichler R, Zimmermann H, "In-line failure analysis on productive wafers with dual-beam SEM/FIB systems," Proceedings of SPIE-The International Society for Optical Engineering 4406, no. In-Line

Characterization, Yield, Reliability, and Failure Analysis in Microelectronic Manufacturing II, p 21-30 (2001).

White HJ, Pu Y, Rafailovich M, Sokolov J, King AH, Giannuzzi LA, Urbanik-Shannon C, Kempshall BW, Eisenberg A, Schwarz SA, Strzhemechny YM, Focused Ion Beam/ Lift-Out TEM Cross Sections of Block Copolymer Films Ordered on Silicon Substrates," Polymer, v 42 n 4, p 1613-1619 (2000).

Yaguchi T, Kamino T, Ishitani T and Urao R., "Method for cross sectional TEM specimen preparation of composite materials using a dedicated FIB system," Microscopy and Microanalysis 5, 363-370 (1999).

Yaguchi T, Kamino T, Sasaki M, Barbezat G and Urao R, "Cross section specimen preparation and observation of a plasma sprayed coating using an FIB/TEM system," Microscopy and Microanalysis 6, 218-223 (2000).

Yaguchi T, Urao R, Kamino T, Ohnishi T, Hashimoto T, Umemura K, Tomimatsu S, "A FIB micro-sampling technique and a site-specific TEM specimen preparation method for precision materials characterization," Mater. Res. Soc. Symp. Proc. 636, Nonlithographic and Lithographic Methods of Nanofabrication: From Ultralarge-Scale Integration to Photonics to Molecular Electronics D9.35/1-D9.35/6 (2001).

Chapter 11

A FIB MICRO-SAMPLING TECHNIQUE AND A SITE SPECIFIC TEM SPECIMEN PREPARATION METHOD

T.Kamino*, T.Yaguchi*, T.Hashimoto**, T.Ohnishi** and K.Umemura***
**Hitachi Science Systems, ** Hitachi High-Technologies, ***Hitachi, Central Laboratory*

Abstract: A FIB micro-sampling technique has been developed to facilitate TEM specimen preparation while allowing samples to remain intact. A deep trench is FIB-milled to remove a portion of the sample containing the region of interest. A micromanipulator is employed for the purpose of lifting out a small portion of the sample, i.e., the micro-sample. FIB assisted metal deposition is used to bond the micro-sample to the micromanipulator. The micro-sample is subsequently lifted out and mounted onto an edge of the micro-sample carrier using FIB assisted metal deposition. The micro-sample is then thinned to the thickness of about 0.1μm for TEM observation. All of the above steps are accomplished under vacuum in the same FIB system. This procedure is a reliable TEM specimen preparation technique when the evaluation or failure analysis of a specific site is required. Both cross sectional and plan view TEM specimen preparations are feasible with this technique. In addition, a technique to prepare TEM specimens from a specific site has also been developed. In this technique, an FIB system equipped with a FIB/TEM(STEM) compatible specimen holder is used for thinning of the samples, e.g., a micro-sample. The compatible specimen holder permits repeated alternating FIB milling and TEM(STEM) observation, enabling TEM specimen preparation from a specific site.

Key words: FIB micro-sampling technique, FIB milling, TEM observation, FIB/TEM(STEM) system, site specific TEM specimen preparation

1. FIB MICRO-SAMPLING TECHNIQUE

1.1 FIB Micro-Sampling for Cross Sectional TEM

The FIB lift-out technique has been developed and applied as a method to lift out a TEM specimens directly from a region of interest in the evaluation of various materials (Giannuzzi et al., 1997, 1998; Stevie et al., 1998). Although the FIB micro-sampling technique resembles the FIB lift-out technique in the respect that the bulk sample remains intact, the method to lift out a micro-sample and the thickness of the lifted out sample are different (Yaguchi et al., 1998, 1999, 2000). The FIB micro-sampling technique is developed to remove a portion of the sample with a thickness of a few micrometers. The process of lifting out the micro-sample, mounting the micro-sample on a special carrier, and thinning of the micro-sample by FIB milling are performed in the FIB system continuously. The procedure for the FIB micro-sampling is schematically illustrated in Figure 1.

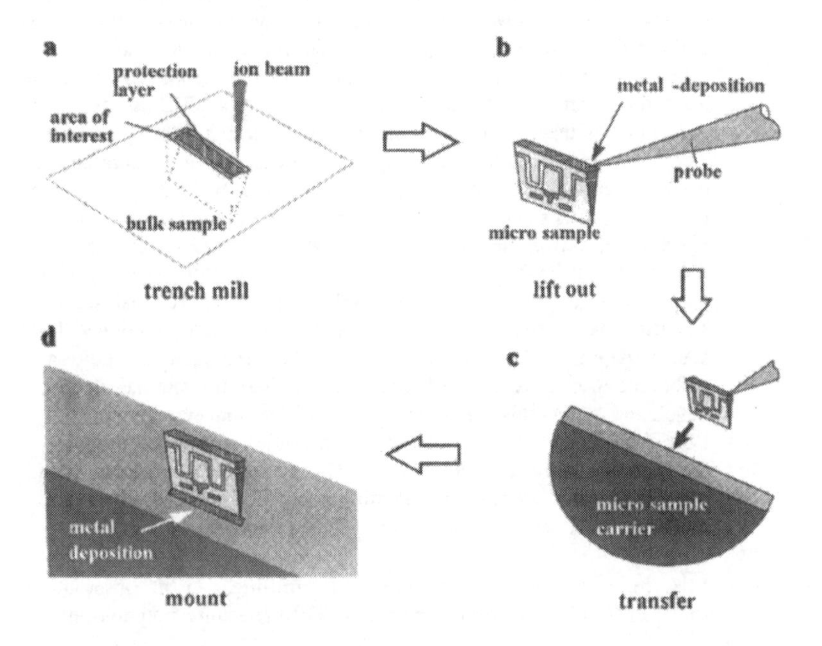

Figure 11.1. Schematic illustration of the FIB micro-sampling technique.

First, the region of interest is covered with a protective layer of W or Pt by FIB assisted deposition. Then the sample is deep trenched by FIB milling to remove a portion of sample (micro-sample) from the region of interest (Fig. 1a). Next, the micro-sample is separated from the bulk sample and lifted out using a micromanipulator (Fig. 1b). The micro-sample is then mounted onto an edge of a micro-sample carrier (Fig. 1c). Finally, the mounted micro-sample is secured by an FIB assisted metal deposition (Fig. 1d). The micro-sample carrier is mounted on an FIB/TEM(STEM) compatible specimen holder so that thinning of the micro-sample can be performed just after securing it to the micro-sample carrier. The time that is required for the whole micro-sampling procedure is dependent on the size of the micro-sample and the elemental composition of the sample. The total preparation time for a 15 μm wide by 15 μm high micro-sample from a semiconductor device is approximately 1 to 1.5 hours.

Figure 2 shows the micromanipulator unit used in the micro-sampling technique. The micromanipulator unit consists of a side-entry stage with a motor driving system and a probe holder with a W probe. The probe holder can be motor driven in three directions (x, y and z) at various rates. This allows for the finely controlled and precisely performed approach of the W probe to the micro-sample. Figure 3 shows the operations of both the micromanipulator and the sample holders in the FIB micro-sampling technique. First, the holder with a bulk sample is inserted into the FIB system and a micro-sample is produced from the bulk sample (Fig. 3a). Next, the micromanipulator is moved into position and the micro-sample is bonded to the W probe (Fig. 3b). Then the micromanipulator with the micro-sample attached is moved to the stand-by position and the bulk sample holder is removed from the FIB instrument (Fig. 3c). Next, a sample holder with a micro-sample carrier is inserted into the FIB instrument (Fig. 3d). Then, the micro-sample is transferred and mounted on the edge of the micro-sample carrier (Fig. 3e). After mounting the micro-sample, the W probe is separated from the micro-sample with FIB milling and the micromanipulator is moved to the stand-by position (Fig. 3f).

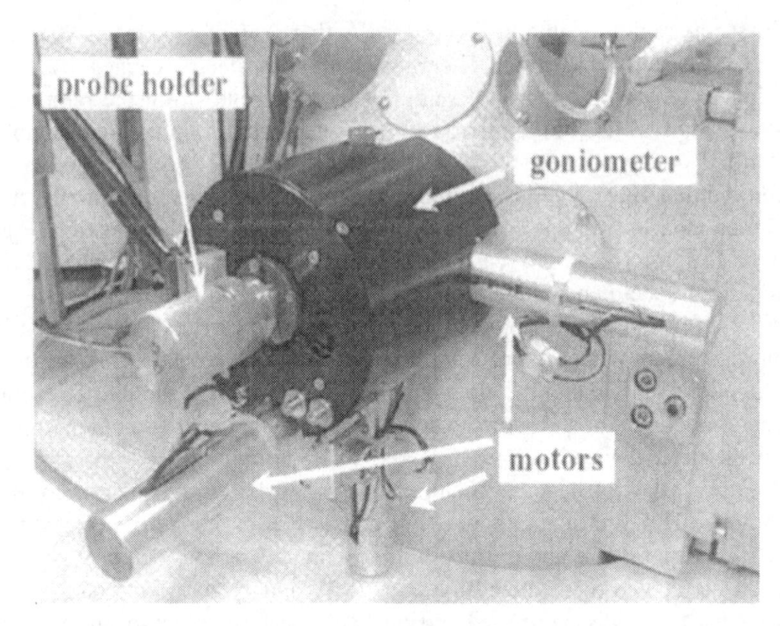

Figure 11.2. Micromanipulator used in the FIB micro-sampling technique.

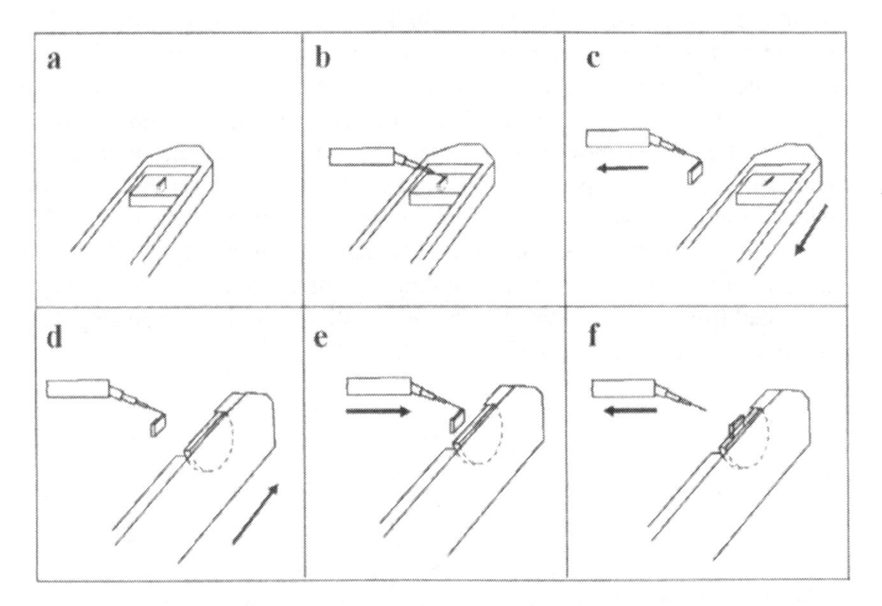

Figure 11.3. Operations of a micromanipulator probe and sample holders in the FIB micro-sampling technique.

A series of FIB images showing the procedure to prepare a micro-sample is shown in Figure 4. First, the area to be examined is localized (Fig. 4a). Then, metal (W or Pt) is deposited as the protection layer over the region of interest (Fig. 4b). The thickness of the deposition layer is normally 0.5 to 1 µm. Next, the front side of the region of interest is deep trenched by FIB milling (Fig. 4c). After that, the back side and both right and left sides of the region of interest are deep trenched leaving a micro-bridge at the upper left corner of the sample (Fig. 4d). The sample is then tilted 60° and the bottom of the sample is milled to separate the bottom portion of the sample from the bulk sample (Fig. 4e). The bulk sample is then tilted back to normal to the ion beam (Fig. 4f), and a micromanipulator probe is bonded to the micro-sample with W deposition. The size of the deposited W bonding layer is about 2 µm x 2 µm and the thickness of the layer is approximately 0.5 µm. After the probe bonding, the micro-bridge is cut off with FIB milling to completely separate the micro-sample from the bulk sample (Fig. 4g). Then, the micromanipulator probe is elevated to lift out the micro-sample (Fig. 4h). The milling and bonding operations are performed using a Ga^+ ion beam at an accelerating voltage of 30kV. The beam currents and the beam diameters of the Ga^+ ion beam used in these operations ranged from 10-20 nA and 400-600 nm, respectively.

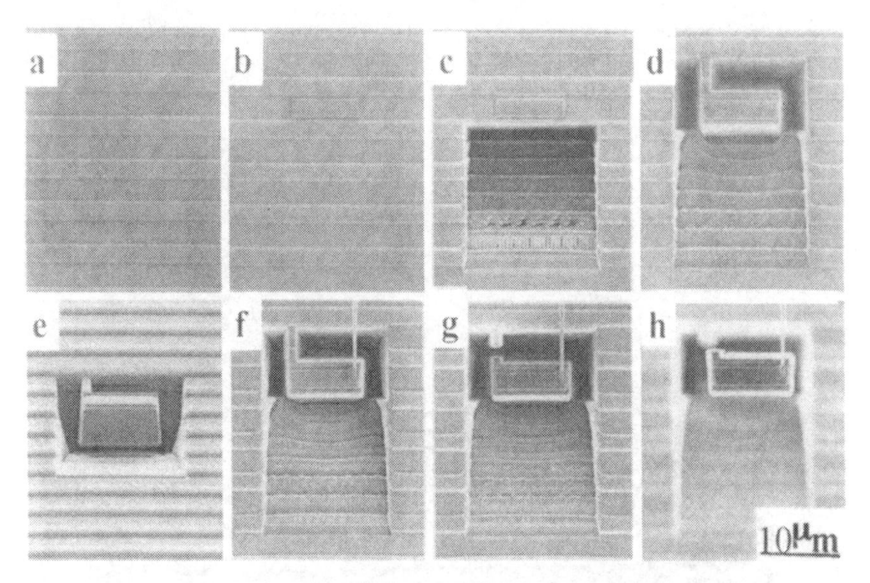

Figure 11.4. Series of FIB images showing the procedure for the FIB micro-sample preparation.

Figure 5 shows SEM images of a micro-sample mounted on the edge of a micro-sample carrier (Fig. 5a) and the micro-sample after preparation for TEM observation (Fig. 5b). A low magnification TEM image of a micro-sample prepared from a Si device is shown in Figure 6. The thin layer on the top of the specimen is the W deposition layer. The piece of dense material on the upper right of the micro-sample is the tip of the W probe used to manipulate the micro-sample to the micro-sample carrier.

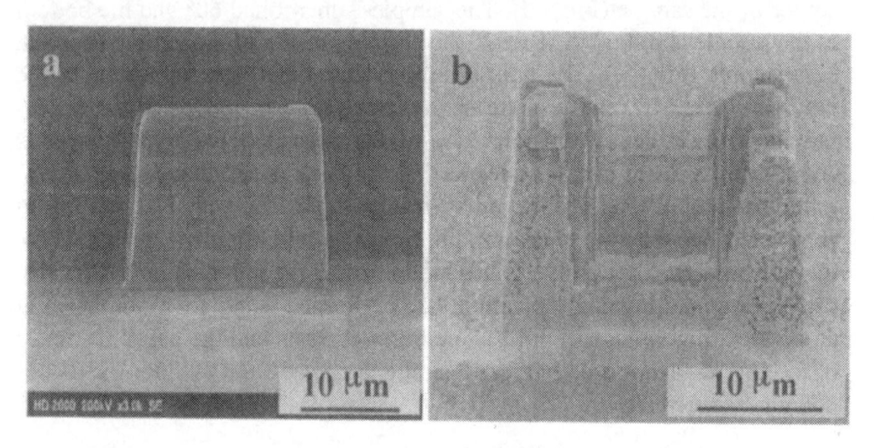

Figure 11.5. (a) Micro-sample mounted on an edge of a micro-sample carrier (b) and a micro-sample after thinning for TEM observation.

Figure 11.6. Low magnification TEM image of a micro-sample prepared from a Si device.

1.2 FIB Micro-Sampling for Plan View TEM

Although the standard FIB technique can be applied to fabricate plan view TEM specimens (Young, 1997), the FIB micro-sampling technique makes the procedure simple and reliable. The procedure for an FIB micro-sampling plan view TEM specimen is shown in Figure 7. The basic operation is similar to the method for cross sectional TEM specimen preparation. The shape of the micro-sample and the tilting angle of the sample used to mill away the bottom for separation of the micro-sample are slightly different. Here, deep trenching is not necessary but a wider surface than used for cross sectional TEM specimens is required (Fig. 7b). The micro-sample is positioned so that the back of the micro-sample faces the edge of the carrier (Fig. 7b). Then the micro-sample is bonded to the edge of the carrier with W deposition (Fig. 7c). After bonding the micro-sample, the carrier is rotated so that the surface of the micro-sample is parallel to the incident ion beam (Fig. 7d). In order to obtain a plan view TEM specimen from the surface layer to be characterized, thinning of the micro-sample is carried out from substrate side (Fig. 7e).

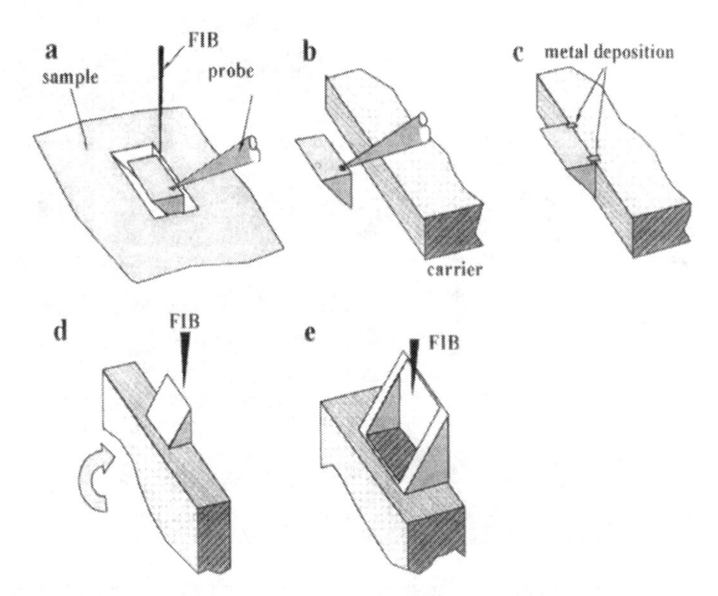

Figure 11.7. Procedure for the preparation of a FIB micro-sample for plan view TEM observation.

1.3 Cross Sectional and Plan View TEM Specimen Preparation from the Same Sample.

The preparation of both cross sectional and plan view TEM specimens from a sample using the FIB micro-sampling technique is possible. This is advantageous when structural evaluation of a sample from two or more directions are needed. An example of the application of this technique is illustrated in Figure 8. Cross sectional and plan view micro-samples were prepared from a magnetic disk (Fig. 8a) and thinned to the thickness of about 600 nm. Figure 8b shows a plan view TEM image of the Cr layer. In this case, both the Co layer and substrate were removed by FIB milling to obtain the thin foil plan view TEM specimen of the Cr layer where the grains of Cr in the layer are clearly observed. A cross sectional TEM image of all of the layers is shown in Figure 8c. (Note that the arrows relating the schematic diagram to the TEM images should be reversed.) Using these specimens, the orientations of the growth of each grain of Cr and Co can be determined and the thickness of each layer can be precisely measured.

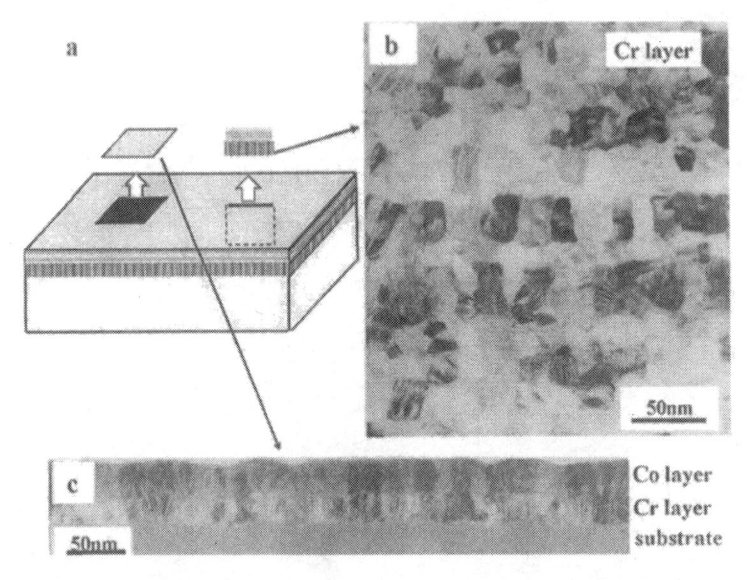

Figure 11.8. Observation of a magnetic disk by the micro-sampling technique. (a) Illustration of the lifting of micro-samples from a magnetic disk for both cross sectional and plan view TEM observation. (b) Plan view TEM image of the Cr layer in a magnetic disk. (c) Cross sectional TEM image of the Co and Cr layers.

1.4 A FIB micro-sampling for Cross Sectional and Plan View TEM Observation of a Specific Site

Preparation of a micro-sample for cross sectional and plan view TEM observation of a specific site is possible. Figure 9 shows the procedure to prepare the specimen for both cross sectional and plan view TEM observation of the same site. First, a micro-sample is prepared from the specific site for plan view TEM observation (Fig. 9a-e). The site is localized and characterized by TEM observation. After the plan view TEM observation, the micro-sample is transferred back to the FIB system for in order to prepare a cross sectional TEM specimen. In the FIB system, The carrier is rotated so the incident ion beam is perpendicular to the micro-sample (Fig. 9f). Then, the micro-sample is separated from the carrier by FIB milling, subsequently transferred to the surface of a flat substrate material (Fig. 9g-h), and secured with FIB assisted metal deposition (Fig. 9i). Finally, the site to be characterized is thinned for cross sectional TEM observation (Fig. 9j).

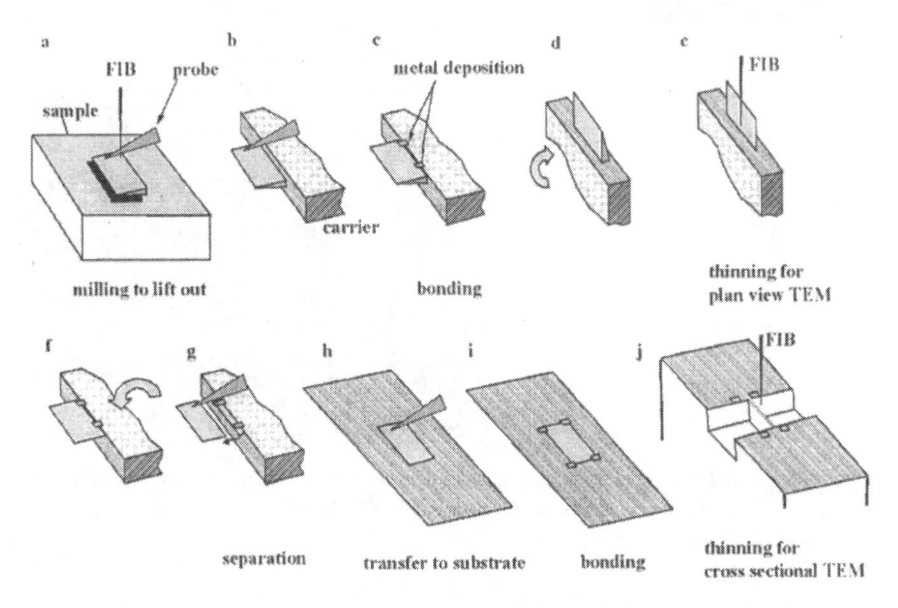

Figure 11.9. Procedure for the micro-sampling of a specific site for cross sectional and plan view TEM observation (see text for details).

An example of the application of this technique for the characterization of the crystal lattice defects in a Si device is shown in Figure 10. Figure 10a shows a plan view TEM image of the crystal lattice defects that are below the trenches of a test elements group (TEG) pattern. The dark lines in the trenches are the crystal lattice defects. Figure 10b shows a cross sectional TEM image of the same site as shown in Figure 10a. It is clearly demonstrated that the incidence of the crystal lattice defects occurred at the Si substrate near the bottoms of the trenches. Figure 10c shows a high resolution TEM image of a crystal lattice defect. It is clearly evident that the lattice defects occurred along the (111) planes of Si single crystal substrate.

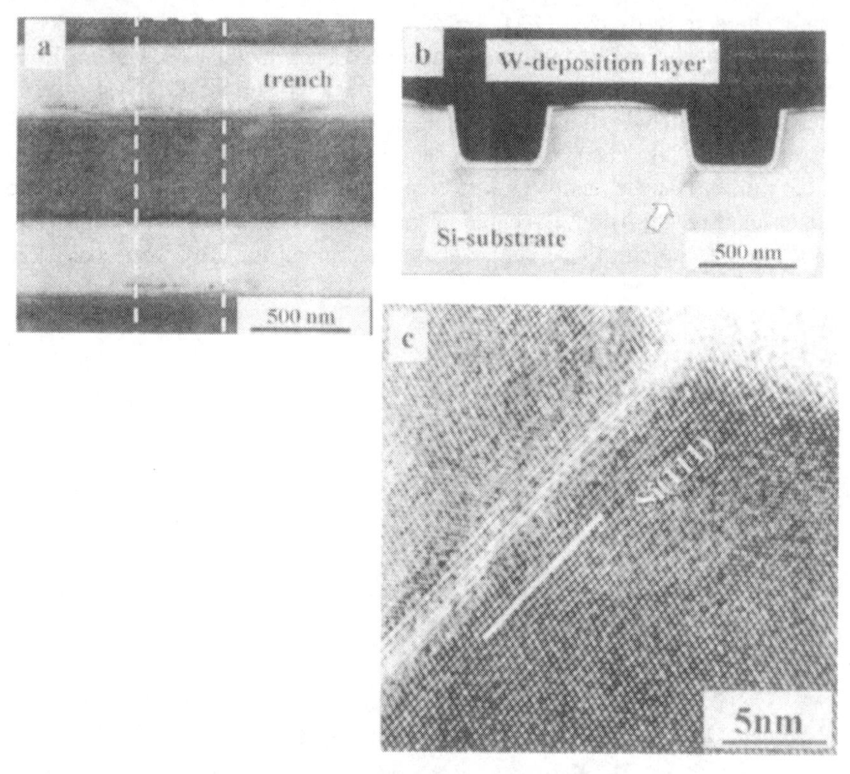

Figure 11.10. Example of cross sectional and plan view TEM images observation of the same lattice defect found in a Si device.

2. SITE SPECIFIC TEM SPECIMEN PREPARATION USING A FIB-TEM SYSTEM:

Although standard FIB techniques are one of the most reliable methods to prepare a specimen from a region of interest (Kirk et al., 1989; Park, 1990; Basile et al., 1992; Bender et al., 1997), the required accuracy for the localization of the specific site for TEM specimen preparation of electronic devices is often beyond the capability of the standard FIB techniques. In this chapter, a method for TEM specimen preparation from a specific site is described. In this method, a dedicated FIB system equipped with a side entry specimen stage and an intermediate-voltage TEM equipped with STEM/SEM units are used. A FIB-TEM compatible specimen holder is also used in the method so that both FIB milling and TEM observation is possible without remounting the specimen (Kitano et al., 1995).

The procedure for site specific TEM specimen preparation is described in Figure 11. First, the sample is FIB milled to the thickness of 3 to 5 μm leaving the site to be characterized almost in the center of the remaining area (Fig. 11a). Then, the sample is transferred to a TEM to observe STEM (Fig. 11b) and SEM images from both sides (Fig. 11c). In this method, the STEM observation is carried for localizing the specific site and the SEM images are utilized for measuring the distances from both cross sectional surfaces to the specific site. After the STEM and SEM image observation, the specimen is transferred to the FIB system again for further thinning. The FIB milling and STEM and SEM image observation procedure is repeatedly carried out until a specimen with a thickness of about 0.1 μm is obtained from the specific site (Fig. 11d).

2.1 Application Of The Site Specific TEM Specimen Preparation For The Characterization Of A Precipitate In A Steel Sample

An application of the site specific TEM specimen preparation method for the characterization of a precipitate in a steel is shown in Figures 12-16. Figure 12a-c shows a STEM image (Fig. 12a) and the SEM images (Fig. 12b,c) of both sides of the cross sectional surfaces of a steel sample after thinning to the thickness of 2 μm. In the STEM image, several precipitates are observed in the area. The precipitate indicated by the arrow is the chosen precipitate for the characterization. The precipitate is faintly visible in the SEM image shown in Figure 12b but is not seen in the SEM shown in Figure 12c at all. This indicates that the precipitate is located closer to the surface shown in Figure 12b, but far from the surface shown in Figure 12c.

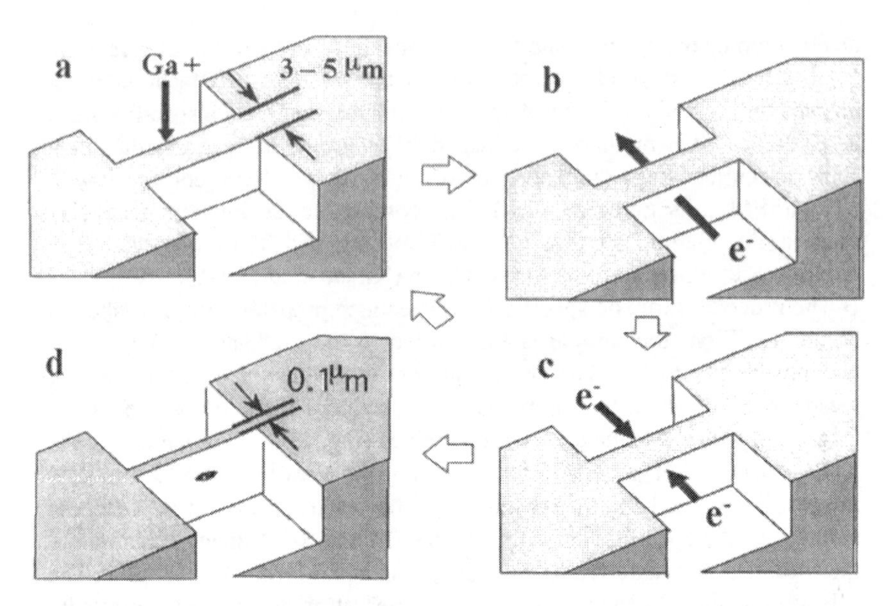

Figure 11.11. A procedure for site specific TEM specimen preparation using a FIB-TEM system.

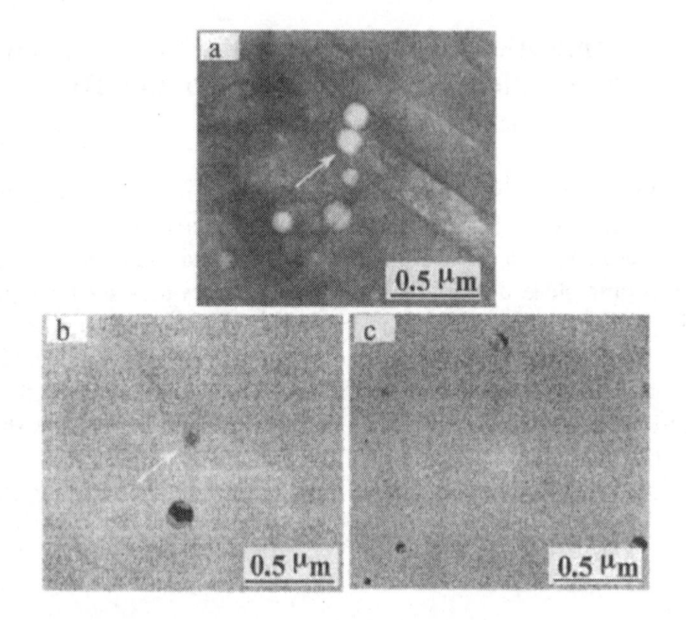

Figure 11.12. (a) STEM and (b,c) SEM images of a steel sample after thinning to the thickness of about 2µm.

Figure 13 shows the STEM and SEM images of the sample after thinning to the thickness of 1 μm. The diameter of the precipitate in Figure 13b is now larger than that in Figure 12b, but is not seen in Figure 13c yet. This means that the center of the precipitate is cross-sectioned at the surface observed in Figure 13b. Therefore, further milling for thinning the sample has to be carried out on the surface observed in Figure 13c.

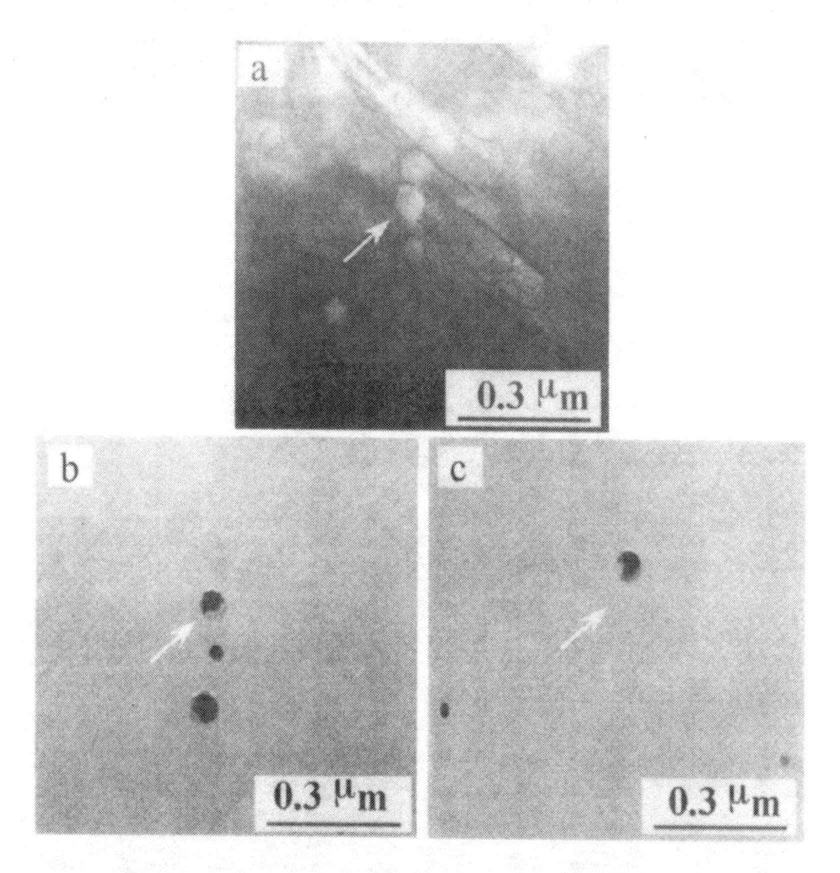

Figure 11.13. (a) STEM and (b,c) SEM images of a steel sample after thinning to the thickness of about 1μm.

Figure 14 shows STEM (Fig. 14a) and SEM images (Fig. 14b,c) of the sample after thinning to the thickness of 0.5 μm. The diameter of the precipitate observed in Figure 14b is similar to the real size of the precipitate shown in the STEM image in Figure 14a. This means the center of the precipitate is now located on the surface shown in Figure 14b, and no additional thinning is necessary on this side. However, the size of the precipitate observed in Figure 14c is still smaller than the real size of the precipitate observed in the STEM image in Figure 14a. This means that the next milling should be done only on the surface observed in Figure 14c. Figure 15 shows STEM (Fig. 15a) and SEM (Fig. 15b,c) images of the sample after thinning to the thickness of 0.1 μm. The size of the precipitate observed in Figure15 b,c are similar to the real size of the precipitate. This indicates that the precipitate was thinned exactly at its center. Figure 16 shows a STEM image (Fig. 16a) and the SEM (Fig. 16 b-g) elemental maps of the sample after thinning to the thickness of 0.1 μm. The precipitate is a combination of MnS and Al, Si, and O as shown by the elemental maps in Figure 16b-f.

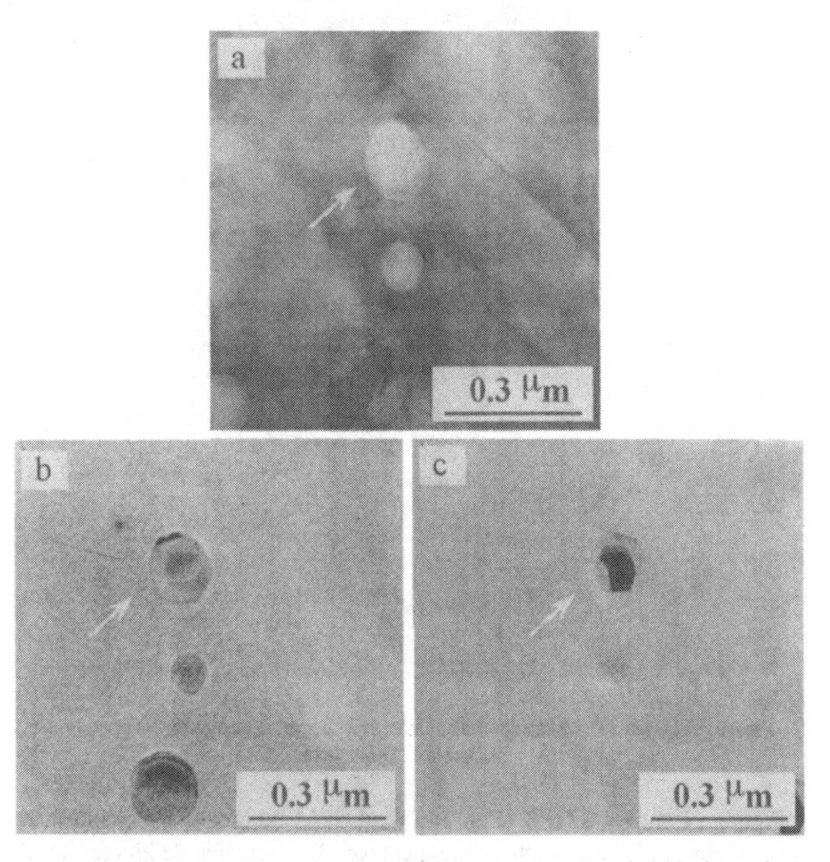

Figure 11.14. (a) STEM and (b,c) SEM images of a steel sample after thinning to the thickness of about 0.5 μm.

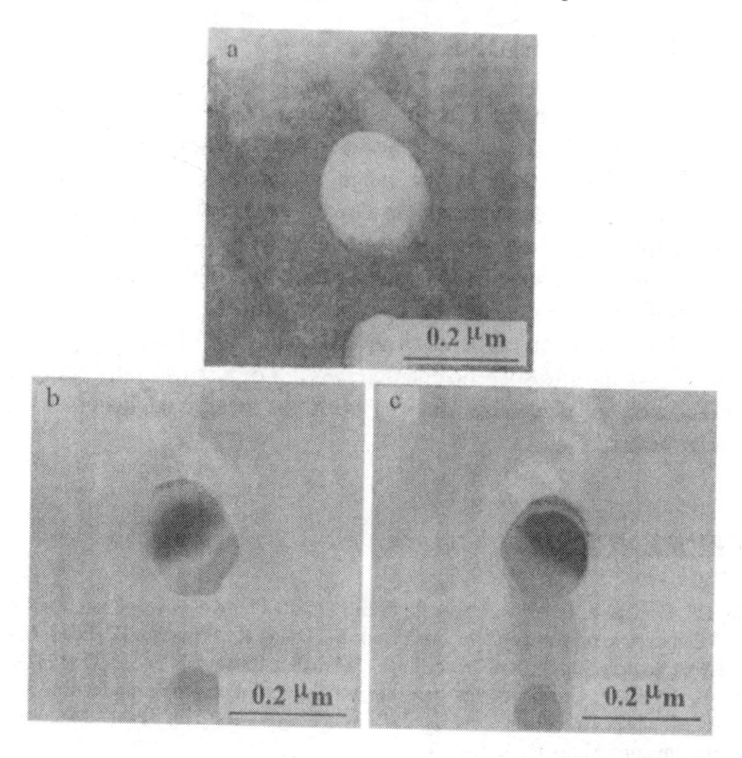

Figure 11.15. (a) STEM and (b,c) SEM images of a steel sample after thinning to the thickness of about 0.1 μm.

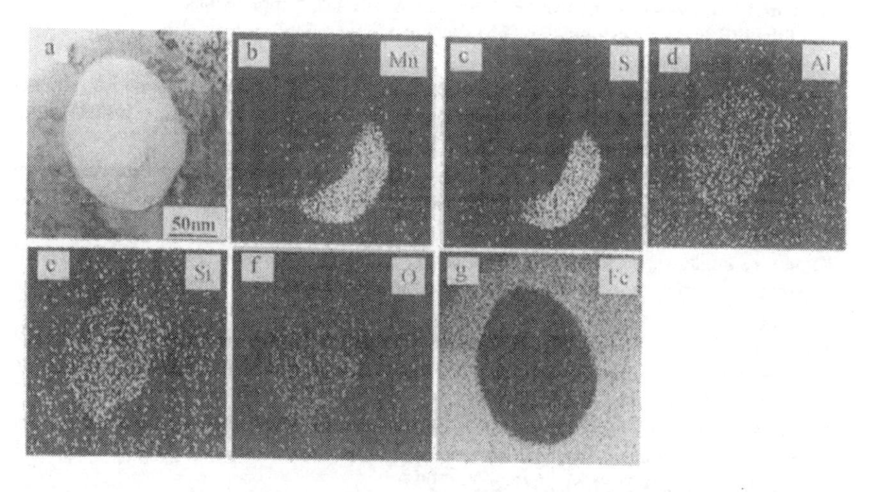

Figure 11.16. (a) STEM and elemental maps of (b) Mn, (c) S, (d) Al, (e) Si, (f) O and (g) Fe of the precipitate.

3. CONCLUSIONS

In comparison with the FIB lift-out technique, the FIB micro-sampling technique is rather time consuming because the thickness of a sample picked up from a bulk sample is in the range of microns, and post-thinning is required for TEM observation. However, it is unique in the pin-point characterization especially when a localization of the site to be characterized is difficult. In a site-specific characterization or failure analyses, the FIB micro-sampling technique can be combined with the newly developed site specific TEM specimen preparation method using an FIB-TEM system. The method allows the site-specific characterization of a structurally complicated materials such as electronic devices with the positional accuracy of 0.1 micron or better.

REFERENCES

Basile, DP, Boylan R, Baker B, Hayes K, Soza D, "FIBXTEM – Focused ion beam milling for TEM sample preparation", In: Anderson, R., Tracy, R., Bravman, B. (Eds.). Materials Research Society Symposium Proceedings, 254. MRS, Pittsburgh, pp. 23-41 (1992).

Bender H, Roussel P, Cross-sectional transmission electron microscopy and focused ion beam study of advanced silicon devices. In: Inst. Phys. Conf. Ser. Proc. Microscopy of Semiconducting Materials X. (1997).

Giannuzzi LA, Drown JL, Brown SR, Irwin RB, Stevie FA, "Specimen preparation for Transmission Electron Microscopy of Materials", Mater. Res. Soc. Symp. Proc. Vol.480:19 (1997).

Giannuzzi LA, Drown JL, Brown SR, Irwin RB, Stevie FA, "Applications of the FIB lift-out technique for TEM specimen preparation", Microsc. Res. Tech. 41:285-290 (1998).

Giannuzzi LA, Prenitzer BI, Drown-Macdonald JL, Brown SR, Irwin RB, Stevie FA, Shofner TL, "Advances in the FIB lift-out technique for TEM specimen preparation: HREM lattice imaging". Microstructural Science 26:26-29 (1998).

Kirk ECG, Williams DA, Ahmed H, "Cross-sectional transmission electron microscopy of precisely selected regions from semiconductor devices", Inst. Phys. Ser. 100, 501-506 (1989).

Kitano Y, Fujikawa Y, Kamino T, Yaguchi T, Sȧka H, "TEM observation of micrometer-sized Ni powder Particles thinned by FIB cutting technique", J. Electron Microsc. 44, 410-413 (1995).

Kitano Y, Fujikawa Y, Takeshita H, Kamino T, Yaguchi T, Matsumoto H, Koike H, "TEM observation of micrometer-sized Ni powder particles thinned by FIB cutting technique", J. Electron Microsc. 44,376-383 (1995).

Overwijk MHF, van den Heuvel FC, Bulle-Lieuwma CWT, "Novel scheme for the preparation of transmission electron microscopy specimens with a focused ion beam", J. Vac. Sci. Technol. B 11(6), 2021-2024 (1993).

Park K, "Cross-sectional TEM specimen preparation of semiconductor devices by focused ion beam etching", In: Anderson, R. (Ed.), Materials Research Society Symposium Proceedings, MRS, Pittsburgh, 199, pp. 271-280 (1990).

Stevie FA, Irwin RB, Shofner TL, Brown SR, Drown JL, Giannuzzi LA, "Plan view TEM sample preparation using the focused ion beam technique", Proc. Int. Conf. American Institute. of Physics, Gaithersburg(Woodbury, New York), Chap.449: 868-871 (1998).

Yaguchi T, Matsumoto H, Kamino T, Ishitani T and Urao R, "A method for cross sectional thin specimen preparation from specific site using combination of focused ion beam system and intermediate voltage electron microscope and its application to the characterization of a precipitate in a steel", Microscopy and Microanalysis 7 (1998).

Yaguchi T, Kamino T, Ishitani T and Urao R, "Method for cross sectional TEM specimen preparation of composite materials using a dedicated FIB system", Microscopy and Microanalysis 5: 363-370 (1999).

Yaguchi T, Kamino T, Sasaki M, Barbezat G and Urao R, "Cross sectional specimen preparation and observation of a plasma sprayed coating using an FIB/TEM system," Microscopy and Microanalysis 6: 218-223 (2000).

Young RJ, Kirk ECG, Williams DA, Ahmed H, "Fabrication of planar and cross-sectional TEM specimens using a focused ion beam", In: Anderson, R. (ed.), Material Research Society Symposium Proceedings, MRS Pittsburgh, 199, pp. 205-216 (1990).

Chapter 12

DUAL-BEAM (FIB-SEM) SYSTEMS

Techniques and Automated Applications

Richard J. Young and Mary V. Moore
FEI Company, Hillsboro, Oregon

Abstract: The dual beam incorporates both a focused ion beam (FIB) and a scanning electron microscope (SEM) in a single system. This combination offers several advantages over a single-beam FIB system, especially for sample preparation and microscopy applications, in which the ion beam can be used for site-specific material removal and the SEM for nondestructive imaging and analysis. Dual-beam system configurations are presented, along with a number of key techniques and applications. These include precision cross-sectioning, TEM sample preparation, and automated 3D process control.

Key words: dual beam, FIB, SEM, failure analysis, process control, 3D metrology, TEM sample preparation, automation, nanofabrication, EBSD, cryo

1. INTRODUCTION

The dual beam incorporates both a focused ion beam (FIB) column and a scanning electron microscope (SEM) column in a single system. This combination is especially useful for cross-section sample preparation using the electron beam to view the cross-section face as the ion beam mills normal to the sample surface. This monitoring allows the milling to be stopped precisely when the feature of interest is exposed. On a single beam FIB, a series of tilting and beam current changes would normally be required to monitor the cross-section face position in this way.

This chapter starts with an overview of dual-beam system configurations, and then discusses key techniques and application areas. As other applications of FIB are discussed elsewhere in more detail, this chapter focuses on the particular benefits of the dual beam and any special considerations with using this type of system.

2. DUAL-BEAM SYSTEM CONFIGURATIONS

A number of different configurations have been proposed for combining FIB and SEM columns on a single system. The first dual beams were SEMs that had a gallium-source FIB column added (Sudraud et al., 1988). Commercial systems began to appear in the early 1990s. The motivation for these systems was to avoid having to use multiple tools for sample preparation and imaging. Currently, a variety of dual-beam systems are available, typically employing a gallium (Ga) FIB column and a field-emission SEM column, which support sample types from small specimens on TEM (transmission electron microscope) grids through to 300 mm wafers.

The typical dual-beam column configuration is a vertical electron column with a tilted ion column. Figure 1 shows such a configuration with the ion beam at 52° tilt to the vertical. In this case, the sample will be tilted to 52° for milling normal to the sample surface. An alternative geometry is with the FIB column vertical and the SEM at an angle. The advantage of this is that sample tilt is not required for milling normal to the sample, potentially simplifying system operation. However, using a tilted SEM column is not a typical configuration for most electron column designs, and so this would need to be considered. The simpler and more compact electrostatic design of FIB columns usually makes it the easier column to mount tilted and still get optimum performance from both columns.

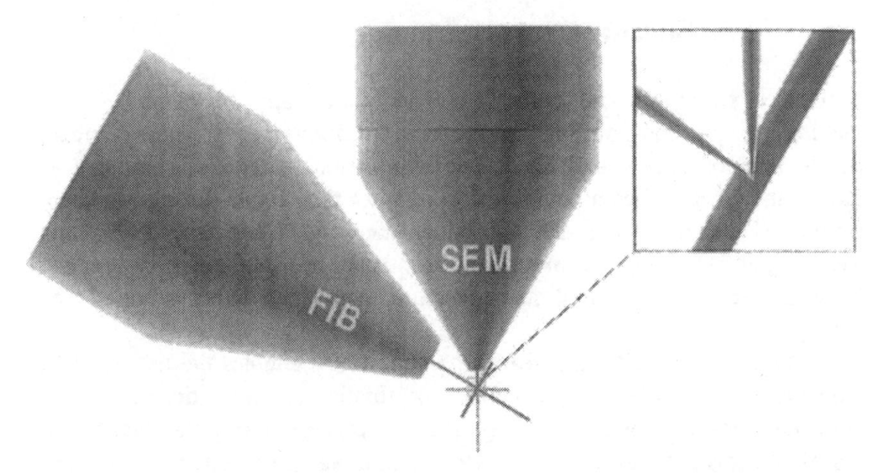

Figure 12-1. A typical dual-beam system configuration. The vertical SEM column and tilted FIB column have a single "coincident" point on the sample. SEM imaging during FIB cross-sectioning enables real-time monitoring of the milling process.

To enable ion milling and electron imaging of the same region, dual beams typically have a coincident point where the two beams intersect with the sample. This is the normal operating position for the system. Options, such as gas injectors and EDS (energy-dispersive X-ray spectrometry), are also aligned for optimum operation at this position. With the sample at the coincident point, the advantages of the dual beam compared to a single-beam FIB become apparent. As Fig. 1 illustrates, the SEM can be used to monitor the cross-section face as the FIB mills. On a single-beam FIB, such milling and imaging would require sample tilt (and perhaps beam current change) between each step, resulting in a much more time-consuming process. The dual beam is therefore extremely well suited for precisely locating cross sections through defect sites or other similar highly localized areas of interest.

3. CROSS-SECTIONING IN A DUAL BEAM

Precision cross-sectioning is used in many applications, including defect analysis, process and materials characterization, and process monitoring and control. The techniques that enable unique contributions with the dual beam for these applications are highlighted in the following sections.

3.1 Site-Specific Cross-Sectioning

Device structures and defect sites are continually shrinking in size, making precise end-pointing on the area of interest ever more critical. Semiconductor defect analysis is one application that takes advantage of having the electron beam positioned to monitor the sample during ion beam milling (Delenia et al., 1995; Zimmerman and Chapman, 1999; and Teshima, 2001). A high-resolution electron beam in a dual-beam system can be used to identify abnormal microstructure, even if the defect is buried (see Fig. 2).

Not only is the SEM useful for imaging, it also enables the use of EDS, which gives compositional (elemental) information about the defect. Having coincident beams considerably speeds up the process. Using the dual beam, real-time decisions can be made about which imaging or analysis techniques to employ to characterize the defect, or whether to do additional milling. Combining these features with accurate stage navigation makes wafer-level defect analysis very efficient on dual-beam systems (Delenia et al., 1995). Wafer dual beams have also been used in-line for 3D structural analysis of wafers that then continue on in the manufacturing process (Weiland et al., 2000). See section 5 for further discussion of the possibilities of automated 3D process control.

The ability to use the electron beam to record images of successively milled slices through a region is a key benefit of a dual-beam system. This "slice and view" technique can also be automated, producing a series of SEM images through the area of interest, allowing rapid characterization of defects and other structures. A slice and view series can be acquired in two ways. The first is to alternate the milling and imaging procedures, so that a high-resolution SEM image is acquired after a slice of a given thickness is milled from the sample. The second method is to use simultaneous milling and imaging, where SEM images are acquired while the milling continues with the ion beam. In this mode, secondary electrons are present from both beams and so the relative sizes of the two beam currents and the detector type and settings need to be considered to get the best SEM image quality.

An automated slice and view series is shown in Fig. 3. Here, 55 images through a defect have been captured, of which 18 are shown. The full set of images can be viewed as an animation or movie, showing progression through the sample. This allows the user to understand better the three-dimensional nature of the defect or sample. It is also possible for the user to monitor the process, stopping it once something of interest is reached and allowing the rest of the defect to be analyzed further with techniques such as EDS.

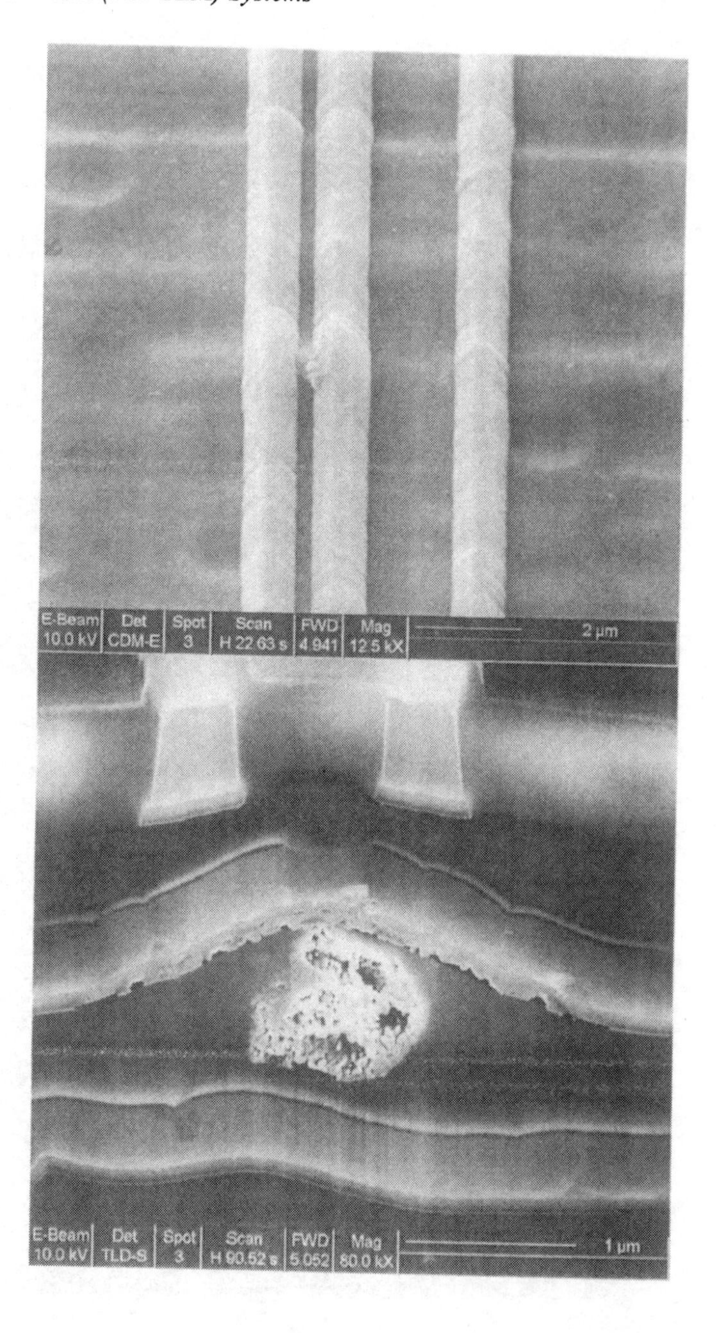

Figure 12-2. Subsurface defect: surface before mill and cross section showing exposed defect.

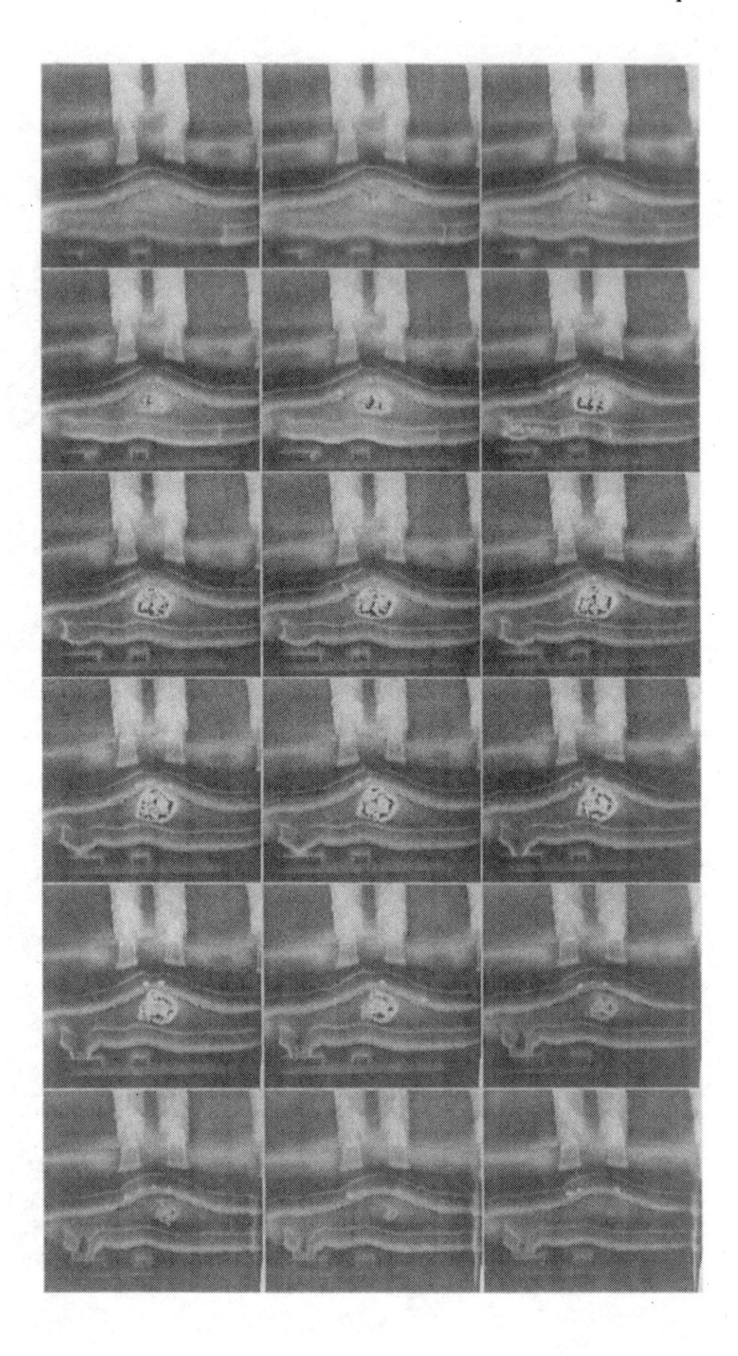

Figure 12-3. Automated slice-and-view series, through the same defect as Fig. 2. Out of the full set of 55 images, 18 are shown here. Each slice is approximately 50 nm thick.

Once an array of images has been created, the three-dimensional volume can be explored in ways other than by just looking at the slices individually or as an animation. For example, sections can be made through the sample space in other directions, as shown in Fig. 4. Such reconstruction techniques rely on careful alignment between images and on compensation for the thickness of the slices, as this is normally several times larger than the image-pixel size in the individual images.

In addition to looking at defects or structures, automated slice-and-view can also be useful as an initial survey through an unknown area to find the precise location of a feature of interest. Once the position is found in one of the images, a cross section or TEM section can be prepared through the equivalent position elsewhere on the sample. Doing an initial survey in this way is a great time-saver for the analyst searching for the correct area to cross-section.

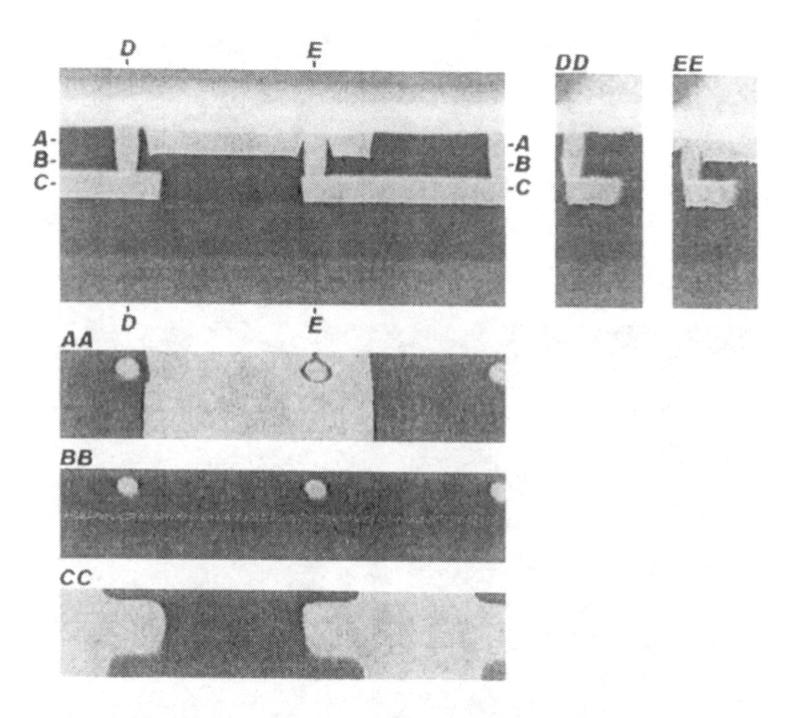

Figure 12-4. Image reconstruction from an automated slice-and-view series of 36 images (for more details on the sample, see Fig. 5). The image at upper-left is one of the original slices from the series. This image shows the positions of the 5 reconstructed slices (AA, BB, etc.). AA, BB, and CC are plan views, while DD and EE are front-to-back views. Main image width is 8.0 μm, each slice approximately 40 nm thick.

The dual beam offers in-situ preparation capability for other analysis techniques, such as electron backscatter diffraction (EBSD) (Matteson et al., 2002). EBSD allows sub-micron grain orientation data to be acquired by recording and indexing a diffraction pattern at each position of interest to build up a map. For EBSD the sample must be highly tilted (typically 70-75°) towards the electron beam and so in the dual beam a pretilted stub is used to allow preparation and imaging of both the sidewall and top surface of the sample (Fig. 5). By combining preparation and analysis in one system any oxidation effects are eliminated and multiple sections can be acquired enabling 3D analysis of a volume of interest.

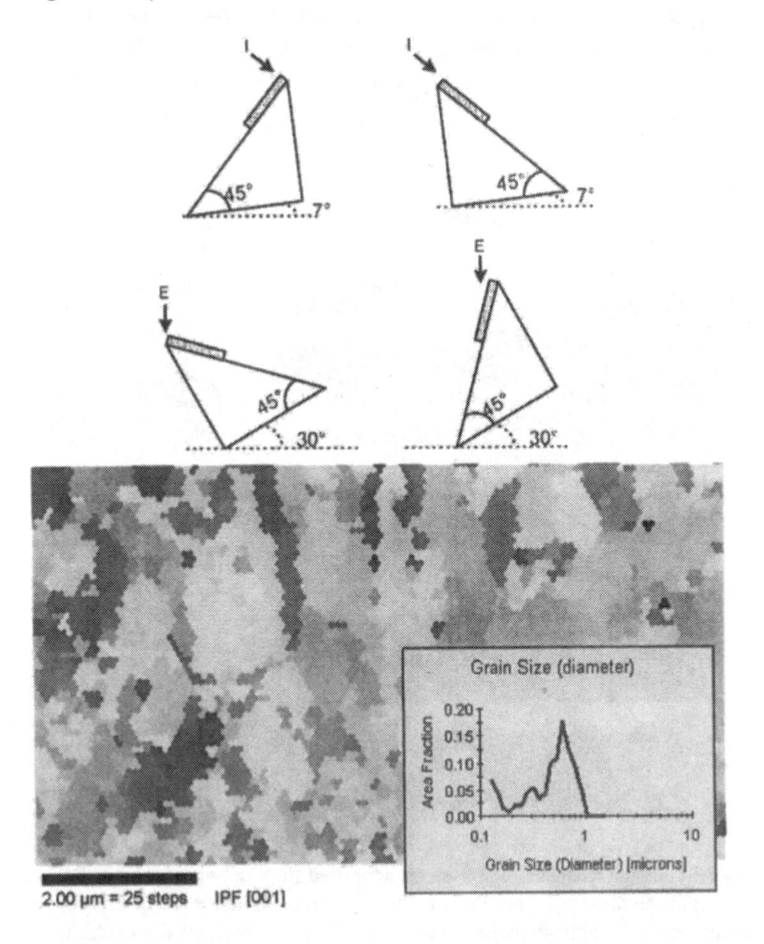

Figure 12-5. EBSD (electron backscattered diffraction) on a dual beam. Using a 45° pretilted stub holder enables ion beam preparation parallel to the sidewall (top-left) or top surface (top-right), and electron beam analysis at 75° in each case (below). Resulting EBSD map and grain size distribution of steel sample with CrC inclusions (bottom).

The combination of a dual beam with a cryogenically cooled stage (Mulders, 2003) also offers exciting possibilities for sample preparation and analysis of biological or other beam sensitive samples. The use of such a "cryo" stage in a SEM is a common way to handle wet or water containing samples, where the low temperature (down to 77 K using liquid nitrogen) keeps the vapor pressure of the sample below the operating pressure of the system. Mulders found that site-specific cross-sections into biological samples under cryo conditions showed very good information that was comparable or better than obtained with non-site-specific cryo fracturing (Fig. 6).

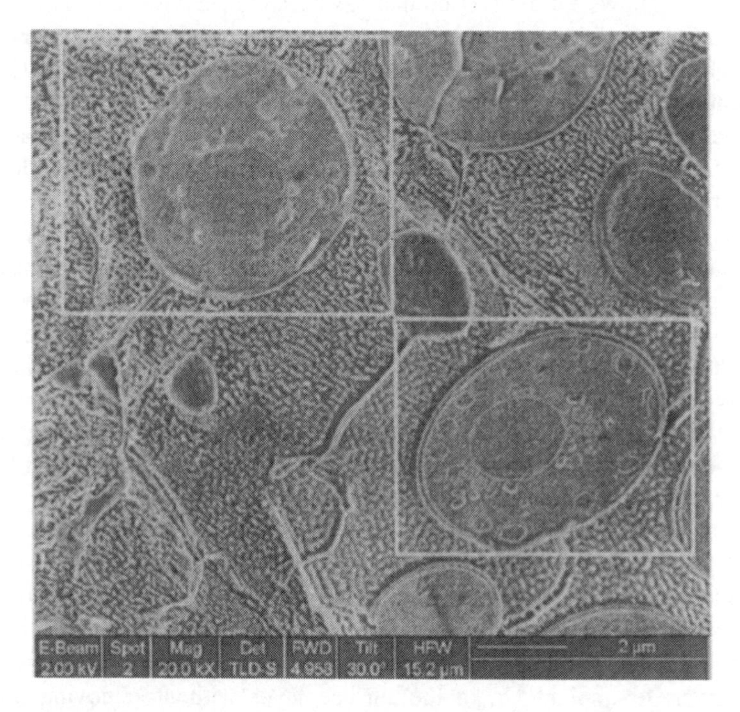

Figure 12-6. Bakers' yeast showing both a cryo fracture (top left box) and a FIB cross-section (lower right) through yeast cells. Cell details, such as membrane and nucleus are clearly visible in each case.

3.2 Complete Cross-Section Preparation and Enhancement

Traditional methods of cross-section preparation normally include some form of wet chemical delineation etch to bring out various layers in the sample more clearly in the SEM. These techniques are also possible with FIB or dual-beam prepared samples, but they involve removing the sample from the system and etching the whole sample before placing the sample in the final imaging tool. Using gas chemistries in a dual beam, complete sample preparation and imaging can be achieved within a single instrument.

Figure 7 shows a cross section that has been etched with delineation etch gas and dusted with a very thin layer of deposited metal in the dual beam, highlighting the low-k dielectric layers. The grain contrast in the Cu is brought out with backscattered electron imaging. The dusting of platinum material expands the choice of kVs (electron beam voltages) for imaging beyond what would otherwise be feasible with the charge-sensitive low-k dielectrics, allowing optimization of overall image quality.

The top-down image of the location was achieved at a higher kV, taking advantage of the increased penetration depth of the electron beam for ease in locating features under passivation layers. For thicker oxide passivation layers, even higher kVs can be used for navigation and location. The ion beam has a much lower penetration depth than the electron beam. While the surface sensitivity of ion imaging can be useful to locate slight perturbations in planarized surfaces, it also can lead to charge buildup if a grounding path is not established. Ga implantation can help spread the charge along the surface, but a ground is required to drain this charge.

Fortunately, a number of charge reduction methods can be employed. In Fig. 7, as mentioned, the electron beam at medium-high kV was used to locate the area of interest. Then, small grounding holes were drilled to short to the substrate. To provide a grounding path from the area of interest to the grounding holes, the area was imaged at low magnification with a medium ion current to implant Ga in the surface layer without removing much material. This step can be iterated if electron charge buildup occurs during SEM imaging. An alternative is to deposit a grounding strap. Sometimes a light dusting of conductive material is all that is needed; this can be accomplished with either beam, but if contamination or surface damage is a concern, electron beam-deposited platinum is most useful.

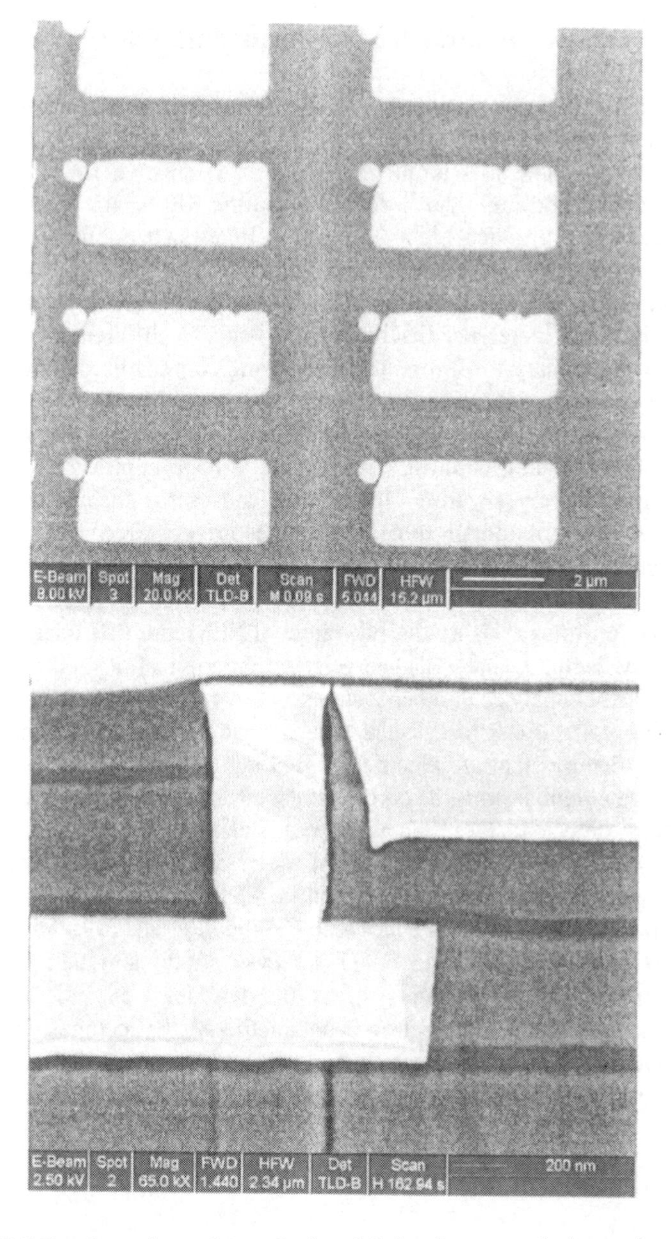

Figure 12-7. Top-down view and cross section of Cu low-k structure. In the top-down view, the electron beam views the Cu through the top insulating layer. The section has been decorated in situ with delineation etch (DE) followed by a dusting of Pt. DE produces differential etching of interlayer dielectrics, and the Pt dusting avoids charging during imaging.

3.3 Ion-Beam and Electron-beam Imaging

Although the electron beam is considered the primary imaging tool in a dual beam, the imaging from the ion and electron beams is often complementary. The FIB is an imaging tool in its own right (Levi-Setti, 1983), and can add important information to the SEM data. While electron beams generally produce higher resolution images, the ion-beam imaging contrast is often strong. Factors contributing to ion-beam contrast include ion channeling effects, passive voltage contrast, sensitivity to surface topography, and materials differences that lead to differential secondary-electron or secondary-ion emissions. The strong contrast in ion-beam images from ion channeling and passive voltage contrast is illustrated in Fig. 8. In this semiconductor sample, the grain contrast is enhanced in the FIB image due to ion channeling contrast. The passive voltage contrast effect, visible only with secondary-electron FIB images, causes the insulating layers to appear very dark. Although most FIB images are acquired using secondary electrons, secondary-ion imaging can provide further image information. For example, more topographical information can often be gleaned from a secondary-ion image. Using the full range of SEM and FIB imaging modes can be very useful in materials characterization applications, eliminating the need for further analysis in other systems.

Passive voltage contrast occurs when, during imaging with an incident ion beam, features that are electrically floating accumulate positive charge. The surface potential prevents secondary electrons from escaping the surface and being detected; these areas appear dark in the ion-beam-induced secondary-electron image. No external electrical potential is required for such results. Passive voltage contrast in FIB images is used in defect detection in semiconductor devices (see Campbell et al., 1995 for technique summary and Luo and Song, 1997 for case study; although both these references use single ion-beam systems, the dual beam expands the options of the analyst to include electron-beam methods). By comparing the ion-beam image to the electron-beam image, electrical connectivity can be often be determined at a glance.

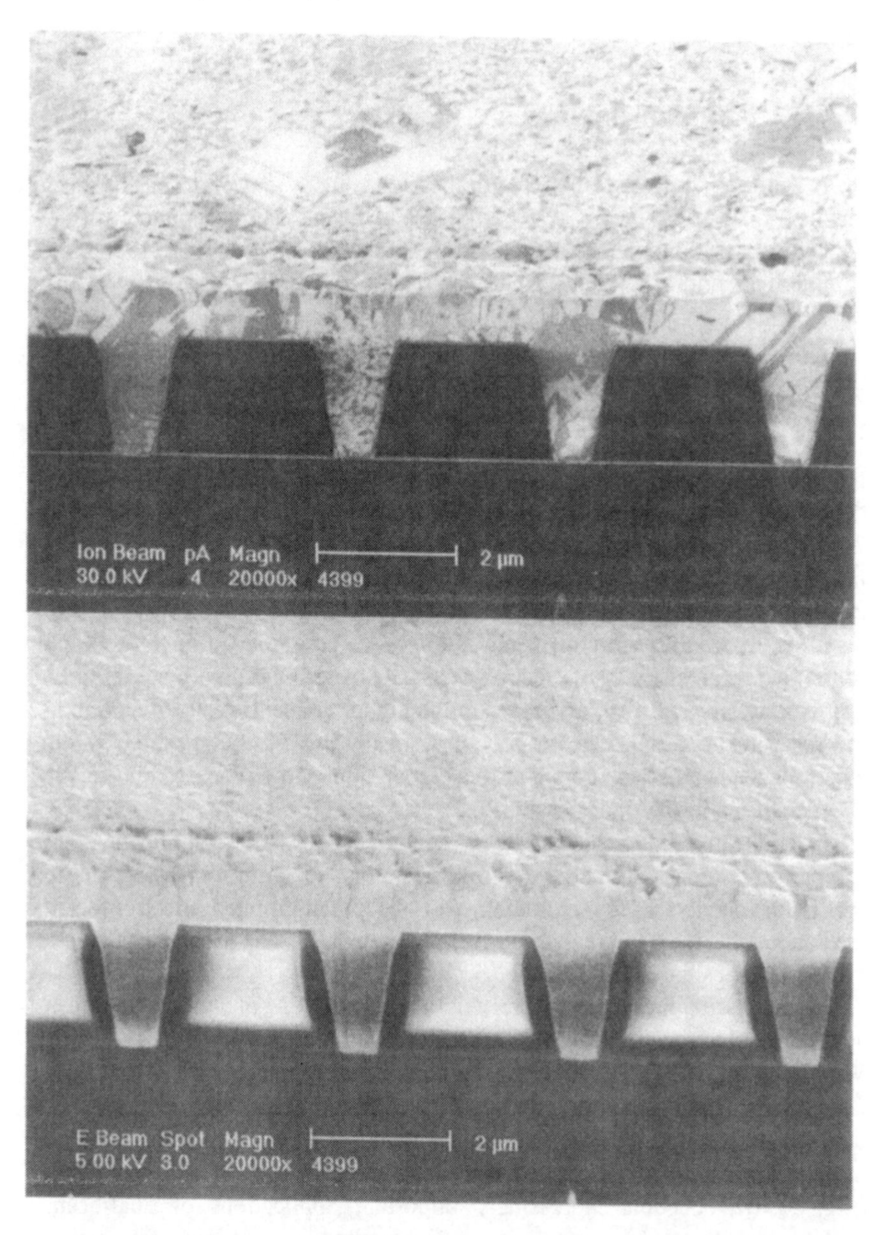

Figure 12-8. Comparison of FIB and SEM imaging. Top: ion beam incident, secondary electrons detected; bottom: electron beam incident, secondary electrons detected. In the FIB image, the grain contrast of the Cu is enhanced by channeling effects, and passive voltage contrast makes the dielectric layers appear dark.

3.4 Electron-Beam Gas Chemistry

As already indicated, the electron beam can also be used to perform gas chemistry. In dual-beam systems this has been mainly applied to the deposition of conductive and insulative layers (Lipp et al., 1996a,b,c; Zimmerman and Chapman, 1999), and also to etching (Gstöttner et al., 2001).

The electron beam can normally use the same gas injectors as the ion beam, so most dual beams are already configured for some gas chemistry. In general, lower beam voltages result in higher deposition rates (Lipp et al., 1996a). This is attributed to secondary electrons being mainly responsible for the deposition, rather than the primary beam itself. Electron-beam deposition for sample protection or charge control is typically performed at 2-3 kV, generally using more beam current than would be used for a normal imaging application. For nanofabrication applications higher voltages (e.g. 20-30 kV) are often used (Weber et al., 1995; Morimoto et al., 1996), resulting in smaller structures but at lower volume deposition rates. Lipp found that the resistivity of electron-beam-deposited platinum material to be higher than with the FIB, attributed to the higher concentrations of carbon in the resulting material. The resistivity of insulating films deposited by the electron beam was also found to be higher than the FIB case, due to the absence of Ga in the final film.

In addition to the electron-beam conductor deposition applications of sample protection and charge control, insulator deposition can also be used for failure analysis. Zimmermann and Chapman applied electron-beam insulator deposition to create high-resistivity pads for electrically isolating a fail down to a single DRAM cell.

Etching with the electron beam is also possible, but has mainly been applied to electron-beam systems (see for example: Nakamae et al., 1992; Winkler et al., 1994). Gstöttner used a dual-beam system to etch Si and SiO_2 using XeF_2, finding that the rate was 50 and 600 times slower, respectively, than the corresponding dose with an ion beam. With the expanding interest in nanofabrication it is expected that electron beam etching and deposition (Fig. 9) will become increasingly important applications of dual-beam systems, to complement the well-established ion beam fabrication capabilities.

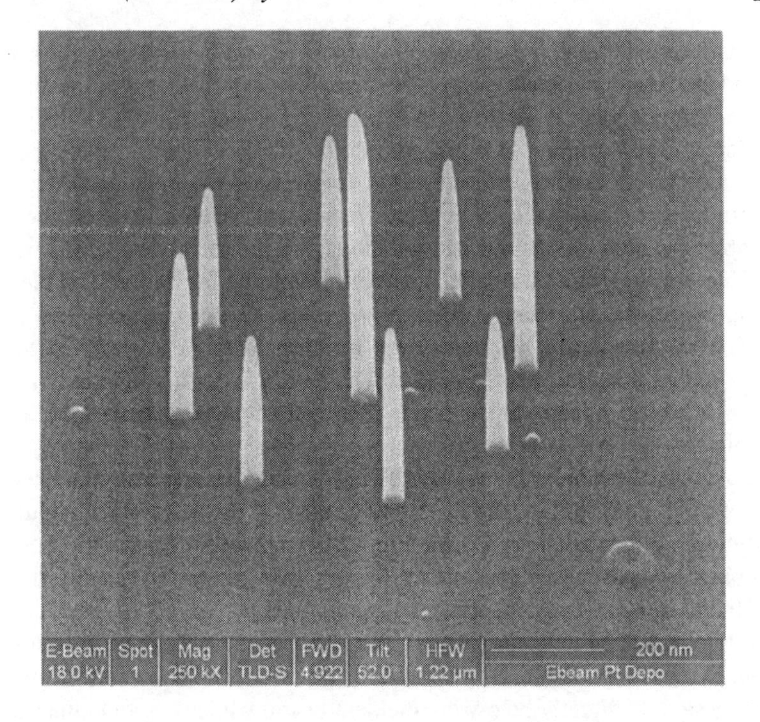

Figure 12-9. Array of pillars created by electron-beam deposition. Pillars are 30 nm in diameter, with end radius < 10 nm.

4. TEM SAMPLE PREPARATION

4.1 Benefits of the Dual Beam

There are a number of benefits to using the dual beam to prepare TEM samples. Most critically, the electron beam can be used to monitor the sample preparation, allowing the section to be precisely positioned through the point of interest (Ma et al., 1999; Xu et al., 2000). In addition, the electron beam can be used to monitor sample thickness and to check that the protective layer on the sample has remained intact.

Ion implantation from initial imaging or deposition of the protective layer can cause a 20-30 nm damage region. If the region of interest is close to the sample surface, then to avoid this damage, electron-beam-induced deposition can be used to put down a 50-100-nm thick protection layer before any ion beam imaging is performed (Lipp et al., 1996c). Ion beam deposition can then be used to deposit the rest of the typically 1 μm thick layer. An alternative technique is to deposit a blanket film over the whole sample before starting the sample preparation; however, this can make locating the area of interest more difficult.

For all TEM samples it is critical that no direct ion-beam imaging of the final membrane occurs; otherwise ion implantation damage will occur. A benefit of the dual beam is that electron-beam imaging of even the final membrane can be carried out without damaging the membrane in this way.

Thin sections have a smaller interaction volume for an incident electron beam. Therefore, this sample geometry is useful to increase resolution in the SEM when interaction volume is the limiting factor, e.g., with EDS. For EDS, spatial resolution below 30 nm is possible with a thinned sample, which is orders of magnitude better than can often be achieved on a bulk sample.

Some dual-beam systems use STEM (scanning transmission electron microscopy) detectors to take further advantage of the increased imaging and analytical information available in thin sections. Such SEM-STEM combinations produce high-contrast images from samples typically thinned to 50-200 nm, at incident electron-beam voltages of up to 30 kV. In a multiregion STEM detector (Hayles and van Veen, 1998), there are different imaging modes available. Bright field images are formed from electrons that exit the sample directly below the incident beam, while dark field images are formed from scattered electrons. The bright field and dark field images produced often show complementary information, with the dark field images displaying strong materials (Z, atomic number) contrast (Fig. 10).

By mounting the sample on a pivoting sub-stage it is possible to orient the sample for FIB thinning and then to position it for SEM-STEM imaging without breaking vacuum, as would normally be required. Such a "flip-stage" arrangement allows for successive thinning and imaging operations to ensure that the region of interest is in the thinned section, and also enables images to be acquired at different sample thicknesses. Complimentary information can be obtained from iterative thinning. In imaging, materials contrast (especially between regions of similar atomic number) is enhanced in a thicker sample, while resolution is optimized in a thinner sample. In compositional analysis by EDS, it might be desirable to obtain signal from the (less efficiently excited) lighter elements in a thicker sample, then analyze the heavier elements after further thinning. Solving more analysis problems in the preparation system also allows the TEM to focus on the samples where atomic level capabilities are really required. Overall, dual-beam sample preparation and SEM-STEM imaging is an excellent combination for structural and compositional analysis, especially when using automated sample preparation, which is discussed next.

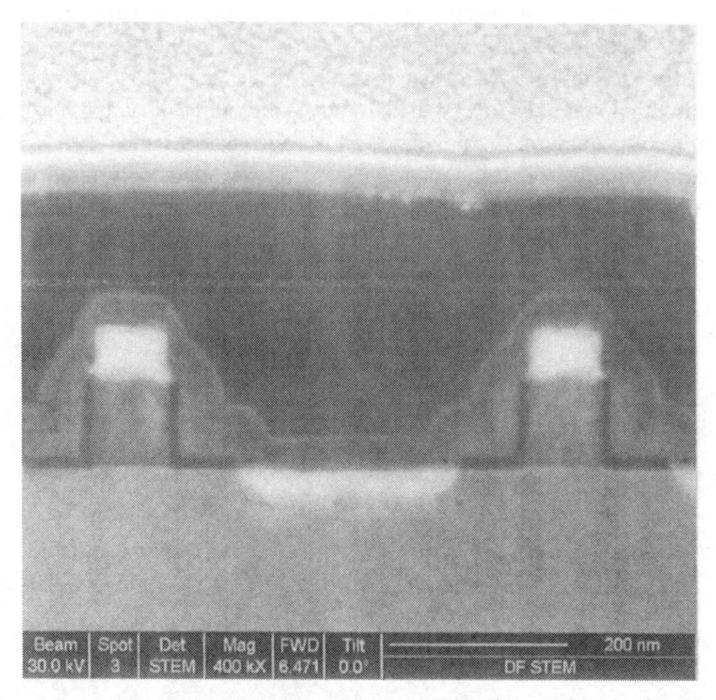

Figure 12-10. High-magnification, dark field SEM-STEM image of a pair of transistors.

4.2 Automation of TEM Sample Preparation

As the use of dual beam and FIB for TEM sample preparation has become more routine, there has also been an increase in the number of samples that need to be analyzed, especially for semiconductor and data storage applications. Automated TEM sample preparation can be used on dual-beam and FIB systems to meet these demands for faster sample preparation and to improve system utilization (Young et al., 1998; Young, 2000; and Moore, 2002).

Preparation of TEM samples consists of a series of imaging, milling, and deposition steps that can be repeated at multiple sites. It is common to use larger beam currents for the initial bulk milling steps, and then to use progressively lower beam currents as the beam approaches the analysis site. Small stage tilts are used during the final milling steps to produce parallel-sided membranes. With the addition of image matching (also known as image or pattern recognition), this multistep process can be fully automated (Fig. 11). These automated routines allow samples with thicknesses < 100

nm to be prepared unattended, with multiple samples prepared in a single session or even overnight. The use of the same recipe from sample-to-sample also enables more consistent results to be achieved than would be expected with a manual process. This removes any variability that may occur with multiple operators. In addition to automation of TEM sample preparation for lab use, this technique can be extended to process control applications, where the large numbers of samples required would preclude a manual process in all but the most limited implementation.

Figure 12-11. Automated TEM sample preparation of a lift-out sample. The crosses milled into the sample are fiducials for the image matching routines used by the automation.

5. AUTOMATED 3D PROCESS CONTROL

Manufacturing processes increasingly rely on metrology to provide the data needed to produce high-yield, reliable products. "Top-down" or "2D" metrology relies on a surface image to determine critical dimensions (CD) on the sample. Increasingly, such images do not have the information required to understand the full 3D structure of complex devices. 3D metrology, in contrast, can provide this process control data by making a section through the precise area of interest, allowing surface features as well as buried features and geometries to be observed and measured.

The first use of the dual beam in an automated process control application was in the data storage industry. These in-line dual beams use automation software that includes a robust pattern-recognition system, which is used to automatically align, position, cross-section, image, and then measure preselected features (Fig. 12). After collecting metrology data from similar structures at several locations across the wafer, the wafer can continue in the production flow. This measurement procedure can be repeated several times at different points in the production flow to keep the process within a tight tolerance for maximum yield and device performance. In addition to 3D-CD information recorded automatically, the images provide information about the quality of the manufacturing process.

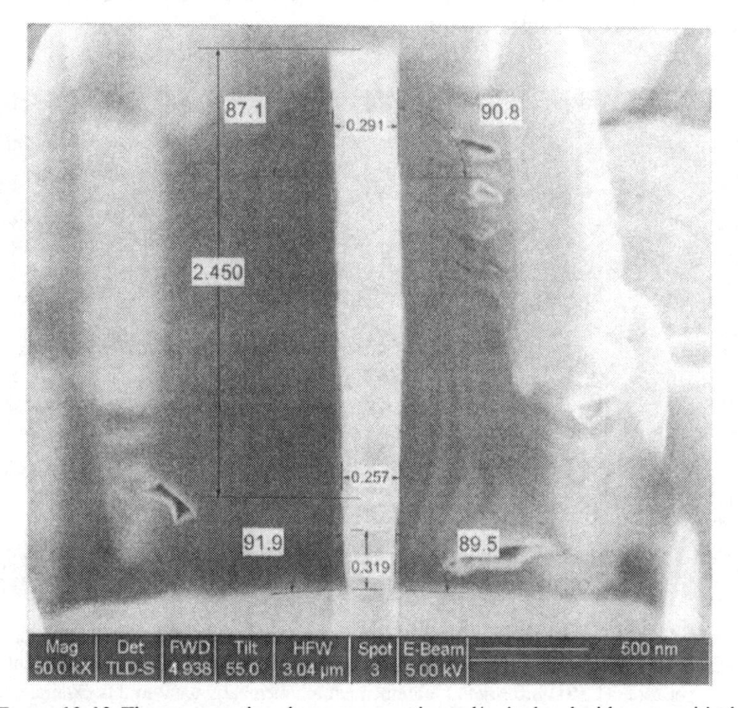

Figure 12-12. The cross section shows a magnetic read/write head with some critical measurements of subsurface features made from the SEM image. Using automated software that executes a cut, look, and measure routine, these precision cross sections and measurements can be obtained rapidly enough to make in-line process control feasible.

Incorporating dual beams into the manufacturing line enables process engineers to identify excursions much earlier in the process. This information can help determine if a previous process step needs to be adjusted and how to disposition nonviable product (rework or scrap, saving dollars either way).

In addition to being used in the production line, dual beams can be used to qualify other process tools. In an example from the semiconductor industry, the use of a dual beam to calibrate a stepper tool is shown in Fig. 13. The cross section, milled with the FIB and imaged with the SEM, is part of a photolithography process characterization study (Berger et al., 2001). As feature sizes decrease, the lithographic process windows narrow, invalidating 2D techniques for qualifying process tools. At the same time, wafer sizes are increasing, making it more attractive to be able to cross-section features without breaking wafers. The ability of the dual beams to rapidly provide full 3D structural information without breaking wafers is key to automated process control applications.

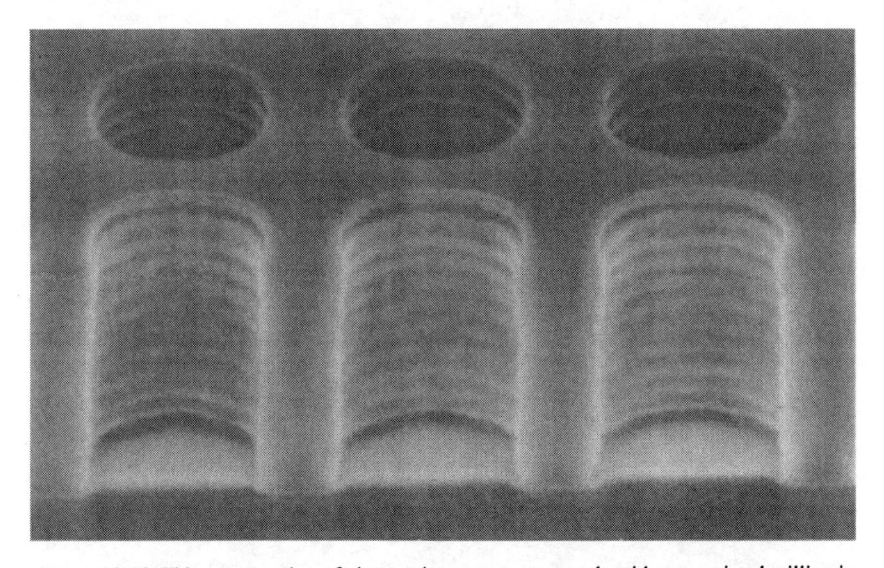

Figure 12-13. This cross section of photoresist contacts was made with gas-assisted milling in the dual beam. The contacts are part of a focused exposure matrix (FEM) needed to calibrate a stepper tool. The 3D information obtained complements 2D data; in this example, the standing wave effect and side wall angle are observable only in cross section.

6. CONCLUSIONS

A dual-beam system allows sample preparation, imaging, and analysis to be accomplished in one tool. This offers great time savings and opens up new applications possibilities.

In the lab, the dual beam provides unparalleled flexibility in 3D structural analysis and increasingly in nanofabrication. While the ion beam is primarily used for precision milling and the electron beam for high-resolution nondestructive imaging, there are additional features of a dual beam system. The ion beam and electron beam complement each other in charge reduction, delineation of cross sections, protective depositions, and imaging information. The electron beam can be used to monitor the ion beam milling to endpoint precisely on the feature of interest, and the presence of the electron beam also enables EDS, which gives elemental compositional information about the sample. Based on this structural and compositional information, milling can continue to examine additional locations. Having all these capabilities in one system means that real-time decisions can be made to produce the sample and information you need, without having to break vacuum to do it.

For rapid wafer-level defect analysis, the dual beam can be used on its own or in conjunction with defect review tools. The dual beam is invaluable for locating and analyzing subsurface defects. The electron beam can be used for relocation of defects from defect review files, and both beams can enhance location of defects using voltage contrast.

For subsurface defect analysis and process control, the dual beam removes the need to break wafers to obtain 3D structural information. The time savings of having both beams in one tool makes high-resolution 3D metrology feasible for in-line process control. With the decreasing dimensions and increasing complexity of modern high technology devices, automated structural process control is a rapidly growing application of dual beams.

ACKNOWLEDGEMENTS

The authors wish to thank our colleagues in FEI for contributing images and information, especially Michael Bernas, Sean Da, Mark Darus, Denis Desloge, Daniel Morse, Hans Mulders, Steve Reyntjens, Terri Shofner, Lyn Swanson, Janet Teshima, and Matthew Weschler.

REFERENCES

Berger S, Desloge D, Virgalla R, Davis T, Paxton T, and Witko D, Proceedings of SPIE, vol. 4344 (2001).
Campbell AN, Soden JM, Rife J, and Lee RG, Proceedings of the 21st ISTFA Conference, 33 (1995).
Delenia E, Tracy B, and Fatemi H, Evaluation Engineering, October (1995).
Gstöttner J, Stepper C, and Schmitt-Landsiedel D, uropean FIB Users Group Meeting, Arcachon, France (2001).
Hayles MF and van Veen GNA, International Patents Pending (1998).
Levi-Setti R, Scanning Electron Microscopy (SEM Inc., Chicago) (1983).
Lipp S, Frey L, Lehrer C, Demm E, Pauthner S, and Ryssel,H Microelectron. Reliab. 36, 1779 (1996a).
Lipp S, Frey L, Lehrer C, Frank B, Demm E, Pauthner S, and Ryssel H, J. Vac. Sci. Technol. B14, 3920 (1996b).
Lipp S, Frey L, Lehrer C, Frank B, Demm E, Pauthner S, and Ryssel H, J. Vac. Sci. Technol. B14, 3996 (1996c).
Luo D and Song X, Proceedings of the 23rd ISTFA Conference, 339 (1997).
Ma Z, Davies B, Brandt J, Baker B, Headley K, and Miner B, Microscopy and Microanalysis, proceedings, vol. 5, supplement 2, 904
Matteson T.L., Schwarz S.W., Houge E.C., Kempshall B.W., and Giannuzzi L.A. (2002) J. Electron. Mat. 31, 33 (1999).
Moore MV, Microscopy and Microanalysis, proceedings, vol. 8, supplement 2, (2002).
Morimoto H, Kishimoto T, Takai M, Yura S, Hosono A, Okuda S, Lipp S, Frey L, and Ryssel H, Jpn. J. Appl. Phys 35, 6623 (1996).
Mulders H, G.I.T. Image & Microscopy 2/2003, 8 (2003).
Nakamae K., Tanimoto H., Takase T., Fujioka H., and Ura K. (1992) J. Phys. D: Appl. Phys. 25, 1681
Sudraud P, Ben Assayag G, and Bon M, J. Vac. Sci. Technol. B6, 234 (1988).
Teshima J, Semiconductor International, July 2001, 171 (2001).
Weber M, Rudolph M, Kretz J, and Koops HWP, J. Vac. Sci. Technol. B13, 461 (1995).
Weiland R, Boit C, Dawes N, Dziesiaty A, Demm E, Ebersberger B, Frey L, Geyer S, Hirsch A, Lehrer C, Meis P, Kamolz M, Lezec H, Rettenmaier H., Tittes W, Treichler R, and Zimmermann H, Proceedings of the 26st ISTFA Conference, 393 (2000).
Winkler D, Zimmermann H, Gessner O, Sturm M, and Formanek H, Proceedings of 20th ISTFA Conference, 19 (1994).
Xu Y, Schwappach C, and Cervantes R, Microscopy and Microanalysis proc., vol. 6, supplement 2, 516 (2000).
Young RJ, Carleson PD, Hunt T, and Walker JFProceedings of 24th ISTFA Conference, 329 , (1998).
Young RJ, Microscopy and Microanalysis proc., vol. 6, supplement 2, 512 (2000).
Zimmermann G and Chapman R, Proceedings of 25th ISTFA Conference, 311 (1999).

Chapter 13

FOCUSED ION BEAM SECONDARY ION MASS SPECTROMETRY (FIB-SIMS)

F. A. Stevie

North Carolina State University, Analytical Instrumentation Facility, Raleigh, NC 27695

Abstract: Focused ion beam secondary ion mass spectrometry (FIB-SIMS) is a material identification technique that combines the high spatial resolution of FIB with the high elemental sensitivity of SIMS. FIB-SIMS has generated important elemental analysis data in the form of ion images for a wide range of applications including minerals, metal alloys, semiconductors, and biological materials. Quantitative imaging is possible by analysis of cross sections of reference samples in which elements are ion implanted or present at known concentrations in the substrate of interest. Elemental information on complex materials can be obtained using the high mass resolution capability of magnetic sector or time of flight analyzers. This technique will provide improved elemental sensitivity as new methods for enhancement of secondary ion yields can be developed.

Key words: Focused ion beam, FIB, secondary ion mass spectrometry, SIMS, imaging, quadrupole, magnetic sector, time of flight, TOF

1. INTRODUCTION

Secondary ion mass spectrometry (SIMS) is a very sensitive analytical technique that can provide parts per million (ppm) to parts per billion (ppb) sensitivity for most elements when relatively large areas of analysis are available. As the analytical volume is reduced, SIMS may not be able to provide the ppm to ppb sensitivity desired for trace identification, but can still provide good detection limits for matrix elements. SIMS analysis of a small area can be comparable with Auger Electron Spectroscopy (AES), depending on the area of analysis, the application, and the elements of interest. Since FIB-SIMS is one of only a few analytical techniques, such as AES and Energy Dispersive Spectrometry (EDS), which are capable of small area elemental analysis, this method deserves consideration.

SIMS has the advantages of high sensitivity and good depth resolution for almost all elements, and is specifically of importance for detection of hydrogen which cannot be analyzed using AES or EDS. SIMS also is isotope specific and can be quantified with the use of standards. SIMS has the disadvantages of a destructive analytical technique, many mass interferences, and difficult quantification. SIMS secondary ion yields vary significantly over the periodic table and are also sensitive to differences in matrix composition.

Effective use of FIB-SIMS requires an understanding of the SIMS process. Sputtering is discussed in another chapter of this book, but forms the basis of SIMS and a brief summary is warranted. The bombarding primary ions lose energy through a collision cascade that takes place on the order of 200 femto-seconds. Sufficient energy is expended in the vicinity of the surface to remove atoms or molecules. Most of the sputtered species are neutrals, but positive and negative secondary ions are also ejected. The sputtered species depart with an energy distribution that has a most probable energy of only a few electron volts. Sputtering yield has been studied as a function of primary ion energy, angle of incidence, and mass, and also as a function of the mass of the target element. (Benninghoven, 1987) Sputtering yield does not change dramatically with primary energy but shows a broad peak at a few tens of keV. At low energy, the total energy into the substrate is reduced and hence the amount of sputtering is decreased, while at high energy most of the collision cascade is too far into the matrix to remove significant amounts of material at the surface. The sputtering yield increases with incidence angle from normal until near grazing incidence is reached. At normal incidence the collision cascade is deeper into the substrate than for more oblique incidence. For incidence greater than 70 degrees from normal, the interaction of the beam with the substrate is significantly reduced. Sputtering yield increases with bombarding mass up to approximately atomic number 30, and then shows only a minor increase for the rest of the periodic table. Sputtering yield varies dramatically with substrate element and shows trends related to location in the periodic table. Variations in sputtering rate of an order of magnitude are often possible for the components of a substrate.

The primary ion beam used in FIB systems is almost universally Ga^+, but In^+ is also commercially available. These sources predominate because they provide a stable beam for a reasonable lifetime (can exceed 1000 hours) and can be reliably and reproducibly manufactured. While the use of a gallium primary beam provides very small beam size (commercially available systems can provide a beam with diameter < 10 nm at the sample) and therefore excellent spatial resolution for site specific micro-structural cross-sectioning, the use of a gallium primary beam for the production of secondary ions provides sensitivity similar to that for Ar^+, which is typically about a factor of 40 less than for O_2^+.

Positive and negative SIMS yields are significantly improved by use of O_2^+ and Cs^+ primary beams, respectively. These beams provide chemical reactions with the surface that increase the ion yield. In SIMS, for a given

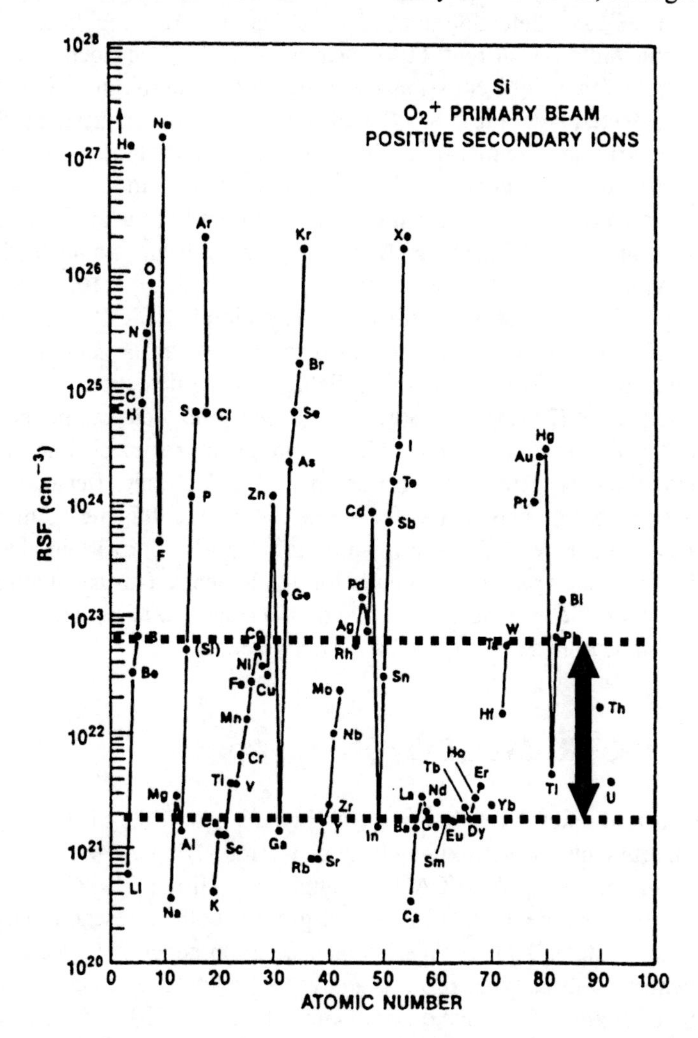

Figure 13-2. Secondary ion yields for oxygen primary beam. Arrow shows increase in secondary ion yield desired for broader range of applications.

primary ion, the secondary ion yield that is generated can vary by orders of magnitude over the periodic table. Figure 1 shows secondary ion yields of the periodic table for O_2^+ bombardment of silicon. The difference in secondary ion yield, and therefore the difference in detection between elements such as Li, Na, K, Al (all readily detected) and the transition metal elements (highly desired but difficult to detect) us about a factor of 30. (Wilson, et al., 1989)

It is important to compare SIMS with EDS and AES for small area analysis. EDS has a detection limit of 1 – 2% atomic for a region as small as 5 nm (achievable on current TEM instruments). AES can detect at about 0.1% atomic for a field emission instrument over approximately a 20 nm diameter detected area. For a SIMS analysis of a 0.1 μm diameter particle in a silicon matrix, and 1 nm/s sputtering rate, the number of atoms/s available is 4×10^5 (5×10^{22} atoms/cm^3 x 7.5×10^{-11} cm^2 x 1×10^{-7} cm/s). A reasonable useful yield (ratio of ions detected to atoms sputtered in the analysis region) for a quadrupole SIMS analyzer with O_2^+ primary beam would be 1×10^{-3}, which would provide 400 counts/s or a detection limit of 2×10^{20} atoms/cm^3 (0.2% atomic) if 1 count is significant (background is typically less than 0.1 counts/sec). For a 0.03 μm (30 nm) diameter particle at the same sputtering rate, there are only 3×10^4 atoms/s available and the detection limit will be 1×10^{21} atoms/cm^3 (2% atomic). However, if the useful yield can be increased by an order of magnitude (achievable with magnetic sector or time of flight analyzers) then the sensitivity will again be 0.2% atomic. Because of the wide range of SIMS sensitivities for the periodic table, this means that FIB-SIMS can compare very favorably with AES and EDS for certain elements, but will not be the optimum technique for all elements. This calculation also presumes that the secondary ion yields for Ga^+ primary beam can be increased to near those for O_2^+ or Cs^+ primary beams.

2. INSTRUMENTATION

The current lateral resolution capability for non-FIB SIMS magnetic sector instruments is approximately 200 nm for O_2^+ (CAMECA IMS-6F) and 50 nm for Cs^+ (CAMECA Nanosims 50). (Hillion, 1994) FIB-SIMS systems have been developed by several groups, both as research tools and as practical systems and provide lateral resolution superior to the non-FIB instruments. All three types of mass analyzers (quadrupole, magnetic sector, and time of flight (TOF)), have been matched with an FIB source. Research systems at the University of Chicago led the way in FIB-SIMS with first a quadrupole and then a magnetic sector analyzer using a Ga^+ liquid metal ion source primary beam. (Levvi-Setti, 1980, 1984, 1994, Chabala, 1996) The magnetic sector provided 4-20% efficiency and mass resolution of 570 based on a M/delta M calculation.

Several generations of FIB-SIMS with quadrupole analyzers were produced by FEI. (Crow, 1995, Dingle, 1997) Instruments with magnetic sector and TOF analyzers have been developed and are currently showing excellent results for lateral resolution and sensitivity. (Tomiyasu, 1998) Currently available TOF-SIMS systems are often configured with FIB sources, but beam broadening associated with the pulsing of the TOF beam results in a lateral resolution not much better than the Cs^+ beam dedicated SIMS systems.

3. APPLICATIONS

There are three main forms in which SIMS data is obtained. Ion images provide lateral elemental distribution. Mass spectra provide an elemental survey over the periodic table and also show isotopic information. Depth profiles indicate the elemental distribution as a function of depth. The imaging applications predominate because FIB-SIMS has been used primarily to improve lateral resolution. Depth resolution has been of less emphasis because the lateral resolution is achieved by use of high primary beam energies (>20 keV) that degrade depth resolution because of increased ion-specimen interaction volume.

3.1 Elemental Imaging

Data from the University of Chicago FIB-SIMS in the 1980's provided elemental maps using Ga^+ probes of diameter 20 – 90 nm. (R. Levvi-Setti 1980, 1984, 1986, Chabala, 1988) This research group investigated a wide range of materials, including minerals, metal alloys, and biological materials. (Berry, 1988, Levvi-Setti, 1993)

An example is shown in Figure 2 where secondary ion images have been obtained form an alloy of Al-Si-Mg-Cu. The Al image on the left shows non-uniform distribution with the Al concentrated in small, mostly circular regions. Higher Al concentration is indicated by a brighter color. The Mg

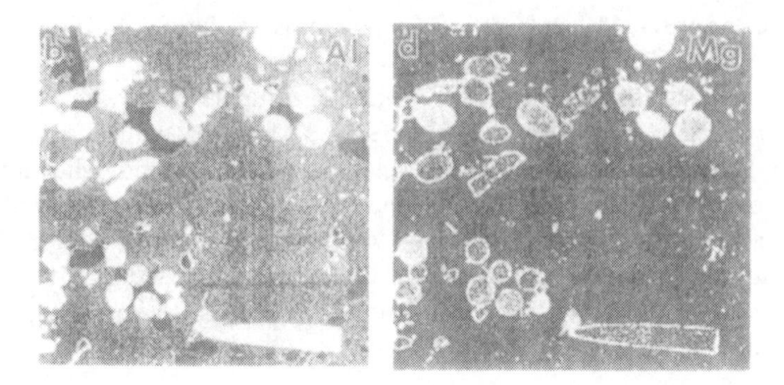

Figure 13.2. Ga^+ SIMS images of a polished section of Al-Si-Mg-Cu alloy with reinforcing Saffil fibers. Left image is $^{27}Al^+$ and right image is $^{24}Mg^+$. From Levi-Setti, 1993, with permission of Scanning Microscopy.

image on the right shows the Mg is coincident with the Al but for most of the localized areas where Mg is detected, the Mg concentration is higher at the edges than in the center. Results such as this provide significant information on the spatial relationship of elements within this matrix.

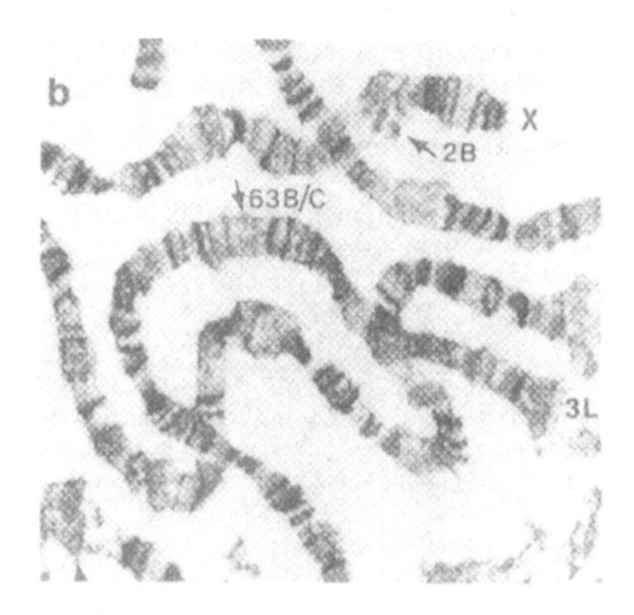

Figure 13-3. CN^- image from labeled chromosome. Image shown in inverted contrast. Labeled features are landmarks used to correlate images. From Levi-Setti, 1994, with permission of Journal of Microscopy and Blackwell Publishing

FIB-SIMS has been used for biological applications and has, for example, been able to show the exact lateral content of a strontium enriched layer in mollusks and location of copper in mysids. (Walker, 1998) The data were normalized by plotting the ratio of strontium to calcium. Because SIMS is isotopic specific, it is also possible to utilize isotopic labeling for biological experiments. (Hallegot, 1990; Levi-Setti, 1994) Figure 3 shows a $^{12}C^{14}N^-$ image from a chromosome. Analysis of labeled chromosomes reveals the areal density of proteins, DNA and RNA. Analysis for Br^- shows a similar distribution. Chromosomal banding information is obtained from the CN^- image and matches optical microscopy of orcein stained chromosomes.

FIB-SIMS elemental mapping has been applied to semiconductor structures. SIMS has excellent sensitivity for Na, and an example of Na detection has been made for random access memory (RAM) devices. (Perrin, 1994) FIB-SIMS was used to study the effects of etching 500 nm semiconductor features in a high density, low pressure plasma reactor. (Downey, 1998) The sample was angled with respect to the beam to expose more side wall surface area. The FIB beam was also used to cut through a line to expose the interior. A polymeric material from photoresist, metal, etch gases, and oxide was found on the side wall. Boron in particular was detected as a result of the BCl_3 etchant and was shown to be removed after subsequent processing. Figure 4 shows the B elemental map, including the FIB exposure cut, and indicates the B is present only on the side walls. FIB-SIMS has also been effectively used for particle and defect identification. (Sakamoto, 1996, 2000)

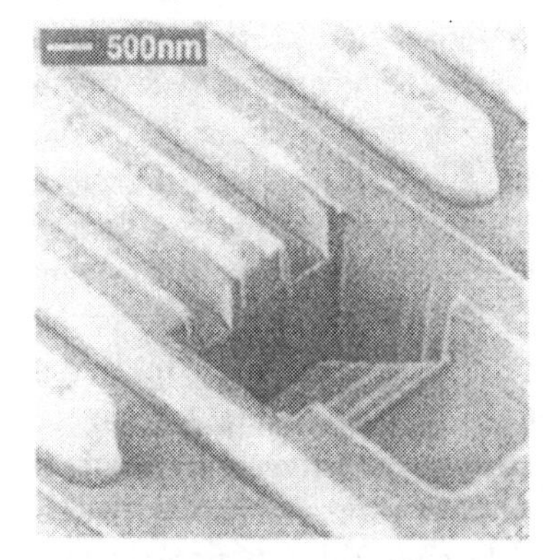

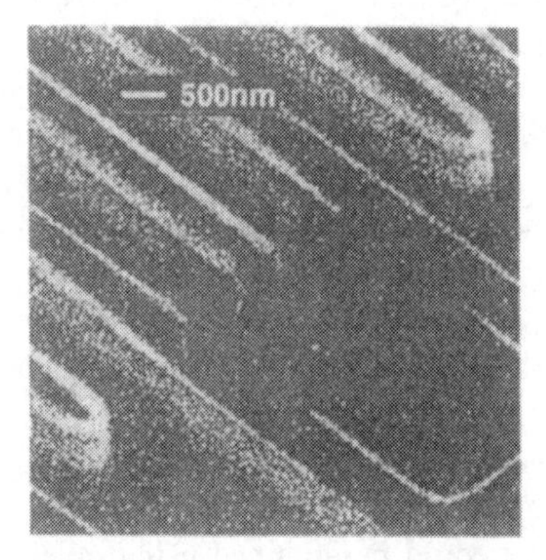

Figure 13-4. Boron residue detected on sidewalls. Figure on left is total secondary ion image and figure on right is $^{11}B^+$ ion image. From Downey, 1998, with permission of Wiley.

3.2 End Point Detection

FIB systems are used to make modifications to semiconductor circuits or to masks used in semiconductor manufacture. FIB-SIMS can help determine when the cutting process of the repair should be stopped. A SIMS depth profile or constant monitoring of a secondary ion species can help determine the sputtering conditions, particularly sputtering time, required to reach the depth of the layer of interest so that the repair can be completed. Using FIB-SIMS, depth profiles can be obtained at a high aspect ratio (ratio of depth to width), especially when gas assisted etching is used. (Abramo, 1994) The SIMS yields on certain metals, such as Al and Cr, are high enough to detect the matrix species. End point detection of Cr has been used for mask repair. (Vasile, 1989) End point detection was used for repairs of semiconductor circuits using B from Borophosphosilicate glass (BPSG), Si and Al. (Crow, 1991) This approach was also used in micro-machining by detection of an Al layer with a quadrupole analyzer system. (Prewett, 1989) The total secondary ion signal can be used to distinguish between a conductive and nonconductive material. (Hill, 1993) Analysis has also been made of a YbaCuO superconducting film on a yttrium iron garnet substrate with $SrTiO_3$ and MgO buffer layers to prevent diffusion. (Montgomery, 1998) The FIB was used to cut stepped trenches in concert with image analysis to convert secondary ion images of cross sections through the superconducting film multilayer into a depth calibrated depth profile.

3.3 Image Quantification

Quantification of images has been achieved, such as for analysis of glasses. (Satoh, 1988) Of substantial interest is the ability to obtain quantitative two and three dimensional dopant profiles for silicon technology. The SEMATECH requirements for dopant mapping are 5 nm spatial resolution, $1E17/cm^{-3}$ sensitivity, and 5% concentration precision. Even though FIB-SIMS will not have the sensitivity to provide the desired detection limit at the stated spatial resolution, it may provide the dopant distribution at a higher concentration which may then be matched to data from another method, such as TEM imaging after preferential etching of the sample to delineate lower concentrations. One approach was to use the depth resolution of SIMS and profile a two dimensional region at right angles to the surface, a technique that requires precise specimen preparation. (Criegern, 1994, 1998) It is possible that information on the region of interest could be obtained using other 3D SIMS approaches (Gnaser, 1996) such as a tomography procedure. In this concept, SIMS would provide a representation of the 2D or 3D dopant profile by using FIB sample preparation and analysis of specimens taken at several different angles. Analysis at normal incidence provides the ability to increase total secondary ion counts on a structure with rectilinear geometry by sputtering into a feature that has some thickness and accumulating counts. (Stevie, 1998, 1999) This structure repeatability is commonly found in semiconductor devices. This approach, when applied to cross sectioned specimens from ion implanted wafers, also provides a method for calibration of images and evaluation of the effectiveness of gases used to enhance secondary ion yields.

Quantitative and 3D analysis and simultaneous detection have been achieved with a magnetic sector analyzer using the shave-off mode to eliminate topographic effects, including edge enhancement. (Nihei, 1997, 1997; Tomiyasu 1995, 1998; Cheng, 1998; Sakamoto, 1999) A fast horizontal sweep is combined with a slow vertical sweep. As a result, knock on mixing and sputter deposition are not significant. A spatial resolution of 50 nm and depth resolution of 5 nm has been achieved. (Tomiyasu, 1998) Relative sensitivity factors (RSF) were obtained for quantification by analysis of elements on glass matrices. This method was used for studies such as determination of contaminants in steel. (Tomiyasu, 1996, 2000)

A method has been developed for focused ion beam reconstruction techniques that, coupled with secondary ion mass spectrometry, allows the three-dimensional reconstruction of geometrical and chemical relationships in a specimen with 20 nm resolution. (Dunn, 1999) This resolution was obtained in a specimen with field of view up to 100 μm^3 and reconstructions that contain up to 10^7 independent "voxels" of information.

If one can add the ability to do peak selection of mass spectra, then the masses of interest can be scanned in a very short time. (Stevie, 1999) This ability may help avoid the problem of sputtering away the region of interest before it can be identified, which can happen if one has to sequentially accumulate ion images. Sputtering during image accumulation of the first few species may remove too much material before acquisition of species at the end of the list.

4. SECONDARY ION YIELD ENHANCEMENT

Several studies have examined secondary ion yields for a gallium primary beam. (Satoh, 1988; Bennett 1991) The difference in secondary ion yields between gallium and oxygen primary ion sources for analysis of SiO_2 is relatively small due to the enhancement of secondary ions by the oxygen in the oxide (see Table 1). Therefore, if the material of interest can be oxidized during analysis, then the FIB-SIMS ion yields may be able to provide sufficient sensitivity. It has been observed that a substantial increase in secondary ion yields using a gallium primary beam can be achieved by the addition of a suitable element or gas to the surface of the region being sputtered. Enhancement of the secondary ion yield from addition of cesium in analysis of arsenic on a time of flight SIMS system has been obtained with use of a cesium beam interleaved with a gallium beam to enhance the yield from gallium bombardment for a time of flight SIMS system. (Schueler, 1998) This system can sputter with one beam and analyze with another. The analysis was made using a 25 keV Ga^+ beam at 600 pA over a 12 μm x 12 μm area within the larger 800 μm x 800 μm raster of a 160 nA Cs^+ beam at 3 keV. At each pixel the surface was exposed to Ga^+ for 35 μs followed by Cs^+ for 45 μs, with a waiting period of 10 μs. The As^- enhancement was approximately a factor of 500.

Table 13-1. Secondary ion useful yields for oxygen, cesium, and gallium bombardment [Schmacher et al., 1992]

Matrix	Species	Primary Ion O_2^+	Primary Ion Cs^+	Primary Ion Ga^+
Si	Si^+	$5.1x10^{-3}$		$6.0x10^{-5}$
	Si^-		$1.6x10^{-2}$	
	B^+	$3.1x10^{-3}$		$6.4x10^{-6}$
	B^-		$2.8x10^{-4}$	
	As^+	$1.3x10^{-5}$		$3.5x10^{-7}$
SiO_2	Si^+	$3.8x10^{-2}$		$1.4x10^{-2}$
	Si^-		$4.9x10^{-3}$	
	B^+	$3.9x10^{-3}$		$1.6x10^{-3}$
	B^-		$5.6x10^{-4}$	

Other studies have been undertaken to explore different ways to enhance the secondary ion yield. Water is used as a source to selectively remove carbon based materials in the presence of the gallium FIB beam. Investigation of water as a SIMS yield enhancer has shown positive results but with some inconsistency. For example, some improvement on secondary ion yields has been noted by use of an oxygen gas or a water source on the FEI system. (Sakamoto, 1994, 1997; Shofner, 2001) The implanted Ga may also be used as an internal standard to monitor enhancement. (Sakamoto, 1997) One of the most important variables is sputtering rate. The surface must be affected by the enhancing species at a rate appropriate to the rate of removal. Current work in this area includes development of a FIB ion source that contains oxygen or cesium. (Guharay, 1999)

5. SUMMARY

FIB-SIMS has demonstrated the ability to provide small area elemental identification for specific elements that exceeds the capability of the limited number of alternatives (AES, EDS). FIB-SIMS also has the ability to obtain isotope specific data. However, this method will not achieve full potential until the secondary ion yield is enhanced by either use of a new gas or development of a gas source. The continued reduction of effective primary beam diameter in TOF-SIMS instruments may facilitate the use of this technique because the TOF analyzer has excellent transmission.

REFERENCES

Abramo M, Hahn L, and Moszkowicz L, ISTFA Proceedings, p. 439 (1994).

Bennett J and Simons D, J. Vac. Sci. Technol. A9, 1379 (1991).

Benninghoven A, Rudenauer RF, and Werner HW, Secondary Ion Mass Spectrometry, John Wiley & Sons, New York (1987).

Berry, JP, Escaig F, Levvi-Setti R and Chabala J, Secondary Ion Mass Spectrometry, SIMS VI, Benninghoven A, Huber AM, and Werner HW, eds., John Wiley & Sons, Chichester, p. 901 (1988).

Chabala JM, Levi-Setti R, and Wang, YL, J. Vac. Sci. Technol. B6, 910 (1988).

Chabala JM, Secondary Ion Mass Spectrometry, SIMS X Proceedings, Benninghoven A et al., eds., Wiley, Chichester, p. 23 (1996).

Cheng Z, Sakamato T, Takahashi M, Kuramoto Y, Owari M, and Nihei Y, J. Vac. Sci. Technol. B16, 2473 (1998).

Criegern R von, Jahnel F, Bianco M, and Lange-Gieseler R, J. Vac. Sci. Technol. B12, 234 (1994).

Criegern R von, Jahnel F, Lange-Geissler R, Pearson P, Hobler G, and Simonescu A, J. Vac. Sci. Technol. B16, 386 (1998).

Crow GA, ISTFA Proceedings, p.401 (1991).

Crow GA, Christman L, and Utlaut M, J. Vac. Sci. Technol. B13, 2607 (1995).

Dingle T, et al., Secondary Ion Mass Spectrometry, SIMS X, Benninghoven A, et al., eds., Wiley, Chichester, p. 517 (1997).

Downey SW, Stevie FA, Colonell JI, Brown S, Dingle T, and Christman L, SIMS XI, Gillen G, et al., eds., Wiley, Chichester, p.163 (1998).

Dunn DH and Hull R, Appl. Phys. Lett. 75, 3414 (1999)

Gnaser H, Surf. Interf. Anal., 24, 483 (1996).

Guharay SK and Orloff J, SIMS Workshop 1999.

Hallegot P, Girod C, LeBeau MM, and Levi-Setti R, Secondary Ion Mass Spectrometry, SIMS VII, Benninghoven A, Evans CA, McKeegan KD, Storms HA, and Werner HW, eds., John Wiley & Sons, Chichester, p. 327 (1990).

Harriott LR and Vasile MJ, J. Vac. Sci. Technol. B7, 181 (1989).

Hill R, Morgan JC, Lee RG, and Olson T, Microelectronic Engineering 21, 201 (1993).

Hillion F, Daigne B, Girard F, Slodzian G, and Schuhmacher M, Secondary Ion Mass Spectrometry, SIMS IX, Benninghoven A, et al., eds, Wiley, Chichester, p. 254 (1994).

Kikuma J and Imai H, Surf. Interf. Anal. 31, 901 (2001).

Levvi-Setti R and Fox TR, Nucl. Instr. Methods 168, 139 (1980).
Levi-Setti R, Wang YL, and Crow B, J. de Phys. 45, C9-197 (1984).
Levi-Setti R, Crow G, and Wang Y L, Secondary Ion Mass Spectrometry, SIMS V, Benninghoven A, et al., eds., Springer Verlag, Berlin, p. 132 (1986).
Levi-Setti R, Chabala JM, Li J, Gavrilov KL, Mogilevsky R, and Soni KK, Scanning Microscopy 7, 1161 (1993).
Levi-Setti R, Chabala JM, Gavrilov K, Li J, Soni KK, and Mogilevsky R, Secondary Ion Mass Spectrometry, SIMS IX, Benninghoven A, et al., eds., Wiley, Chichester, p. 233 (1994).
Levvi-Setti R, Chabala JM, Smolik S, Journal of Microscopy 175, 44 (1994).
Montgomery NJ, McPhail DS, Chater RJ, and Dingle T, Secondary Ion Mass Spectrometry, SIMS XI, Gillen G, et al., eds., Wiley, Chichester, p. 631 (1998).
Nihei Y, Tomiyasu B, Sakamoto T, and Owari M, J. Trace and Microprobe Techniques 15, 593 (1997).
Nihei Y, J. Surface Analysis 3, 178 (1997).
Perrin D and Seifert W, ISTFA Proceedings 13, 409 (1994).
Prewett PD, Marriott P, and Bishop HE, Microelectronic Engineering 10, 1 (1989).
Sakamoto T, Tomiyasu B, Owari M, and Nihei Y, Surface and Interface Analysis 22, 106 (1994).
Sakamoto T, Owari M, and Nihei Y, Jpn. J. Appl. Phys. 36, 1287 (1997).
Sakamoto T et al., Bunseki Kag 45, 479 (1996).
Sakamoto T, Kuramoto Y, Cheng ZhH, Takanashi K, Wu H, Owari M, and Nihei Y, Secondary Ion Mass Spectrometry, SIMS XII, Benninghoven A., et al., eds., Elsevier, p. 217 (2000).
Satoh H, Owari M, and Nihei Y, J. Vac. Sci. Technol. B6, 915 (1988).
Schuhmacher M, Migeon HN, Rasser B, Secondary Ion Mass Spectrometry, SIMS VIII, Benninghoven A, et al., eds., Wiley, Chichester, p. 49 (1992).
Schueler B, Gemmill J, Reed DA, Reich F, and Smith NS, Secondary Ion Mass Spectrometry, SIMS XI, Gillen G, et al., eds., Wiley, Chichester, p. 759 (1998).
Shofner TL, Stevie, FA, McKinley JM, and Lomness J, ISTFA 2001 Proceedings, p. 281 (2001).
Stevie F. A., Downey S. W., Brown S. Shofner T., Decker M., Dingle T., and Christman L., Microscopy and Microanalysis 1998 Proceedings, p. 650 (1998).
Stevie FA, Downey SW, Brown SR, Shofner TL, Decker MA, Dingle T, and Christman L, J. Vac. Sci. Technol. B17, 2476 (1999).
Tomiyasu B et al., Tetsu Hagan 81, 977 (1995).
Tomiyasu B et al., Bunseki Kag 45, 485 (1996).
Tomiyasu B, Fukuju I, Komatsubara H, Owari M, and Nihei Y, Nucl. Instru. Meth. Phys. Research B136-138, 1028 (1998).
Tomiyasu B, Owari M, and Nihei Y, Bunseki Kagaku 49, 593 (2000).
Walker JF, Robinson J, Peck LS, Garnacho E, and Pugh PJA, Electron Microscopy 1998, ICEM 14 Proceedings, p. 315 (1998).
Wilson RG, Stevie FA and Magee CW, Secondary Ion Mass Spectrometry, Wiley (New York) 1989.

Chapter 14

QUANTITATIVE THREE-DIMENSIONAL ANALYSIS USING FOCUSED ION BEAM MICROSCOPY

D.N. Dunn, A.J. Kubis and R. Hull
University of Virginia, Department of Materials Science and Engineering

Abstract: In this chapter, we review progress in quantitative three-dimensional chemical and microstructural volume analysis using focused ion beam tomography. A brief survey of techniques previously developed for solid state tomographic reconstructions is presented and the relevance of focused ion beam tomography to these techniques is discussed. Methods for quantitatively reconstructing three-dimensional spatial and chemical maps using focused ion beam microscopy are then discussed in detail. Finally, areas for future technique development and improvement are discussed.

Key words: Focused ion beam microscopy, tomography, shaped based interpolation, quantitative volume reconstruction.

1. INTRODUCTION

The great majority of materials characterization techniques currently available are limited to one- or two- dimensional representations of materials properties (here we take the term "properties" to include structure and chemistry). For example, spectroscopic and X-Ray or neutron diffraction techniques typically measure the average properties of relatively large volumes of material (although micro-spot modes for two-dimensional mapping or near surface measurement modes often exist). High resolution imaging techniques, such as electron microscopy, generally produce two-dimensional representations either of surface or near-surface properties (such as in secondary electron imaging), or of an average property measurement projected through a sample thickness (such as in transmission electron imaging). Scanning probe microscopy techniques, such as atomic force microscopy or scanning tunnelling microscopy, may be regarded as providing "two and a half dimensions" in that that they have provide high lateral resolution (down to atomic dimensions) perpendicular to a surface normal, and provide ultra-high resolution (down to a small fraction of an Angstrom) parallel to the surface normal. However, these scanning probe techniques are limited to surface measurement and cannot provide sub-surface information, although combined etching / probe microscopy techniques have been used to infer three dimensional structures (Magerle, 2000) in a similar fashion to the sequential FIB sputtering and imaging techniques described in this chapter.

Three-dimensional imaging techniques have of course long been enormously important in the medical field. Magnetic resonance imaging (MRI) and computerized axial tomography (CAT) techniques provide essentially non-destructive (critical in human care!) methods for three-dimensional imaging of internal organs. The resolution of such techniques is of order of a fraction of a millimeter, which couples well with the length scales inherent to human organs and most flaws within them. However, improvement of resolution, signal, reconstruction and interpretation algorithms remains a highly important medical research field.

Several other techniques have been developed for three dimensional imaging of (primarily) inorganic materials, as will be discussed later in this article. Perhaps foremost among these is X-Ray micro-tomography which has a resolution on the order of a micron, is non-destructive, but has limited chemical sensitivity. Another important technique is the position sensitive atom probe, which uses field ion desorption from an ultra-high curvature tip, coupled with time-of-flight mass spectrometry and a position sensitive detector, to reconstruct the three dimensional structure and chemistry of the tip with atomic resolution. The field of view of this technique is limited to less than a cubic micron, and requires that samples be prepared using specialized geometries. Furthermore, samples must also be conducting in most cases.

These techniques leave an important part of dimensional space inaccessible, corresponding to length scales of tens of nanometers to tens of microns. As the field of nanotechnology becomes increasingly important, these length scales become increasingly critical. Further, the ability to engineer on the nanoscale (i.e. on length scales of order tens to hundreds of atomic dimensions) implies control (and therefore measurement) of properties in three dimensions. It is to this important regime that focused ion beam tomography is ideally suited.

The essence of focused ion beam tomography is the serial acquisition of images at different depths in a structure. These images may be created by secondary electrons, by secondary ions, or by mass-filtered secondary ions to create chemical maps. The resolution of such images is determined by a combination of the primary ion beam diameter at the sample surface, the secondary electron / ion escape depth, and the implant depth / lateral spreading of the primary ions in the sample. For a state-of-the-art primary ion beam diameter of 10 nm, and a beam energy of 30 keV, this translates into a practical spatial resolution of order 20 - 30 nm for most inorganic materials, both parallel and perpendicular to the sample surface normal, as will be described later in this article. The set of images recorded at different depths from the original sample surface are then aligned and concatenated in the computer using appropriate interpolation algorithms between slices. This forms a three dimensional representation of the sample which can contain up to 10^7 or more individual voxel elements, each of which are separately addressable in the reconstructed image. Three-dimensional structural and chemical relationships between components in the structure may thus readily be investigated.

A major issue in the generation of such three-dimensional reconstructions is signal to noise. The secondary electron yield (number of secondary electrons emitted per incident ion) is relatively high, generally greater than one for conducting or semi-conducting materials. Thus, even at 10 nm resolution there are high numbers of secondary electrons to detect - for a yield of 2, and a (10 nm^3) voxel volume, there will be of order 10^6 secondary electrons generated per voxel. Secondary ion yields are many orders of magnitude lower, as discussed in the chapter on secondary ion mass spectroscopy (SIMS) in the focused ion beam by F.A. Stevie in this volume. Collection efficiencies through quadrupole mass spectrometers typically are also relatively low. This means that chemical mapping using mass-filtered secondary ion species may often be signal limited. Compounding this issue is the fact that there is only one opportunity to detect a sputtered ion. This is in contrast to techniques such as electron energy loss spectroscopy in the transmission electron microscope, where low interaction cross-sections between the primary electron and the appropriate chemically-sensitive inelastic scattering event in the sample may be overcome by acquisition of spectra from literally billions of primary electrons incident on the same volume of sample. This "one shot" aspect of FIB-SIMS reconstructions means that efficient strategies for secondary ion (and to a lesser extent, electron) detection are crucial. Despite these limitations, it is our experience that extremely useful chemical reconstructions may be obtained for inorganic materials at local atomic concentrations of order a few percent or greater, and at spatial resolutions of order tens of nanometers.

1.1 Micro-Tomography Techniques

Tomography has become increasingly important over the past ten years for imaging and reconstructing three-dimensional volumes of many materials. Many techniques used for obtaining two-dimensional images and elemental maps have been modified or combined with sectioning techniques to reconstruct three-dimensional renderings from two-dimensional information. This allows investigators to establish three-dimensional microstructural and chemical relationships directly. In addition, these three-dimensional relationships provide insight into structure property relationships and also aid in materials design.

One of the most common probes for micro-tomography are x-rays. X-ray tomography records x-ray attenuation as a function of angle of rotation to produce a two-dimensional image (Copley et al. 1994). A series of these images is collected as a function of distance along an object and concatenated into a three dimensional volume. The most important difficulties with x-ray attenuation measurements are signal to noise and signal localization. One method to overcome these problems, Cone-Beam X-Ray Tomography (CBXRT), has recently been suggested to improve signal to noise (Defrise 2001). In CBXRT, two –dimensional x-ray attenuation data are collected in parallel using an array detector, thus improving signal to noise and, in some cases, removing the need for interpolative methods during volume reconstruction (Copley et al. 1994). A difficulty with CBXRT is however, that collected data need to be processed with complex deconvolution algorithms. Whether this technique will find applications in materials analysis is yet to be seen.

X-ray tomography can be used to reconstruct the three-dimensional morphology and, in some cases, chemical distributions within a material. Elemental distribution can be inferred in some cases because x-ray attenuation coefficients depend upon chemical constituents within a volume (Owens et al. 2001). Since x-ray attenuation is dependent on the density of a material and as well as elemental constituents, care should be taken in interpreting the data.

As with most experimental methods, ultimate resolution is an important consideration in x-ray tomography. Many different types of x-ray sources can be used to perform x-ray tomography, but to get to high spatial resolution a synchrotron source is used. Using synchrotron radiation, spatial resolution of up to 1 μm can be obtained. In addition to micron scale spatial resolution, x-ray tomography is also a non-destructive technique. This allows the sample to be used for other experiments and enables the evolution of microstructural and chemical changes in materials to be investigated as a function of time (Babout, 2001; Lin, 2000; Lu et al., 2001; Lin et al., 2000; Martin et al., 2000; Guvenilir et al., 1997; Ludwig and Bellet, 2000).

Electron probes have also found widespread use in microtomography (Mangan and Shiflet, 1997; Soto et al., 1994). In conjunction with serial sectioning, both Scanning Electron Microscopy (SEM) and Transmission Electron Microscopy (TEM) have been used to reconstruct material volumes. Two different methods have been used to produce SEM tomographic reconstructions. In the first, samples were serially sectioned using standard polishing techniques and SEM imaging. In this technique, a sample is removed from the chamber and polished between images. This allows slices to be taken at 200-300 nm depth increments.

A second, non-destructive method for SEM tomography is to collect backscattered electrons as a function of primary electron energy (Soto et al., 1994). Since backscattered electron emission depth varies as a function of incident beam energy, a sample can be imaged as a function of depth by increasing the incident beam energy. Noise from a particular depth can be reduced by energy filtering of backscattered electrons.

TEM has also been used in conjunction with serial sectioning methods to reconstruct volumes in biological materials. Sections 1-2 μm in thickness were taken from a structure of interest and each slice was analyzed in the microscope. In addition, three-dimensional reconstructions of biological molecules have been done using TEM without using serial sectioning, but by varying defocus and tilt in samples suitable for phase contrast imaging (Niedrig and Rau, 1998; Ruprecht and Nield, 2001). In both SEM and TEM reconstructions, lateral resolution is limited by microscope contrast transfer functions, while depth resolution is limited by sectioning technique or minimum defocus and tilt increments.

Two techniques that allow very high spatial resolution are Scanning Force Microscopy (SFM) (Magerle, 2000) combined with plasma etching and Atom Probe Field-Ion Microscopy (FIM) (Magerle, 2000; Miller, 2000; Deconihout et al., 1999; Blavette et al., 1993). Proof of concept for SFM was done using block copolymer samples and by sectioning with plasma etches techniques. An advantage of this technique is that changes in topography due to preferential etching can be followed directly with SFM during data collection. Lateral resolution for this technique is of order 10 nm.

FIM atomic scale reconstructions of three-dimensional elemental positions are performed with the aid of Time of Flight Mass Spectrometry (TOFMS) and a position sensitive detector. A sample is formed into a probe with a tip radius of less than 50 nm and a taper angle of less than 10 degrees. Monolayers of atoms are field desorbed in a pulse mode and filtered spatially and by mass using TOFMS. This information is then reconstructed into a three-dimensional volume with near atomic resolution. Since a sample has to be formed into a conductive probe of very fine dimensions, this technique is limited to applications for which samples can be fabricated.

A number of approaches have been developed to use primary ion beams for reconstructing three-dimensional volumes from both images and elemental maps. For example, a non-destructive method using Particle Induced x-ray Emission (PIXE) and Scanning Transmission Ion Microscopy (STIM) has been developed using Focused Ion Beams (Jamieson, 1998; Malmqvist, 1995). An ion beam, usually a focused proton beam, is rastered over the surface of a sample and emitted x-ray intensity is recorded as function of position. Since x-ray attenuation is a non-linear process, local x-ray attenuation factors are calculated from STIM density reconstructions to account for differences in x-ray production cross sections (Malmqvist, 1995). By using accurate modeling and advanced algorithms for concatenating experimental data, volume reconstructions can be obtained with spatial resolutions as high as 0.5 μm (Schofield, 1995). This technique has been used to study both biological samples as well as metallurgical samples (Sakellariou et al., 1997; Ng et al., 1997).

SIMS is another technique where primary ion beams are used to obtain elemental information (Benninghoven et al., 1987). The primary ion beam (usually Cs^+ or O^-) is used to sputter material from a surface and sputtered ions are extracted into a mass filter where their yield as a function of mass and position is recorded. Since the primary beam is sputtering the surface away, images are collected at given time intervals that correspond to different depths in the sample (Hutter and Grassebauer, 1992). Slices can then be reconstructed into a three-dimensional volume with approximate lateral resolutions of 2 μm and an approximate vertical resolution of 20 nm (McIntyre et al., 1992; Lu et al., 1994; Hutter et al., 2001; Gammer et al., 2001). One drawback of this method is that sputtering efficiency varies dramatically with both composition and crystallographic orientation (Ng et al., 1997), thus, uncertainties due to differential sputtering increase as a function of sputtered depth. In some cases, uncertainties due to differential sputtering can be compensated for through Atomic Force Microscopy data or by milling the surface perpendicular to the imaging plane, but in most cases this involves removing the sample from the ion microscope and can lead to contamination problems (Wagter et al., 1997).

Focused ion beam microscopes are particularly well suited to tomographic reconstructions because samples can be sectioned and data collected in the same chamber. Several groups attempted to use FIB for three-dimensional analyses but were limited by the capability of computers of the time to handle the amount of data generated (Steiger et al., 1993). With higher capacity computers images could be acquired at increasing depths and the elemental distributions compared as a function of depth. This type of methodology has the same inherent topography problem encountered in ion microscopes but has better lateral resolution. Since a FIB is an ion mill and material can be removed with nanometer precision in-situ, serial sectioning can be performed. This is done by first milling a flat surface and then rotating the sample to image the newly formed surface (Satoh et al., 1991; Nihei, 1997; Nihei et al., 1997; Tomiyasu et al., 1998; Montgomery et al., 1998). This process is repeated to obtain images at various depths in the sample. These images can then be compared to look at changes in both microstructure and changes in composition. Focused ion beam microscopy has been successfully used to examine multi-layer semiconducting devices, embedded particles, alloys, fracture surfaces as well as other materials of interest to the materials science community (Satoh et al., 1991; Nihei, 1997; Wang et al., 1999; Takanashi et al., 2000; Sakamoto et al., 2000).

Volume reconstructions for medical applications have led to extensive research into interpolation algorithms as well as algorithms for noise reduction (Lohmann et al., 1998; Herman et al., 1992; Raya and Udupa, 1990; Leroy, 1998; Drebin et al., 1988). Many of these algorithms can be adapted for rendering volumes of FIB generated data (Dunn and Hull, 1999). For example, volumes have been reconstructed from both secondary electron images and ion maps so that the distribution of elements in semiconductor via structures could be compared. In addition, a more advanced algorithm using shape-based interpolation has been used to look at metal precipitates with irregular features. These reconstructions can be combined with animation routines so that complex structures can be easily visualized. While limits to resolution come from both beam interactions and redeposited material during milling (Rudenauer and Steiger, 1988), lateral resolution of approximately 10 nm can be obtained. Detailed discussion of experimental and reconstruction techniques for FIB tomography now follows.

2. QUANTITATIVE THREE-DIMENSIONAL VOLUME RECONSTRUCTIONS USING FOCUSED ION BEAMS

Focused ion beam tomography is a conceptually simple technique that is capable of yielding quantitative three-dimensional insight into problems in materials science. An essential assumption implicit in focused ion beam tomography is that the geometry and chemistry of a feature can be accurately reconstructed in three-dimensions from discrete two-dimensional slices. Ideally, one might like to disassemble a feature of interest atom by atom using a continuous sputter process, sort atoms by species and position, and then display this information graphically for analysis. Unfortunately, continuous sputter processes yield relatively poor spatial resolution for most materials because differential sputtering and re-deposition seriously degrade resolution and chemical sensitivity. Problems due to differential sputtering and re-deposition can be overcome by discretely sampling a three-dimensional feature of interest as a function of depth and then interpolating geometric and chemical information between data samples. Data sets in this case can be thought of as slices through a sample at different depths.

For each slice, image information is collected using either secondary electron or ion images. Chemical information is collected for each sample using secondary ion mass spectroscopy, energy dispersive x-ray analysis, Auger electron spectroscopy or any other spectroscopic technique capable of producing two-dimensional elemental maps. These slices are then concatenated and the sampled volume is reconstructed using interpolative algorithms. As such, focused ion beam tomography can be thought of as a two-step process, data collection and volume reconstruction. In this chapter, we explain the process of focused ion beam tomography using secondary electron images and SIMS elemental maps. The reader should keep in mind that these are only two of many possible signals that can be used to generate tomographic reconstructions. The volume reconstruction methods discussed below are useful for a variety of imaging and spectroscopic signals.

2.1 Data Collection

Data collection is the most crucial step in focused ion beam tomography because it determines the ultimate accuracy and resolution of three-dimensional reconstructions. A typical data collection algorithm used for focused ion beam tomography is as follows. Secondary electron images and secondary ion mass spectroscopy (SIMS) elemental maps are collected as a function of depth in a structure of interest. Each image and elemental map is divided into pixels, where pixel size is scaled to the probe size. A probe of appropriate size is then scanned over the image area and held at each pixel for a user selected dwell time, typically ranging from 20-40 μSec for secondary electron images and up to 4 mSec for elemental maps. At each pixel, secondary electron and SIMS signals are collected and stored. The feature of interest is then sputtered down to a specified depth with the beam parallel to the surface of interest, and another set of images and elemental maps is collected. To a first approximation, slice thickness is determined by the amount of material removed by the beam when collecting secondary electron images and SIMS elemental maps. In addition, fiducial marks are cut into the periphery of the first slice so that subsequent slices can be aligned with the first slice. These alignment markings are re-cut at several depths to avoid slice misalignment. Secondary electron images and SIMS elemental maps are iteratively collected as a function of depth until a volume of interest has been sampled.

There are several aspects of this procedure that warrant further explanation. First, it is very important that a feature of interest be sectioned with the beam parallel to the surface for data collection because this minimizes deviations from a planar section due to differential sputtering. If one removes material between slices using normal incidence sputtering, surface undulations due to differential sputtering may progressively increase in magnitude and severely degrade depth resolution.

Second, it is important that fiducial marks used for alignment remain sharp and undistorted in each slice. Typically, at least three square fiducial marks are cut into the periphery of each slice so that both rotational and translational drift can be corrected.

Third, care must be taken to ensure that accurate slice depth can be measured. The most straightforward method for measuring slice depth is to measure sputtered depth, edge-on using secondary electron images. In particular, the depth from the original surface is measured with the beam parallel to the sputtered surface. Inherent in this method is the assumption that the image plane is flat and not inclined along the beam. A second method for determining slice depth is to measure the final sputtered depth using scanning probe microscopy. The total depth is then divided by the number of slices to estimate the depth of each slice.

Finally, it is important to choose a probe size that does not over or under sample image and elemental map information in a particular slice. Since focused ion beam imaging and SIMS are inherently ion processes, one has to take account of the normal and lateral extents of signal generation. A typical FIB image or elemental map is divided into voxels with an image plane area determined by the digital-to-analog characteristics of beam scan coils. Since only those ions within 0.5 nm of the surface have enough energy to escape (Sigmund, 1969), a more important consideration is the lateral range of ions once they have impacted a sample surface. Estimates of lateral range for most materials can be made using a typical spreadsheet and approximations to Linhard, Scharff and Schiott (LSS) theory (Schiott, 1970). Using this information, an appropriate probe size can be chosen to ensure that signals measured do in fact arise from within voxel bounds.

The ultimate resolution of FIB tomographic reconstructions is directly determined by instrumental factors and the physical aspects of sputter events. Since we are discussing a sputter based technique, both lateral and depth resolution must be considered.

Lateral resolution is controlled by two factors, the smallest ion probe achievable in a given FIB and the lateral range of probe ions in a feature of interest. The smallest achievable probe for a given FIB system is determined by the FIB column contrast transfer function and coulombic interactions of ions within the beam (Orloff, 1993). Using modern FIB instrumentation, it is possible to achieve ion probes 10 nm in diameter. High current ion probes smaller than 10 nm in diameter are susceptible to perturbations due to columbic interactions (Orloff, 1993).

Since ion probes are used to sputter surfaces, the effect of ion-sample interaction has to be taken into account when considering the ultimate resolution of image and SIMS elemental maps. Because secondary electron images and SIMS elemental maps are incoherent, lateral resolution is directly proportional to effective probe size. As was mentioned earlier, the effective probe size is increased by the lateral range of probe ions in features of interest. To estimate the increase in probe size due to lateral straggle, one can use a variety of techniques. The simplest method is to use LSS theory. Shown in table I are estimates for 30 keV Ga ions in Al, Ti and Si.

Table 14-1. Estimated Lateral Ranges for 30 keV Ga Ions

Sample Material	*Lateral Range (nm)*
Al	7
Ti	7
Si	3

The effective probe size can then be estimated by adding twice the lateral range of probe ions in sample materials to the smallest achievable incident probe diameter. Using the lateral ranges in table I and a 10 nm incident probe, the ultimate lateral resolution of images and elemental maps from these materials are thus approximately 24 nm for Al and Ti and less for Si. As was mentioned earlier, lateral range is strongly dependent upon sample material and incident ion species and energy and should thus be calculated on a case-by-case basis.

Depth resolution in FIB tomography is affected by differential sputtering and damage profiles produced as ions traverse sample materials. To minimize differential sputtering, it is important to section a feature of interest with the ion beam parallel to the surface for data collection. This minimizes differential sputtering during sectioning.

Damage profiles along the beam direction during data collection directly affect the minimum distance between slices for a feature of interest. Knock-on displacement events can distort sample features, so the minimum depth between slices is limited by depth of maximum damage for a given sample material. For 30 keV Ga ions, this depth ranges from a few nanometers to tens of nanometers depending upon sample material.

2.2 Volume Reconstruction Algorithms

Once data from a feature of interest has been collected, it has to be processed to reconstruct three-dimensional image and chemical information. There are several methods that can be used to reconstruct volumes from these data, but we will focus on linear intensity interpolation and shape based interpolative methods.

2.3 Linear Intensity Interpolation

In linear interpolative schemes, collected data are concatenated as a function of depth to produce a three-dimensional set of discretely sampled data. Volume elements (voxels) between slices are then interpolated using linear interpolation algorithms. In linear interpolative schemes, secondary electron intensities or SIMS intensities are used as voxel elements thus, signal intensities are interpolated from slice to slice.

Linear interpolative schemes are accurate and useful for objects that can be generated by extrusion operations such as columnar grains and vias in multilevel interconnect structures used in integrated circuits. For example, Figure 1 shows an Al grain from a metal line in an integrated circuit. This volume reconstruction was done using secondary electron image data from 4 slices through this grain.

Using the same interpolative methods, three-dimensional chemical maps can be reconstructed using linear interpolative schemes. Shown in Figure 2 is a set of Al vias reconstructed using linear interpolation. These vias are cut into SiO_2 interlayer dielectric, lined with Ti and then filled with Al.

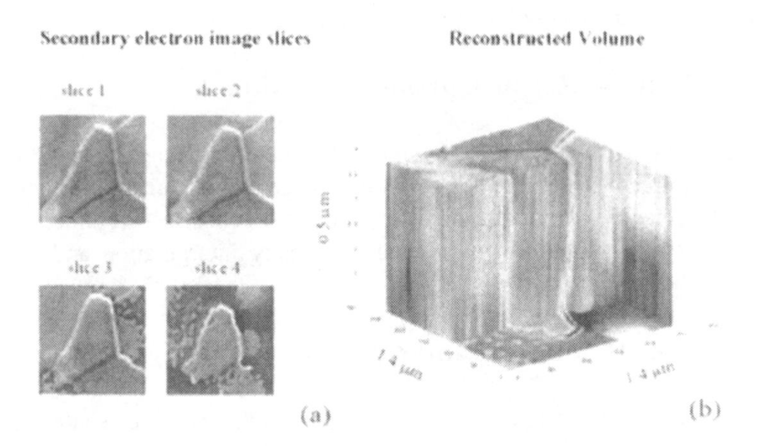

Figure 14-1. Shown is a secondary electron image volume reconstruction of an Al grain. (a) Secondary electron image slices from an Al grain used to reconstruct the volume in b. (b) Volume reconstruction of a single Al grain. Since secondary electron yield tends to be large at surfaces, grain boundaries are accentuated. In this reconstruction, secondary electron intensities are mapped from blue to red, with blue being lowest intensity and red indicating highest intensity (shown as greyscale images here).

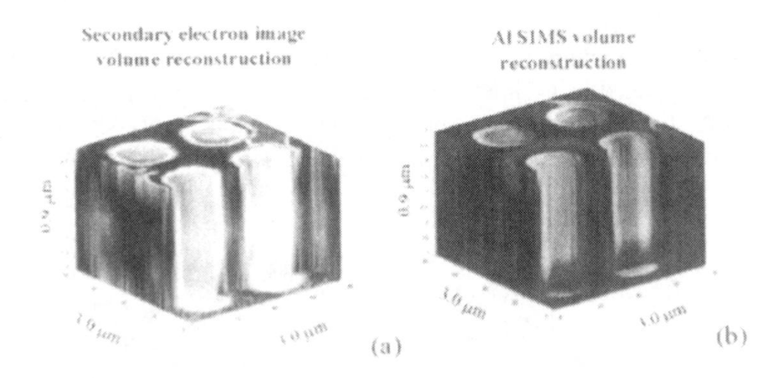

Figure 14-2. Shown are image and SIMS chemical maps reconstructed from secondary electron images and SIMS elemental maps. (a) Secondary electron image reconstruction calculated using linear interpolation from 5 secondary electron images. (b) Al elemental reconstruction calculated using linear interpolation from 5 SIMS Al elemental maps. In this figure, orange and yellow indicate the highest Al signal, while dark-red and black indicate low Al intensity (shown as greyscale images here).

It is clear from Figure 2 that linear intensity interpolation is capable of accurately reconstructing via structures. However, as was mentioned previously, this type of interpolation does not accurately reconstruct features that cannot be generated by extrusion operations. If for instance, these vias contained voids, void edges would be blurred during the interpolation process. This would make it impractical to make quantitative estimates of void volume, surface area, connectivity or morphology.

While linear interpolative schemes are limited to extrusion geometries, they can be used to achieve high spatial resolutions in semiconductor structures. In fact, using multilayer specimens, lateral resolutions of order 22 nm and depth resolutions of order 10 nm have been obtained using linear interpolative schemes (Dunn and Hull, 1999).

2.4 Shape Based Interpolation

Shape-based interpolation is a scheme developed for medical tomographic imaging that uses interfaces present in three-dimensional data sets to reconstruct complex features. Shape-based interpolation is more accurate than intensity based reconstruction techniques because the shape of the object, rather than image intensity, is being interpolated, thus inaccuracies and edge blurring often observed in intensity interpolation are avoided (Rayaand and Udupa, 1990; Herman et al., 1992). Since FIB microscopy images and elemental maps consist primarily of spatially varying intensities, shape-based interpolation can be readily adapted to three-dimensional tomographic reconstructions of geometric and chemical data from FIB microscopy data. Shape based volume reconstructions can be used to not only establish three-dimensional chemical and geometric relationships, but also to obtain quantitative information such as the sharpness of interfaces, surface area, volume and volume fraction of features of interest.

Shape-based interpolation is an interpolative method to reconstruct three-dimensional objects from two-dimensional slices. In this method, the shortest distance of a voxel to the edges of a feature is calculated using either Euclidean or city-block distances (Rayaand and Udupa, 1990) as is shown in Figure 3. If a voxel is inside the edges of a feature, this distance is entered into a voxel as a positive distance; if a voxel is outside the edges of a feature, its closest distance to feature edges is entered into a voxel as a negative distance. Voxels that fall on the edges of a feature, by definition, have zero distances. The determination of whether a voxel is inside or outside of a feature is done using standard inside-out tests (Haines, 1994).

Edge Distances For Shape Based Interpolation

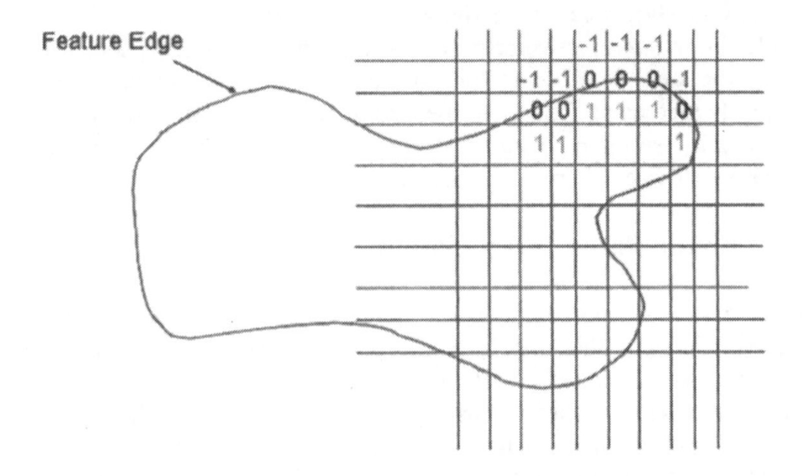

Figure 14-3. Shown is a schematic representation of distances to edges in a slice used for shape based interpolation. The closest distance to an edge is calculated and stored in a pixel. If a pixel is outside an edge, its value is recorded as the negative distance to the edge and if it is inside an edge, its value is recorded as the positive distance to the edge. If a pixel is on an edge, it is given a value of zero.

Once each slice has been processed and edge distances recorded, distances for slices in between recorded data slices are interpolated using bi-linear interpolation (Rayaand and Udupa, 1990). After interpolation has been completed, voxels that have negative edge distances are turned off (set to zero) and the resulting volume is represented by a volume of voxels that have zero or positive distances to feature edges.

SIMS elemental maps can be processed by shape based interpolation using the same method described for image data. In some cases however, SIMS elemental maps have a low signal to background ratio making it difficult to find edges using typical edge detection methods. In these cases, a combination of shape based interpolation methods and intensity interpolation are used to analyze these data. In particular, SIMS data are concatenated into a three-dimensional array and volume between slices is linearly interpolated to produce a three dimensional map of SIMS intensities (Dunn and Hull, 1999). These data are then further processed to determine the chemical distribution of selected elements within or outside the edges of features of interest. Since the volumes described above have voxel values of 1 for voxels inside a feature and zero for those outside of a feature, these volumes can be used to select SIMS intensities inside a given feature by logical AND operations. In addition, the chemical distribution outside of selected features can be selected by NAND logical operations. These logical operations can then be used to investigate changes in chemistry across the interfaces of selected features.

As an example, a typical geometric and chemical reconstruction for Cu-15In cellular colonies is shown in Figure 4. Figure 4 shows a montage of reconstructed volumes and three-dimensional In SIMS elemental maps taken from cellular colonies. Shown in Fig. 4a is a volume reconstruction of several cellular structures of interest that have been reconstructed using shape based interpolation. This reconstruction clearly shows that these structures are not plates as has been assumed in the past. In addition, Figure 4a shows that some of these structures are connected to one another, which would be difficult to tell from a random two-dimensional section. The volume fraction of cellular structures in this view is 0.114 and corresponds to a total reconstructed volume of 2.9 μm^3.

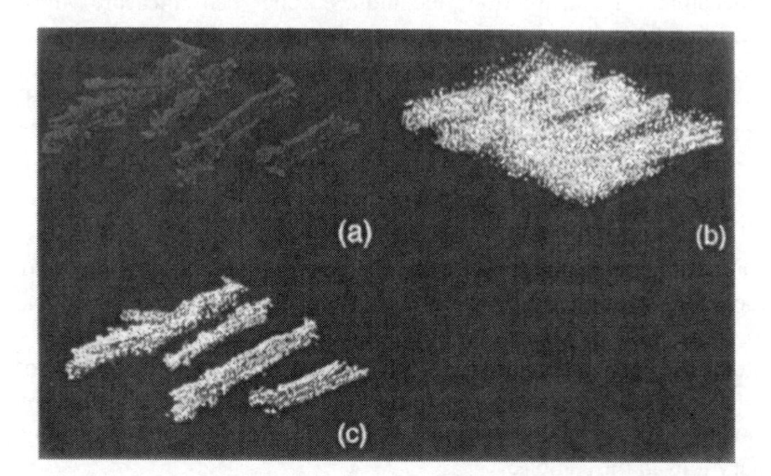

Figure 14-4. (a) An In rich lamellar colony reconstructed using shape based interpolation. (b) Three dimensional In elemental map showing In concentrations throughout the matrix and lamellae for the colony shown in (a). This elemental map was reconstructed using linear interpolation. (c) Logical AND operation of the volume reconstruction shown in (a) with the In elemental map shown in (b). This operation yields a three-dimensional map of In concentrations within lamellae (Sample courtesy of G. Shiflet, University of Virginia).

Figure 4b is a three-dimensional In elemental map showing the spatial variation of In in both the lamellae and the matrix. This chemical reconstruction was done using linear interpolation of In signal between slices. If the elemental map shown in Fig. 4b is operated on using a logical AND operation with the volume in Fig. 4a, the spatial variation of In within lamellae is obtained as shown in Fig. 4c. This operation shows not only the variation of In concentration within lamellae but also yields the position of interfaces between lamellae and the surrounding matrix. Using this position, one can extract data across a specific interface and investigate the abruptness of interfaces both geometrically and chemically.

Three-dimensional reconstructions using FIB microscopy and shape based interpolation can provide critical input for problems in materials science and condensed matter physics. In problems involving solid state phase transformations and reaction kinetics, it is important to ascertain the connectivity of component phases and their respective volume fractions and surface areas. Typically, this type of information is obtained through stereological analyses. One draw back to stereological analyses is that feature geometry must be assumed. Features are typically assumed to take on model shapes such as plates, cylinders, spheres, oblate and prolate spheroids, and then methods of geometric probability are used to infer mean feature dimensions, volume fractions, surface areas, etc. By using FIB three-dimensional reconstruction techniques, one can measure these quantities directly without having to assume model feature shapes. In addition, this technique can be used to test fundamental stereological results that can be used to investigate samples that are not amenable to FIB microscopy.

2.5 Future Prospects and Summary

Focused ion beam tomography is emerging as a very powerful technique for nanoscale three-dimensional structural and chemical reconstructions. The major capabilities of this technique are:

a) A lateral (i.e. perpendicular to the surface normal) spatial resolution of order 20 - 30 nm for most inorganic materials. This is defined mainly by the primary beam diameter and by lateral spread of the implanted primary ions in the sample.
b) A vertical (i.e. parallel to the surface normal) spatial resolution of order 10 nm. This is defined mainly by the implant depth of the primary ions (a function of the implant ion energy and species, and the atomic number of the target sample), and by the escape depth for secondary electrons and ions from the sample surface. Mixing of specimen feature within the primary ion implantation depth also has to be considered. Finally in

collection of low intensity SIMS signals, vertical resolution may effectively be limited by the thickness of the slice that needs to be sputtered to generate a sufficiently strong signal.

c) Reconstructions may contain a million (i.e. 1024 x 1024) pixels or more per slice, depending on the beam raster software and digital-to-analog conversion hardware used, and up to dozens of slices, depending upon available computation time and power. This creates reconstructions containing up to tens of millions of voxels, each of which corresponds to a measured piece of data.
d) Secondary electron reconstructions provide strong signal to noise for each voxel element, for all but insulating materials. Secondary electron contrast provides some chemical differentiation (via differing secondary electron yields), particularly between materials of differing classes (e.g. metals vs. insulators). Secondary electron reconstructions can also be used to map grain topologies in polycrystalline materials, due to orientation dependences of channeling of the primary ion beam and secondary electron emission yields, and to enhanced secondary electron emission at grain boundaries.
e) Mass-filtered secondary ion images can be used to produce chemical-specific reconstructions. The primary challenge here is signal-to-noise in individual voxels, arising from low ionization yields and collection efficiencies.
f) Reconstruction (i.e. slice interpolation) algorithms have been developed that can deal with surface topologies that are of high radius of curvature and even those containing re-entrant features. These reconstruction techniques borrow heavily upon existing algorithms developed for the medical tomography and computer graphics fields.

Areas for future development include:

a) Spatial resolution. Higher lateral resolution will require combinations of smaller primary ion beams and lower lateral spread of primary ions in the target sample (necessitating lower energy and/or higher mass primary ions). The latter approach will also improve vertical resolution, and reduce ion-induced mixing effects in the sample. Substantial decreases (i.e. by a factor of two or more) in the primary ion beam diameter appear to be unlikely in the foreseeable future, at least using conventional liquid metal ion source technology because of the virtual source size. Emission of gas sources cryogenically cooled onto a field emitting tip (Jousten et al. 1982, Thompson et al. 1996) offers the possibility of much smaller virtual source sizes and primary ion beams down to 1 nm, but such sources are at an early stage in their development. Use of higher mass or lower energy primary ions, while maintaining the focused ion beam diameter, create major additional challenges for ion optical design, as in both cases lower ion velocities in the source and imaging system will create additional chromatic spreading. In summary incremental resolution advances appear to be possible, but improvements of resolution below 10 nm appear to be unlikely when primary ions are used for imaging. Hybrid techniques, such as use of field emission electron sources for the imaging component of the process offer the possibility of resolution improvements down to the several nm level. Sample damage during the imaging process is also greatly reduced, and image acquisition times may be greatly enhanced, although secondary electron emission yields are significantly reduced compared to use of primary ion beams for imaging. Dual field emission ion and electron sources have been commercially available for several years. Other approaches that may be considered include use of reactive ion etching processes to reduce damage during material removal.

b) Chemical sensitivity. Several strategies may be used to improve secondary ion ionization yields and collection efficiencies during FIB-SIMS. Techniques we have used that yield measurable but not dramatic (i.e. in the range of 2-3 x) enhancements in overall collection efficiencies include variation of the angle of the primary ion beam with respect to the surface normal, biasing the sample stage, and introduction of reactive ion species (oxygen or iodine) during sputtering. Substantially greater improvements would likely require development of stable sources with much higher ionization efficiency, such as cesium or oxygen. The highly reactive nature of the former has inhibited stable source development, and the lack of a high intensity stable gas ion source has prohibited the latter. Another approach is post-sputtering ionization using high energy photon or electron beams, as been previously demonstrated for SIMS techniques (Ma et al., 1995).

c) Reconstruction algorithms. The medical tomography and computer graphics communities have developed extensive sophisticated algorithms for accurate interpolation between complex shapes. Most of the limited examples of focused ion beam tomography published to date have used simple linear intensity interpolation algorithms between successive slices. Such techniques do not work well for structures with high radius of curvature or discontinuous, branching or re-entrant surfaces. Further development of appropriate algorithms for slice interpolation in FIB tomography is of great importance.

Together these approaches offer the promise for development of FIB tomography into a widely applied technique for three-dimensional structural and chemical resolution, with resolution extending down to the nanometers level. Such a technique would be a crucial component of the developing nanotechnology field.

ACKNOWLEDGEMENTS

The authors would like to acknowledge G. Shiflet (U. Virginia), A. Evans (Princeton) and G. Gilmer (Bell Laboratories) for provision of samples used to generate tomographic reconstructions presented in this article. Important discussions with G. Shiflet, L. Giannuzzi (University of Central Florida), F. Stevie (Agere) and J.F. Walker (FEI Company) are gratefully acknowledged. Funding from IBM, the Defense Advanced Research Projects Agency (Virtual Integrated Prototyping Program), and the National Science Foundation (Materials Research Science and Engineering Center) is gratefully acknowledged.

REFERENCES

Babout L, Maire E, Buffiere JY, Fougeres R, Acta. Mater. 49, 2055-2063 (2001).

Benninghoven A, Rudenauer FG, Werner HW, Secondary Ion Mass Spectrometry Basis Concepts, Instrumental Aspects, Applications and Trends, John Wiley and Sons, New York (1987).

Blavette D, Bostel A, Sarrau JM, Deconihout B, Menand A, Nature 363, 432-435 (1993).

Copley DC, Eberhard JW, Mohr GA, Journal of Metals 46(1) 14-26 (1994).

Deconihout B , Pareige C, Pareige P, Blavette D, Menand A, Microsc. Microanal. 5, 39-47 (1999).

Defrise M, Comput. Med. Imag. Graph. 25, 113-116 (2001).

Drebin RA, Carpenter L, Hanrahan P, Comput. Graphics 22(4) 65-74 (1988).

Dunn DN, Hull R, Appl. Phys. Lett. 75(21) 3414-3416 (1999).

Gammer K, Musser S, Hutter H, Appl. Surf. Sci. 179, 240-244 (2001).

Guvenilir A, Breunig TM, Kinney JH, Stock SR, Acta. Mater. 45(5) 1977-1987 (1997).

Haines E, in *Graphics Gems IV*, (Academic Press, New York) pp 24-46 (1994)

Herman G, Zheng J and Bucholtz CA, IEEE Computer Graphics & Applications, May, 70, (1992).

Herman GT, Zheng J, Bucholtz CA, IEEE Comput. Graphics and Appl. 12(5) 69-79 (1992).

Hutter H, Nowikow K, Gammer K, Appl. Surf. Sci. 179, 161-166 (2001).

Hutter H, Grasserbauer M, Mikrochim. Acta 107, 137-148 (1992).

Jamieson DN, Nucl. Instrum. Methods Phys. Res. B 136/138, 1-13 (1998).

Jousten K, Bohringer K, Borret R and Kalbitzer S, Ultramicroscopy 26 301 (1988).

Levoy M, IEEE Comput. Graphics and Appl. 8(5) 29-37 (1988).

Lin CL, Miller JD, Chem. Eng. J. 77, 79-86 (2000).

Lin, CL, Miller JD, Chem. Eng. J. 77, 79-86 (2000).

Lohmann K, Gundelfinger ED, Scheich H, Grimm R, Tischmeyer W, Richter K, Hess A, J. Neurosci. Meth. 84, 143-154 (1998).

Lu SF, Mount GR, McIntyre NS, Fenster A, Surf. Interface Anal. 21, 177-183 (1994).

Lu D, Zhou M, Dunsmuir JH, Thomann H, Mag. Res. Imaging 19, 443-448 (2001).
Ludwig W, Bellet D, Mater. Sci. Eng. A281, 198-203 (2000).
Ma Z, Thompson RN, Lykke KR, Pellin MJ, and Davis AM, Rev. Sci. Inst.66, 3168 (1995)
Magerle R, Phys. Rev. Lett. 85(13) 2749-2752 (2000).
Malmqvist KG, Nucl. Instrum. Methods Phys. Res. B 104, 138-151 (1995).
Mangan MA, Shiflet GJ, Scripta Mater. 37(4) 517-522 (1997).
Martin CF, Josserond C, Salvo L, Blandin JJ, Cloetens P, Boller E, Scripta Mater. 42, 375-381 (2000).
McIntyre NS, Davidson RD, Weisener CG, Taylor KR, Gonzalez FC, Rasile EM, Brennenstuhl AM, Surf. Interface Anal. 18, 601-603 (1992).
Miller MK, Mater. Charac. 44(1-2) 11-27 (2000).
Montgomery NJ, McPhail DS, Chater RJ, Dingle T, Secondary Mass Spectrometry SIMS X I, Gillen G, Lareau R, Bennett J, Stevie F (ed.), John Wiley and sons, New York, 631-634 (1998).
Ng YK, Orlic I, Liew SC, Loh KK, Tang SM, Osipowicz T, Watt F, Nucl. Instrum. Methods Phys. Res. B 130, 109-112 (1997).
Niedrig H, Rau EI, Nucl. Instrum. Methods Phys. Res. B 142, 523-534 (1998).
Nihei Y, Tomiyasu B, Sakamoto T, Owari M, J. Trace and Microprobe Techniques 15(4) 593-599 (1997).
Nihei Y, J. Surf. Anal. 3(2) 178-184 (1997).
Orloff J, Review of Scientific Instruments, 64, 1105-1129,(1993).
Owens JW, Butler LG, Dupard-Julien C, Garner K, Mater. Res. Bull. 36, 1595-1602 (2001).
Raya SP, Udupa JK, IEEE Trans. On Med. Imag. 9(1) 32-42 (1990).
Rayaand SP and Udupa J, IEEE Transactions on Medical Imaging, 9 , 32-42, (1990).
Rudenauer FG, Steiger W, Ultramicroscopy 24, 115-124 (1988).
Ruprecht J, Nield J, Progress Biophys. Mol. Bio. **75**, 121-164 (2001).
Sakamoto T, Takanashi K, Cheng ZH, Ono N, Wu H, Owari M, Y. Nihei, Inst. Phys. Conf. Ser., No. 165, 9-13 (July 2000).
Sakellariou A, Cholewa M, Saint A, Legge GJF, Nucl. Instrum. Methods Phys. Res. B 130, 253-258 (1997).
Satoh H, Owari M, Nihei Y, J. Vac. Sci. Technol. B 9(5) 2638-2641 (1991).
Schiott HE, Radiation Effects, 6,107-113 (1970).
Schofield RMS, Nucl. Instrum. Methods Phys. Res. B 104, 212-221 (1995).
Sigmund P, Physical Review Review, 184, 383-416 (1969).
Soto GE, Young SJ, Martone ME, Deerinck TJ, Lamont S, Carragher BO, Hama K, Ellisman MH, Neuroimage 1(3) 230-243 (1994).
Steiger W, Rudenauer F, Gnaser H, Pollinger P, Studnicka H, Mikrochim. Acta Supp. 10, 111-117 (1983).
Takanashi K, Wu H, Ono N, Cheng ZH, Sakamoto T, Qwari M, Nihei Y, Inst. Phys. Cof. Ser., No. 165, 9-13 (July 2000).
Thompson W, Armstrong A, Etchin S, Percival R, Saxonis A, Ion-Solid Interactions for Materials Modification and Processing. Symposium. Mater. Res. Soc., pp.687-93. Pittsburgh, PA, USA (1996).
Tomiyasu B, IFukuju I, Komatsubara H, Owari M, Nihei Y, Nucl. Instrum. Methods Phys. Res. B 136/138, 1028-1033 (1998).
Wagter ML, Clarke AH, Taylor KF, van der Heide PAW, McIntyre NS, Surf. Interface Anal. 25, 788-789 (1997).
Wang YZ, Revie RW, Phaneuf MW, Li J, Fatigue Fract. Engng. Mater. Struct. 22, 251-256 (1999).

Chapter 15

APPLICATION OF FIB IN COMBINATION WITH AUGER ELECTRON SPECTROSCOPY

A Practical Guide With Examples

E. L. Principe

Carl Zeiss SMT, Inc., Nano Technology Systems Division 56, D-73447 Oberkochen, Germany

Abstract: This chapter highlights practical applications that combine Auger electron spectroscopy (AES) and focused ion beam (FIB) technology. The examples presented are derived exclusively from the semiconductor industry but the methods described are generally applicable to a variety of industries. The analytical power resulting from the combination of FIB and AES stem from simple concepts; FIB technology provides a controlled site specific method to create a surface (i.e., x-sections and thin membranes) while AES is a surface analytical technique that can yield elemental, and to a varying degree chemical, information from the same *near-surface* region (~10-100Å) of the FIB cut face at moderate lateral spatial resolution (~80-250 Å). Practical operational tips, as well as a discussion of limitations and artifacts to avoid support the examples. Several of the examples presented in this chapter convey the power of combining AES and FIB on an intuitive graphical level.

Key words: Focused Ion Beam, FIB, Auger Electron Spectroscopy, AES, volume reconstruction, electron beam interactions

1. INTRODUCTION

Details related to FIB theory, hardware and operation have already been covered in earlier chapters within this text. Auger electron spectroscopy has a history, scientific scope and range of applications too broad to capture entirely in a single brief chapter. There are also volumes of background information related to electron beam theory, scanning electron microscopy, and x-ray micro-analysis that are relevant to discussions in this chapter that also cannot be covered entirely.

However, a certain amount of theory and related background is critical to have a coherent understanding of the materials presented in this chapter. The most important theory that will enable a clear understanding of AES spectra and the examples presented here relates to the nature of electron beam interactions with solids. This subject accounts for the Auger signals generated, signal-to-background, depth resolution, lateral spatial resolution and source of artifacts such as beam damage. Therefore, the first section summarizes the essentials of electron beam interactions with solids.

Following a theory section the instrumentation utilized for Auger is described briefly. The chapter culminates with a series of examples combining FIB and Auger and a discussion of artifacts.

2. THEORY

The multiple signals that manifest in Auger spectra include Auger electrons, inelastically scattered electrons and plasmons, which all can be utilized for qualitative or quantitative analysis. Theory will also help the reader to compare Auger with larger volume analyses, like x-ray energy dispersive spectroscopy (XEDS) that arise from the same primary electron beam. The theoretical treatment is necessarily selective and abridged. While the coverage is not intended to be comprehensive the core materials are summarized in a fashion suitable for an introduction to the uninitiated, but with content that is still a useful reference to an experienced user. The selected references at the end of the chapter lead the interested reader to a more extensive treatment of these important subjects (Goldstein et al., 1992; Williams and Carter, 1996; Feldman and Mayer, 1986).

2.1 The Auger Electron Transition (Vs. X-Ray Fluorescence)

Auger electron spectroscopy derives its name from Pierre Victor Auger (1899-1993), a French scientist who discovered the tertiary electron scattering process that bears his name in 1923 while investigating energy transitions in gas molecules (Auger, 1923). The Auger transition can be described with reference to Figure 1.

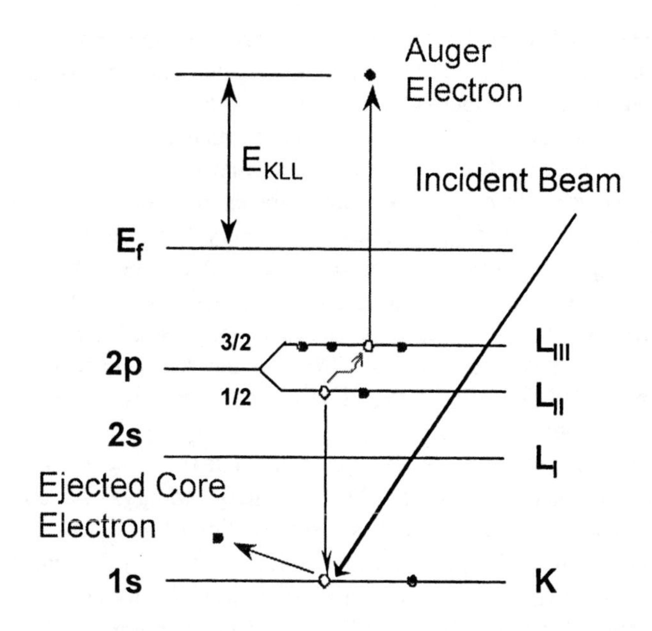

Figure 15-1. Schematic diagram of the Auger electron transition process.

The process leading to an Auger transition begins with the ejection of a core level electron, typically from a K, L, M, or O level shell induced from an incident electron, x-ray, or energetic ion. Following the core level electron ejection a spontaneous relaxation process is initiated and another electron from the same shell (but different sub-shell), or more commonly from another outer shell, fills the initial core vacancy. An Auger transition occurs if all or part of the relaxation energy released as the core vacancy is filled is transferred directly to yet another outer shell electron, and that electron gains sufficient energy to be ejected into vacuum. The final ejected electron is termed an Auger electron and the discrete kinetic energies from those electrons form the spectral line(s) that provide a unique identification for all elements with an atomic number of three and higher. To a first approximation the energy of the outgoing electron is:

$$E_{xyz} \cong E_x(Z) - E_y(Z+\Delta) - E_z\ (Z+\Delta) - \phi_s \ \textit{where}\ 0 \leq \Delta \geq 1 \quad (1)$$

The equation is the difference in binding energy of the core electron and the two outer shell electrons (i.e., E_K-E_{L1}-$E_{L2,3}$) with a delta term to account for the higher binding energy felt by the two outer shell electrons following removal of the core electron modeled as an atom with a higher nuclear charge. The final term is a spectrometer work function that is typically a few electron volts. More exact quantum calculations are possible, but in practice published handbooks with experimental spectra and tabulated binding energy values are compiled for most elements and many compounds (Childs et al., 1995). It is important to note that the kinetic energy of any Auger electron transition is independent of the excitation beam voltage.

Alternatively, instead of the relaxation energy from the first outer shell electron being transferred to a second outer shell electron, the energy released during the relaxation could convert directly into a photon (x-ray). This alternative radiative process produces x-rays that are employed wavelength dispersive spectroscopy (WDS) and x-ray energy dispersive spectroscopy (XEDS). The latter is also known by two interchangeable acronyms, EDS and EDX.

A second distinct process also generates x-rays. The second process forms a continuous spectrum of electromagnetic energy. As an electron approaches a solid surface it may loose any fraction of its kinetic energy due to deceleration caused by the opposing coulombic field of the matrix atoms. The random kinetic energy loss induces emission of photons over a broad energy range, forming the well-known Bremsstrahlung (German for "braking radiation") background. The background radiation extends from close to the energy of the primary incident electrons, down to the infrared.

The radiative x-ray process and the non-radiative Auger electron emission process are competitive and complementary, as only one of the two can occur for a given core hole. For this reason, the probability for the yield of Auger electrons and the probability for fluorescent x-ray yields are also complementary. Auger electron emission is the dominant process for K-shell vacancy binding energies below 2keV, which is why AES can be a favorable technique to detect low Z elements relative to XEDS. As mentioned above, the initial core level electron ejection can be induced by either an electron beam or a photon (x-ray). Therefore, Auger electron spectra also appear in another popular surface analytical technique, x-ray photoelectron spectroscopy (XPS). Excitation of x-ray induced Auger electrons are limited by the energy of the x-ray source, the two most commonly employed being monochromated Al-Kα(1486.3eV) and Mg-Kα (1253.6eV).

2.2 Nomenclature

Auger transition nomenclature is an *x-ray notation* that indicates the three shells involved in the transition. For each type of series (KLL, LMM,

MNN, Coster-Kronig, etc.), there are several possible final state electron configurations of the atom (i.e., six spectral lines and nine final state electron configurations are possible in a KLL series). The final state configuration of the atom is expressed in *spectroscopic notation*. For instance, the final electron state configuration of a KL_1L_1 transition would be expressed as $2s^02p^6$, indicating two vacancies formed in the 2s (L1) shell and all six electrons left in the 2p ($L_{2,3}$) shell. Likewise, a $KL_1L_{2,3}$ transition would yield a final state electron configuration of $2s^12p^5$.

2.3 Electron Beam Interactions with Solids, Interaction Volume and Analytical Volume of AES

Auger electron spectroscopy involves the interaction of electrons with solids and it is certainly appropriate to discuss this particular interaction. But it is also very important to be aware of a host of interactions that occur when an electron beam interacts with a solid. Some interactions are of diagnostic benefit in Auger spectra and others are the source of artifacts the analyst must recognize. The interactions of interest are summarized with reference to Figure 2, depicting Auger electrons, secondary electrons, back-scattered electrons and emitted x-rays. First a qualitative description of the processes, then numerical estimates for the interaction volumes. The analyst may consider entering the equations provided in this section into a spreadsheet in order form a "picture" of the interaction volume and origin of various signals generated.

As the incident electrons move through the sample they are directly scattered by the nuclei as well the other electrons that form the electromagnetic field in the material. The majority of the electron interaction volume occurs over a region that forms a "tear-drop" shape when viewed in cross-section. The tear-drop shape is actually a special case corresponding to high voltage in a low Z matrix, but it serves as a device to illustrate prominent features. The neck of the tear-drop is the region where the interactions are primarily elastic, which means relatively few of the individual electrons exchange energy through singular scattering events while passing farther into the solid. Singular and plural scattering events escalate as the electrons begin to transfer energy and the collective random trajectories of the individual scattered electrons form the bulb shaped portion of the interaction volume.

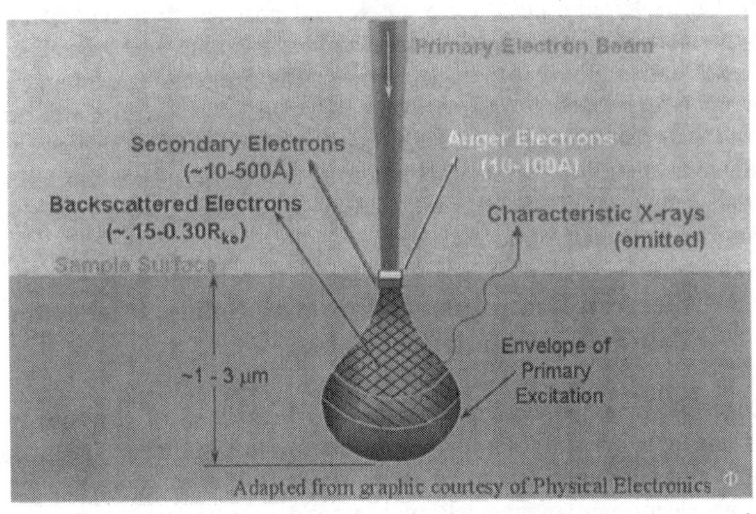

Figure 15-2. Schematic depicting the primary electron beam interactions occurring in bulk solids. The inelastic mean free path of Auger electron energies restrict the analytical volume to a range of ~10-100Å. Other electron beam interactions lead to phonons that are responsible for heat conduction and plasmons that also appear in Auger spectra and have diagnostic applications. X-rays are generated and emitted from a much greater depth, typically 1-3um depending upon the material and accelerating voltage.

One of the numerical expressions to estimate the radial range of electron interaction volume is known as the Kanaya-Okayama Range, R_{KO}, and is given by:

$$R_{KO} = \frac{0.0276 A E_o^{1.67}}{Z^{0.89} \rho} \cos(\theta) \qquad (2)$$

where Z is the atomic number, Eo is the beam voltage in KeV, ρ is the density in g/cm^3, A is the atomic weight in g/g·mole and θ is the angle with respect to the surface normal. Monte Carlo simulations are also routinely employed to estimate the electron interaction volume and other detailed studies related to electron beam interactions. Commercial software, freeware, as well as source code, is available that is specifically designed to aid micro-analysis. Example output from one such commercial software product, Electron Flight Simulator, is shown in Figure 3.

Panel A) of Figure 3 illustrates the electron interaction volume for 20kV electrons at normal incidence in a silicon oxide matrix. The software also estimates the depth from which x-rays are generated (green) and the depths from which x-rays are emitted (red). The interaction volume under these conditions is approximately 7μm (R_{KO}=6.5μm) while the x-rays emanate from ~4μm. In panel B) the 20kV result in silicon oxide is re-scaled to show top 1000Å of the surface. This region models the zone where elastic scattering dominates and beam spreading ensues from single scattering events. An equation to estimate the beam spread, *b*, in the single scattering regime (neck of tear drop) as a function of sample thickness is provided below (after Goldstein and Reed):

$$b = 0.0721 \frac{Z}{E_o} \left(\frac{\rho}{A} \right)^{\frac{1}{2}} t^{\frac{3}{2}} \tag{3}$$

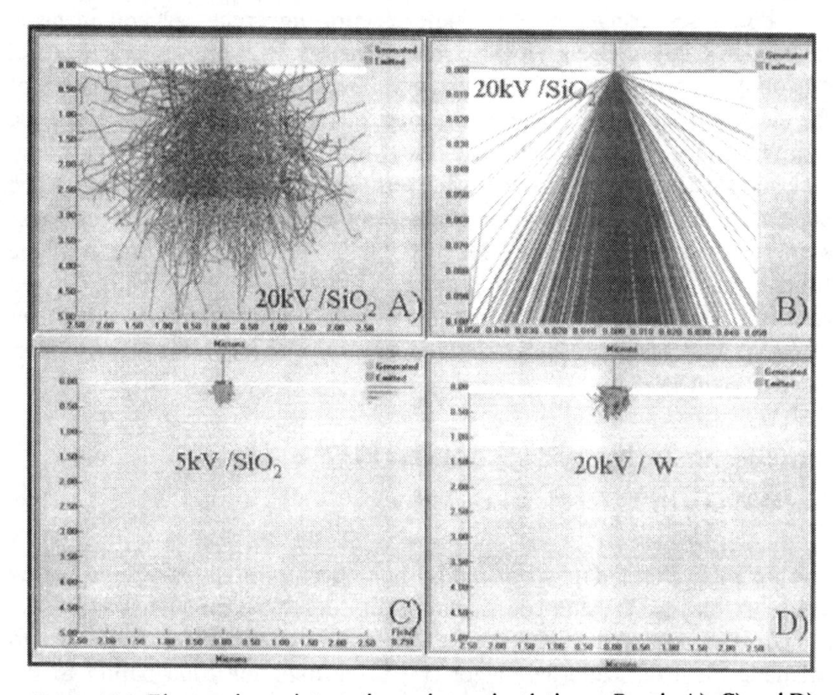

Figure 15-3. Electron beam interaction volume simulations. Panels A), C) and D) show the same vertical scale of 5μm. Panel A) is a Monte Carlo simulation of a 20kV electron beam in a SiO_2 substrate. Panel B) expands the top 1000Å of Panel A) to illustrate beam spreading. Panel C) is a Monte Carlo simulation of a 5kV electron beam in a SiO_2 substrate. Panel D) is a Monte Carlo simulation of a 20kV electron beam in a W substrate. Note the variation of electron interaction volume as a function of voltage and material.

In this form of equation 3 the quantities t and b are in angstroms while the other quantities are as defined above. Applying this equation to 20kV beam in a silicon oxide matrix and using weighted averages for the atomic number and atomic weight we estimate a beam divergence of approximately 11Å in the first 100Å and over 950Å after 2000Å, in reasonable agreement with our Monte Carlo simulation. It is important to note the beam divergence is inversely proportional to the incident beam energy and has a power dependence with density, atomic weight and membrane thickness. In panel C) the electron interaction volume is modeled for a 5kV primary beam in a silicon oxide matrix, with an interaction volume of a little less than 0.55μm (R_{KO}=0.6μm). Panel D) models the interaction volume for a 20kV beam in a tungsten matrix, which yields a volume of a little over 0.7μm (R_{KO}=0.8μm).

Plural scattering means a single electron undergoes several inelastic interactions, loosing a varying amount of energy with each interaction and randomly altering the electron trajectory. Because of the random nature of the electron trajectories some of the electrons are directed back toward the sample surface. Those electrons emitted from the surface following plural inelastic electron scattering events are termed back-scattered electrons. The percentage of the total number of incident electrons that result in back-scattered electrons is a strong function of the average atomic mass of the solid, but is not a strong function of beam voltage. An empirically derived equation to estimate the back-scatter coefficient, n_b, the fraction of incident electrons that result in back-scattered electrons, is provided below (after Love & Scott 1978):

$$\eta_b(Z) = -5.23791\times10^{-3} + 1.5048371\times10^{-2} Z - 1.67373\times10^{-4} Z^2 + 7.16\times10^{-7} Z^3 \quad (4)$$

The weighted average may be used for homogenous mixtures. The equation above yields a backscatter coefficient of 0.175 (17.5%) for silicon and 0.482 (48%) for tungsten at 20KeV. The backscatter coefficient is important to calculate the backscattering factor, r^A, to account for contribution to the Auger electron flux from core ionization induced by backscattered electrons of sufficient energy. The backscattering factor for Auger can be estimated from the following equation (after Reuter, 1972):

$$r^A = 2.8\left(1 - 0.9\frac{E_w}{E_p}\right) \times \eta_b(Z) \tag{5}$$

where Ew is the core level ionization energy, Ep is the primary beam energy, and the backscatter coefficient and numerical estimate as a function of Z was defined above. This equation is said to be a low estimate for core level energies below 500eV and more involved calculations can be applied.

Auger electrons are *generated* throughout the interaction volume. However, because the energy of Auger electrons are typically 2000eV or less, only electrons within approximately the top 100Å or less are emitted without energy loss. The remaining Auger electrons that are generated deeper in the interaction volume continue to loose kinetic energy to the matrix up to the thermal limit and never escape the surface, or contribute to the inelastic background. The fundamental material parameter that dictates the depth from which Auger electrons are emitted is known as the inelastic mean free path (IMFP), given the designation λ^o (this is *not* a wavelength!). The IMFP is defined as the average distance that an electron with a given energy may travel between successive inelastic collisions. The IMFP limits the depth from which Auger electrons can be emitted without energy losses and is the reason AES is a surface sensitive technique. The intensity distribution as a function of depth, *t*, of those electrons emitted with no energy loss follows an exponential decay according to the Beer-Lambert Law such that:

$$I(t) \propto I^o \exp\left(\frac{-t}{\lambda^o \sin(\theta)}\right) \tag{6}$$

where θ is the emission angle of the electrons with respect to the sample plane. An example of the relation is provided through Figure 4 corresponding to 1383eV electrons (XPS Si2p transition) traveling through elemental silicon at normal emission (θ=90°) and having an IMFP defined as 34Å. According to the definition, 63% of the total signal intensity emanates from a depth equal to 1 IMFP length. By convention, the analytical information depth is considered as 3λ, and accounts for 95% of the total intensity. For the purposes of mathematical modeling a depth of 5λ is preferred, which accounts for 99.3% of the electron intensity emanating from the solid. The IMFP is function of the initial kinetic energy of the electron and the properties of the matrix through which the electron travels. The IMFP varies according to the elemental composition, density and chemical bonding of the matrix. The value of λ may be determined from experiment (i.e., TEM correlation), empirical formula, or some combination. The NIST reference at the end of the chapter offers free software to generate IMFP data, as well as effective attenuation length data, from theoretical models or from built-in database compilations (NIST database 82). This information is extremely useful to approximate the analytical volume of AES (or XPS) signal intensity and for numerical calculations.

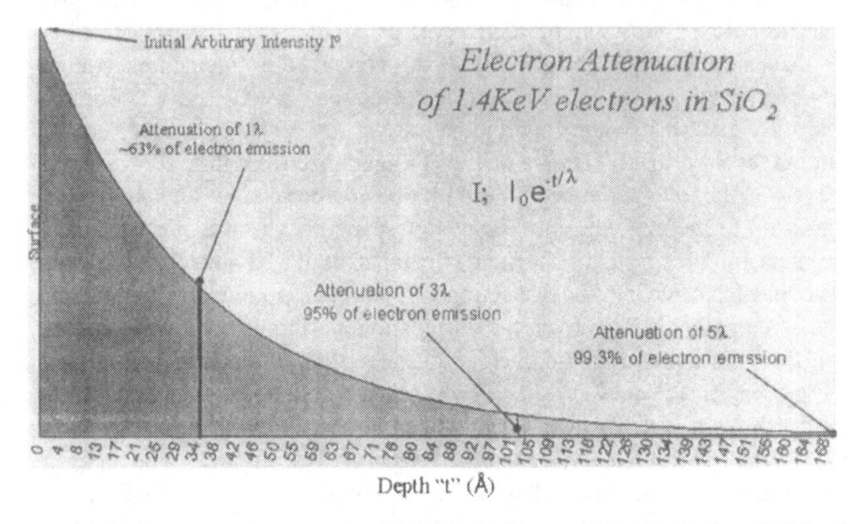

Figure 15-4. Electron attenuation accounts for the surface sensitivity distribution of AES and follows an exponential decay as function of depth according to a Beer-Lambert relation. Approximately 95% of all emission will originate from within 3λ, which is commonly regarded as the analytical depth.

Two other interactions not diagrammed in Figure 2 are important for this discussion, phonons and plasmons. Both processes can be described as collective oscillations. The majority of energy deposited into a bulk sample not emitted through one of the above mechanisms is eventually converted to heat through a phonon mechanism, which is a collective vibration of the atoms in the lattice network. Heat generation through the phonon mechanism occurs in both crystalline and amorphous solids. Each phonon event results in a vibration that accounts for only ~0.1eV energy loss, but the phonon scattering mechanism occurs over a relatively large solid angle. The cross-section (probability) for phonon scattering has a $Z^{1.5}$ dependence. The phonon mechanism leads directly to the sample heating associated with the primary electron beam, which is usually of relatively small magnitude in the electron flux range used in AES. An estimate of specimen heating due to a static electron beam is given below (after Rölle):

$$\Delta T = \frac{E_o I_o}{\pi r_o k_s}, \tag{7}$$

where the product of the primary beam voltage and the primary beam current in the numerator gives the incident power in watts, r_o is the radial extent of the electron interaction volume (i.e., R_{KO}) and k_s is the thermal conductivity of the substrate{watts/(meter.kelvin)}. Assuming a 20kV primary beam at 10nA, a thermal conductivity of 148 (W. m^{-1}.K^{-1}) for Si and 1(W. m^{-1}.K^{-1}) for SiO_2, the equation yields an approximate value for ΔT of 0.1and 10 Kelvin, respectively. For high conductivity metals the temperature rise is even lower. However, the primary source of beam damage in inorganics is due to ionization of matrix atoms and less strongly related to heat generation, so "low" heat generation does not mean low beam damage.

In a classical mechanical interpretation, plasmons represent a longitudinal wave vibration where the vibrating medium is the density of free electrons (electron gas) in the solid having an oscillation frequency ω_p. Consider a spherical volume $4/3\pi r^3$ of electrons with a differential quantity $\delta n = 4\pi r^2 \delta r$ electrons in this volume surrounding a positive ion core, where n is the free electron density and r is the radial distance from the core. A sudden radial contraction, δr, of the free electron gas from its equilibrium radius induces an electric field, $\delta\mathcal{E}$, setting up a retarding force of magnitude $\delta\mathcal{E}e^-$. The classical solution for the fundamental frequency of this forced harmonic oscillator model is given by:

$$\omega_p = \sqrt{\frac{ne^2}{\varepsilon_o m}}, \tag{8}$$

where e is the electron charge, ε_o is the permittivity of free space, and m is the electron mass. The plasma frequency can interpreted as the natural frequency of the electron–ion system and is therefore sensitive to atomic and electronic structure. At the terminating surface of the solid there is an analogous collective oscillation, but in the form of a transverse oscillation wave known as a surface plasmon, designated as ω_{sp}, and related to the bulk plasmon by:

$$\omega_{sp} = \frac{\omega_p}{\sqrt{(1+\varepsilon)}}, \tag{9}$$

where ε is the dielectric constant of the material. Plasmon oscillations are strongly damped but harmonics of the fundamental frequency are also excited. In the very non-classical quantum mechanical sense, the energy transfer associated with this oscillation is quantized and only discrete energy packets of magnitude ω_p may be transferred to the matrix from any interacting electron. Since a plasmon oscillation is an inelastic scattering process with a relative high cross-section, an Auger electron may loose a fraction of its characteristic energy through this process in multiples of ω_p and ω_{sp} and these are known as plasmon losses. By this mechanism, bulk plasmon, surface plasmon, and their respective set of harmonic loss peaks appear in Auger (and XPS) spectra. The harmonics appear with successively diminished intensity. Typical plasmon energies range from 4-20eV, depending upon the material. The intensities of plasmon loss peaks are much stronger for metals due to a higher free electron density and are notably reduced or nearly absent (particularly the harmonics) for oxides and insulators. In terms of practical AES applications plasmon loss peaks have diagnostic benefits and it is relatively easy to qualitatively discern a clean metal or semi-metal from an oxidized metal by the presence and intensity of plasmon loss peaks, even in a rapid and low energy resolution survey scan. It is also evident that detailed high-energy resolution spectra from plasmon loss peaks can reveal characteristics of the free electron density and dielectric constant of the solid.

Secondary electrons are defined by convention as those electrons having energy of 50eV or less but more the 90% have energy of less than 10eV. The secondary electrons are generated when the primary incident electrons and back-scattered electrons transfer energy to weakly bound conduction band (metals) or valence band electrons. Depending upon the material composition and accelerating voltage, 50% or more of the emitted secondary electrons generated can be due to energy transferred from back-scattered electrons. One might assume that since secondary electrons have such low electron energy the signal is exclusively from the near-surface. In one sense this is true, but there can be considerable contribution from back-scattered electrons to the secondary electron intensity and this means the interaction volume of back-scattered electrons influences the secondary electron signal intensity. This is particularly true at higher accelerating voltages and for materials with an average higher atomic number. For this reason, the signal depth of secondary electrons is generally considered to be in the range of 10-500Å. At primary beam energies of 5kV or less the volume from where secondary and backscattered electrons are generated is similar.

2.4 Quantification of Auger Electron Spectra

The integrated intensity of Auger electrons over a thickness t can be related to the concentration X of element i, with reference to equation 5), as follows (Briggs and Seah 1990):

$$I_i = \frac{I_i^o}{\lambda_i^o} \int_0^{\infty} r_i^A(t) X_i(t) \exp\left(\frac{-t}{\lambda_i^o(t)\sin(\theta)} \right) dt \tag{10}$$

In practice, AES quantification is determined through application of relative sensitivity factors that are provided by the equipment manufacturer or determined experimentally. Using this approach quantification is straight forward and given by:

$$X_i = \frac{I_i / S.F._i}{\sum^{n} I_i / S.F._i} \tag{11}$$

where S.F. is the relative sensitivity factor for element i. Accuracy of the final calculation depends upon the accuracy of the S.F., signal-to-noise statistics and matrix effects. Due to moderate or strong matrix effects in AES, it is a good practice to determine your own sensitivity factors from standards or cross-referenced measurements (i.e., RBS or EPMA) for compounds in those applications where improved accuracy is desired.

It fairly simple to provide a first order practical estimate of the detection limit *for a given acquisition condition* using equation 11. Assume a detection limit corresponding to an intensity of one to three times the measured rms noise level at the energy in the spectra where the Auger peak would be expected. This measured intensity can then be substituted into the numerator of equation 11 for the element of interest to yield a practical detection limit (the contribution in the denominator can often be ignored). Essentially, this amounts to quantifying some multiple of the noise to establish the detection limit. Typical AES detection limits range from 0.1-1.0at% over the ~100Å analytical volume and is approximately an order of magnitude poorer than XPS for most elements. The particulars of the detection limit depend upon the element, the analyzer transmission and analysis conditions, where a longer acquisition will enhance signal to noise and a higher probability for Auger emission will yield higher sensitivity. Note that the detection limit can equate to a sub-monolayer quantity if the species is concentrated at the surface.

An example of Auger spectra acquired on a copper surface at 3ekV is shown in Figure 5. Largely due to historical convention stemming from early generation electron analyzers that had a relatively poor signal to noise ratio as compared to modern instrumentation, empirical sensitivity factors for quantification of Auger spectra are based upon the peak-to-peak height of the signal peak in the differentiated spectrum, EdN(E)/dE. Quantification based upon the undifferentiated signal is possible but far less common at this time and must use different area-based sensitivity factors in combination with suitable background subtraction.

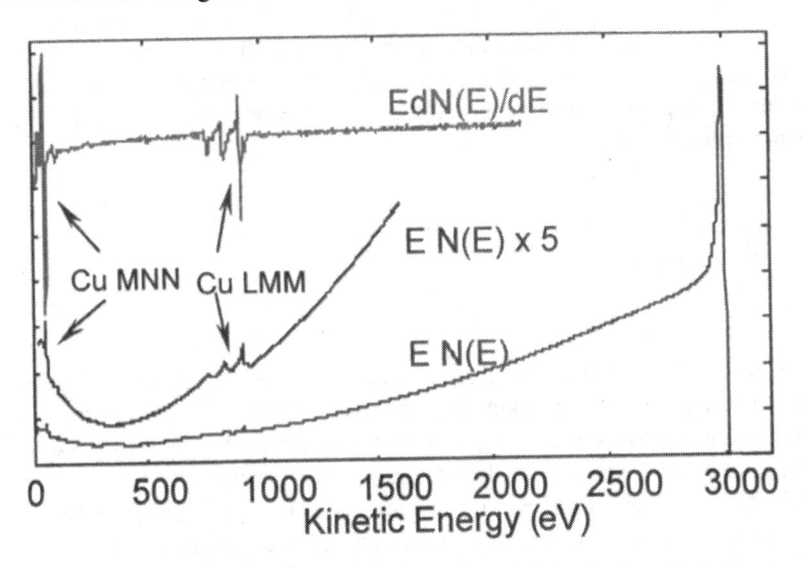

Figure 15-5. An example of Auger electron spectra acquired on a copper surface using a primary electron beam energy of 3KeV. The bottom spectra represents the full scale undifferentiated spectrum. The middle spectrum is an expanded view of a portion of the undifferentiated section. The upper spectrum is the differentiated signal, which removes the background and creates the peak-to-peak features commonly used to quantify Auger data.

2.5 Auger Electron Spectroscopy Instrumentation

This section describes generically and in brief the type of equipment commonly employed in modern commercially available Auger analytical systems. In terms of basic system components, a modern AES system incorporates a vacuum system, a scanning electron gun, a secondary electron detector, and an electron analyzer to measure the kinetic energy of the Auger electrons.

Most AES systems also incorporate a variable energy ion gun that can be used to modify the surface through sputter ion etching. The variable energy ion gun is an important component to a surface analytical tool as it provides a means to sputter etch the surface and generate an elemental or chemical depth profile. Modern ion gun systems can produce ion energies ranging from less than 10eV up to, typically, 5KeV. Thus it is possible to sputter in-situ at etch rates on the order of 1 angstrom per minute, as for example the case of 250eV ions in SiO_2.

In-situ low ion energy etch capability is also useful in FIB applications, including final thinning of membranes for TEM. An example of in-situ low energy sputter ion milling of a FIB membrane is provided in Figure 6. The membrane was initially prepared in a dual beam FIB and transferred to the Auger spectrometer. Low energy ion milling was completed for 20 minutes using 500eV ions rastered over 2X2mm at a current of 540nA. To obtain the highest quality HRTEM images it is critical that in addition to being suitably thin, the samples have minimal damage and minimal surface roughness (Kisielowski et al. 2001, Principe et al. 2002). Sample preparation can be achieved by a number of standard methods but the quality of the final surface is the most important result and when ion milling is employed, is often dictated by the sputter conditions. Due to the relative high gallium ion energies in a dual beam FIB, extremely thin samples can suffer significant amorphous layer damage. A dual beam FIB integrated with a low energy ion gun and an in-situ extractor/manipulator should allow increased flexibility, and improved quality of TEM sample preparation by FIB.

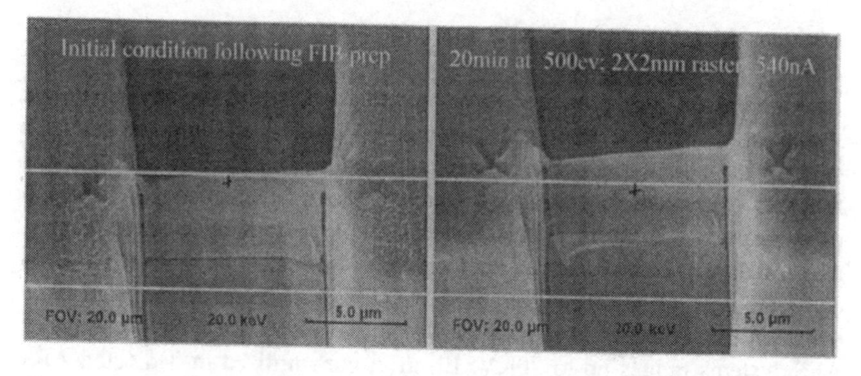

Figure 15-6. In-situ low energy (500eV) Ar^+ ion sputtering of a thin membrane produced by FIB. This capability allows SEM monitoring of final thinning of a TEM membrane and reduced sample damage relative to 10kV or 30kV Ga^+ ions.

There are other interesting possibilities that arise from AES applied to thin membranes. In the theory section of this chapter the background in

AES was attributed to inelastic electron beam interactions within the solid. Beam spreading in the elastic regime was also related to sample thickness. Finally, beam damage is also associated with inelastic scattering events originating deeper in the sample. Thus, the AES spectra background changes as a function of the membrane thickness, the extent of beam spreading may be limited by the membrane thickness, and beam damage can be reduced when AES is performed on a thin membrane. The backscatter contribution to the Auger intensity will also be reduced significantly. It is also conceivable that the intensity and background of AES spectra on a thin membrane can be applied as an end-point detection scheme to determine membrane thickness in-situ.

2.5.1 Ultra-High Vacuum Systems

AES is most effective when conducted in an ultra-high vacuum environment. Since AES is most sensitive to the top few monolayers of a surface the presence of random atoms and molecules in the analytical chamber's ambient environment that can actively adsorb onto the surface of interest before, during and after the analysis are generally undesirable. In modern commercial systems the sample introduction sequence is automated through a load lock and requires 5-15 minutes. In practice, vacuum levels range from $1X10^{-8}$ to $1X10^{-9}$ torr, which is quite acceptable for the majority of AES analytical studies, but of course depends upon the analysis requirements.

2.5.2 Electron Guns in AES Systems

An electron gun produces ionizing radiation, which in any analytical system may serve a manifold purpose. Virtually all commercial AES systems have scanning coils and secondary electron detectors and thus can function equivalent to a standard scanning electron microscope (SEM) in most respects. Because Auger systems employ scanning electron beams, Auger data can be collected in spot mode, as line scans or as area maps. A common type of electron gun used in AES is a thermally-assisted field emission gun (FEG). The practical voltage available on most commercial AES systems ranges up to 20KV. Electron guns utilized in dedicated AES systems produce currents in a range of 10-30nA while collecting spectra. The relatively high current is needed to promote a sufficient flux of Auger electron since, as a simple rule of thumb, approximately 1 Auger electron is produced for every 1000 secondary electrons. This nominal current range may be compared to the 1-2nA level employed in bulk XEDS analysis and the 1-10pA range typical for optimal SE imaging conditions.

The relatively high current required for AES can lead to sample charging problems on insulators. In some instances conducting masks can be placed on the sample surrounding the region of interest to provide a conduction path. Another option to eliminate charging effects on insulating layers less than a ~1.5 microns is to employ a higher acceleration voltage, such as 20KeV, in order to "punch through" the insulating layer and provide a conduction path through the substrate. Based upon the electron beam interactions described above, higher voltages have additional benefits when AES is applied to analysis of particles and thin layers. Electron beam damage due to inelastic scattering interactions at the *near-surface* is reduced at higher voltage, for a given current. Also, the emission of spurious Auger electrons induced by back-scatter electrons from the area immediately surrounding the feature of interest, such as a particle, are reduced at the *near-surface* for a higher voltage at a given current. It is commonly stated that lower voltages produce less beam damage but this is actually a matter of the location in the depth of the sample that inelastic damage is concentrated. In terms of AES, it is preferred to minimize beam damage at the near-surface from where the Auger electrons are emitted. A third option to address charging effects is to use the FIB to produce local holes through the insulation to the conductive substrate to dissipate the surface. This method can also be augmented with electron beam induced gas-assisted metal deposition in the FIB to deposit a metal "mask" near the region of interest. However, care must be taken not to affect the surface to be examined.

Modern commercial AES systems have an image resolution typically in the range of 80-100Å for optimal imaging conditions, which is much poorer than the 30Å or less in a modern dedicated SEM. At currents typical for Auger, the beam size is on the order of ~250 Å. Image resolution limitations are due in part to challenges in the integration and design of a high-resolution electron column that is compatible with a UHV environment that also does not interfere with the operation of electron analyzers (magnetic lenses can cause problems). Industry is demanding better image and spatial resolution, even from dedicated AES systems, so it is likely improvements will be made within the next few years.

2.5.3 Auger Electron Energy Analyzers

Two types of electron kinetic energy analyzers are commonly used in commercial AES instrumentation, cylindrical mirror analyzers (CMA) and hemispherical mirror analyzers (HMA). The historical advantage of a CMA over early generation HMA spectrometers was higher transmission (efficiency and count rate). The fixed retard ratio (FRR) operation mode of a CMA results in an energy resolution, ΔE, that is a function of the kinetic energy of the Auger electron ($\Delta E/E$ = constant, typically 0.4-0.8%) and therefore the energy resolution of the spectra decreases as kinetic energy increases. The inherent line widths of Auger spectra range from ~0.25-1eV (Z<30) and a CMA is quite adequate to resolve major features of core level Auger spectra. However, modern commercial HMA systems provide counts rates that can exceed that of today's commercial CMA spectrometers. An HMA also provides higher energy resolution capability, which can benefit AES studies designed to extract detailed peak shape data and chemical state information. The energy resolution of a HMA is also constant across the spectrum, due to fixed analyzer transmission (FAT) operation mode. Further, modern spectrometers employ channeltrons for electron detection and the systems can be operated in a "snapshot mode" wherein the entire energy window is acquired in parallel across the available independent channels. The snapshot mode is commonly applied for elemental map acquisition in both AES mapping and imaging XPS systems. One late generation HMA commercial spectrometer has 112 independent channels dedicated for energy detection. The higher channeltron density permits relatively rapid acquisition with an energy resolution less than ~0.5eV for a typical 20eV energy window (actual energy resolution is also function of pass energy), and thus can be suitable for mapping and depth profiles requiring moderate energy resolution for chemical separation.

The particular AES system employed to produce the results in this chapter is a SMART Tool, manufactured by Physical Electronics, Inc. In addition to an electron gun and an 8 channeltron CMA, the system also incorporates a focused ion beam (FIB) and an XEDS analyzer. The sample handling allows analysis of 200mm and 300mm full wafers as well as capability for small parts and wafer pieces. Some of the FIB work was also completed on a FEI 1265 dual beam equipped with an XEDS. The 1265 dual beam can also handle 200/300mm full wafers.

3. APPLICATIONS OF FIB AND AUGER

This first example contrasts the much smaller analytical volume of AES relative to XEDS. Panel a) in Figure 7 is a side-view of illustrates a defect located on the edge of an aluminum metal line on a patterned semiconductor wafer. Initial analysis of the defect structure by XEDS at both high and low accelerating voltage indicated aluminum, titanium, oxygen, nitrogen and silicon. Since the film stack consists of these same elements the XEDS analysis was merely suggestive, not conclusive, as to the root cause of the defect. To appreciate the ambiguity that can arise, consider the XEDS volume of 7keV overvoltage needed to excite the Ti_{KLL} line with reasonable intensity. A FIB cut was made to remove the front portion of the defect structure, as shown in panel a). The initial Auger spectral analysis and Auger map (not shown) of the surface shown in panel a) also did not pinpoint the root cause. A series of FIB cuts were made while monitoring the cut face until most of the defect was consumed, as shown in panel b). A number of spot mode analyses were made at different locations in the image in order to determine the elements present and to prepare the energy windows for the Auger map. The resulting high resolution Auger map is illustrated in panel c). From the Auger map it can be observed that the titanium (TiN) layer "wraps" around a silicon defect. Subsequent deposition and etch processing steps on top of the original defect, resulted in the final defect structure shown in panel a). The Auger analysis clearly identifies the root cause.

Several points can be related from this example. It was necessary to continue to cross-section through the defect until a critical point when the elemental distribution determined by the Auger analysis revealed the defect structure and location. Because of the surface sensitivity of Auger it is necessary to conduct the analysis at the appropriate plane of the defect structure. Since it is often not possible to know apriori the optimal location of the FIB cut for Auger analysis, the analyst must be prepared to be patient and willing to make several careful cuts and possibly iterate between the Auger analysis and FIB.

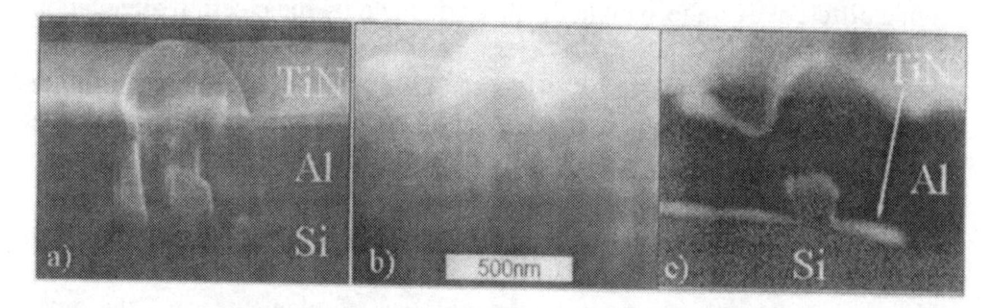

Figure 15-7. FIB-Auger analysis of a defect on a multi-layer patterned wafer. Panel a) depicts the SEM image of the initial defect structure. Panel b) is the defect following FIB sectioning and panel c) is an Auger map of the structure shown in panel b). The titanium (TiN) layer "wraps" around a silicon defect. Subsequent etch and deposition processing steps resulted in the final defect structure shown in panel a)

It should also be mentioned that XEDS can still be very useful. An experienced XEDS analyst may be able to make a very reasonable conclusion in this situation that the root cause was associated with silicon, even without FIB cross-sectioning, by varying the voltage and with careful comparison with a reference area. Likewise, it was not absolutely necessary to acquire an Auger map to determine the root cause, as a few spot mode analyses after the final FIB cut were all that was required for the experienced analyst. However, it is often the non-expert customer that must be convinced of the analytical result, and a single impelling graphic representation can often achieve that objective most effectively. There is also a possibility an Auger map will reveal an unexpected distribution or a feature the selective spot analysis did not pick up. Remember, strong element contrast may not be present in the SEM image, depending upon materials and SEM conditions, and the subjective choices made by the analyst to determine the Auger spot mode analysis locations are guided by the SEM image contrast.

There are a few practical tips for successfully acquiring a moderately high resolution (256 X 256 pixel) Auger map at high magnification, as shown in panel d) of Figure 7. First, the element contrast in a map acquisition on this type of instrument is based upon the intensity difference in the peak height and the background next to the peak in the *undifferentiated* spectrum. A higher accelerating voltage will yield a lower background at higher electron energy, and in some cases this can lead to a better elemental contrast in the Auger map. However, there is a voltage and electron energy dependence on the analyzer transmission function and there is often more than one set of peaks at different electron energies that can one can choose for analysis. Peaks at different electron kinetic energies come from a different average sample depth, as dictated by the electron attenuation length. So one should check to determine the optimal voltage and optimal peak(s) for a given analysis. If a focused spot analysis is required, or high spatial resolution mapping is a goal; a higher voltage will also produce a smaller beam size for a given current while also reducing *near-surface* electron beam induced sample damage.

It is also necessary to use image registration because of sample drift during the acquisition, which can take 20-60 minutes to complete a high-resolution map depending upon the number of elements and the map resolution. A lower resolution map can be complete in 10-15 minutes but still may require image registration. If possible, it is best to set up an image registration target away from the actual region where the Auger map is being acquired. This is because the image registration is based upon the SEM image contrast and the image can change during the map acquisition to the point where the image registration will fail. Selecting an image registration feature can be an issue, since there may be no suitable feature close enough to the analysis area to serve the purpose. However, the FIB can be used to drill a precise etch pattern in the sample next to the analysis region at a location of the analyst's choosing. One or more simple FIB holes provide an excellent dark-bright and constant contrast feature for the image registration.

Further, the small dimension of a FIB hole means the image registration can be completed at high magnification, which ensures a well-registered Auger map. Note that image registration can also be necessary for spot mode analysis and line scan acquisitions and this tip applies to those situations as well. Finally, a higher current will produce better signal to noise for a given analysis setup, so use the highest current practical. A current up to 20nA or even 30nA may be used but one should be careful to consider beam damage effects prior to choosing a very high current, so if it is a new system be conservative (10nA or less) or make a test run. Note that a higher current will typically produce a larger beam size for a given voltage.

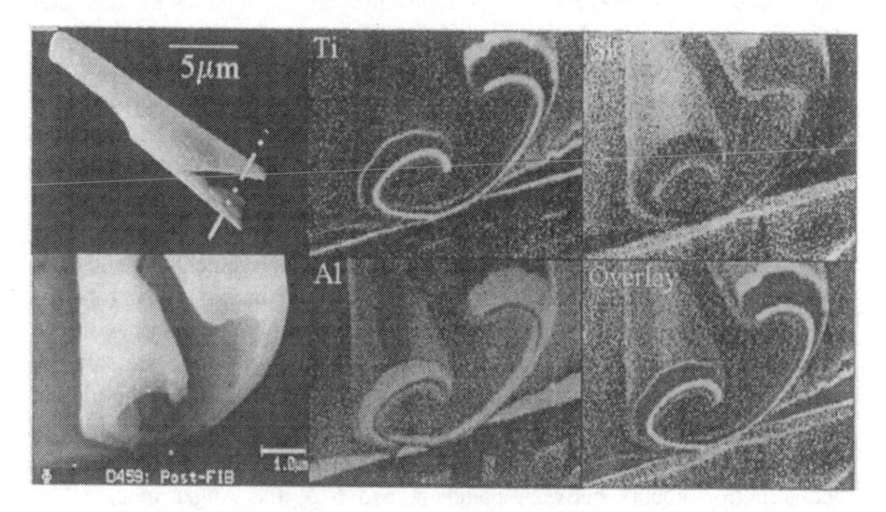

Figure 15-8. FIB-Auger Map on complex defect structure. Panel a) in the upper left is an SEM image of the defect prior to FIB cross-section. Panel b) immediately below is an SEM view following FIB cross-section through the region of the defect indicated by the dashed-dot line in panel a). Panels c) through e) show the Auger maps of Ti(N), SiO2 and Al, respectively.

In Figure 8 a pair of SEM images depicting a defect before and after a FIB cut to expose the cross-section is shown in panels a) and b). The remaining panels display elemental Auger maps that reveal the structure in detail. The valuable information conveyed through this example is related to defect history and structural detail. For instance, it is possible to determine that this complex defect structure was established on the surface prior to, or early into, the first TiN layer deposition process. The entire defect structure was later coated by silicon oxide during the last blanket process step. The ability to distinguish the surface composition, or outer shell component of a defect as compared to the interior, is a very common application of Auger-FIB in complement to XEDS and can often indicate the process step the defect was first introduced. The multi-layered structure of the defect itself is a forensic clue to the origin and mechanism of the defect formation. The detail and spatial resolution in the Auger elemental maps in Figure 8 is impossible to match with an XEDS analysis.

A more sophisticated application of FIB and Auger is discussed next with reference to Figure 9. The subject of interest is a spherical feature approximately 200nm in diameter. The as-received SEM image is shown in the lower left image panel. The particle defect is sitting on top of a layered film stack consisting of elements A and B, where top surface film is composed of layer B and the layer beneath is composed of element A. The total thickness of the layers A+B is approximately the same as the diameter of the spherical defect. An initial XEDS analysis was unable to determine if the spherical feature was composed of element A, element B, or both. Initial Auger surface analysis indicated the exterior of the defect consisted of element A but was resting on the surface of element B. The spherical particle was coated with platinum in a dual beam FIB, transferred to the Auger/FIB SMART Tool and a sequence of eight slices were made while cutting through the 200 nanometer feature. Auger maps were acquired after each FIB slice. The eight part Auger map series of elements A and B is shown in Figure 9 in panels 1) through 8). It can be observed from the Auger map series that the defect structure consists of *both* elements A and B, where element A coats the exterior element A running through the core. A partial volume reconstruction model was then produced of the defect structure from the Auger maps. The result of the reconstruction is shown in the lower right panels of Figure 9. The discontinuity in the volume reconstruction occurs between panels 4 and 5 of the Auger maps. The reconstruction was produced using a custom script and the built-in image analysis capabilities in MatLab®. The final reconstruction model can be rotated, sectioned, viewed from any angle and even animated. This example combines high magnification, image registration, high-resolution Auger elemental mapping and volume reconstruction of a nanostructure. All the operational tips described earlier were applied in this example.

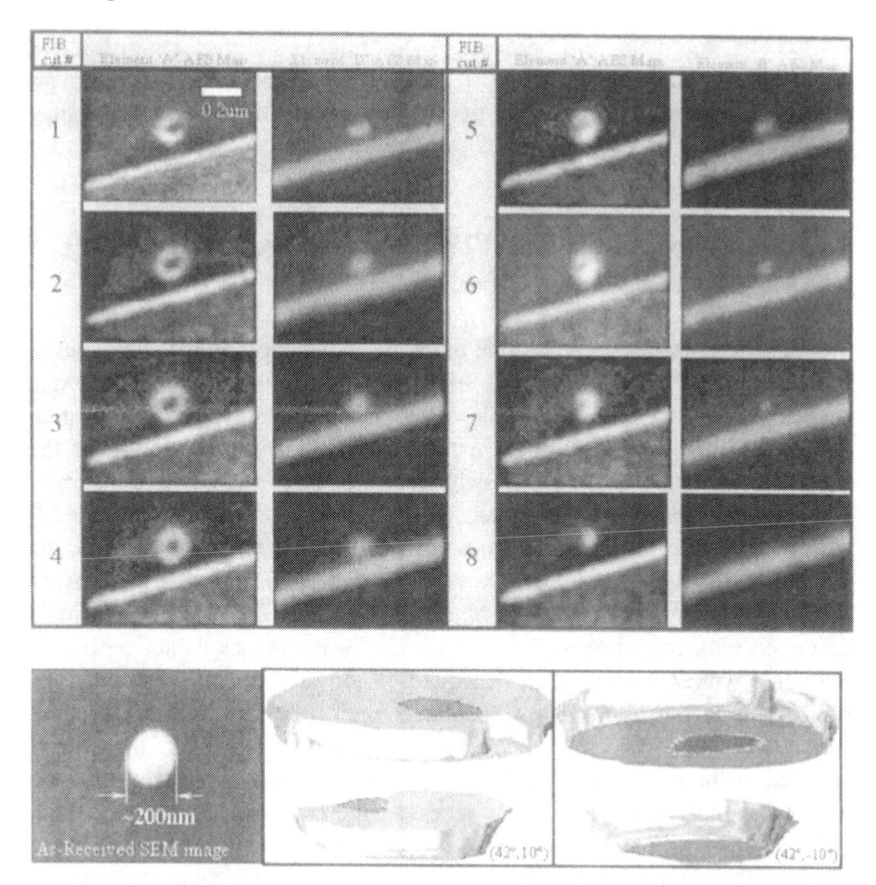

Figure 15-9. A FIB-Auger map multi-slice series and volume reconstruction. The top panel shows a series of eight FIB cuts on a 200nm particle composed of two components, 'A' and 'B'. sitting on top of a layered film stack also of elements 'A' and 'B'. The bottom panel shows an SEM image of the original particle sitting on the film stack next to two isometric views of the volume reconstruction produced from the eight FIB-Auger map series.

Knowing the structure of the defect and the film stack, it is easy to understand the XEDS analysis was problematic due to the significantly greater analytical volume of x-ray analysis. The FIB cuts were made while simultaneously imaging with the SEM, so the progress could be monitored carefully. Each of the FIB cuts required less than ten seconds, on average, and each series of Auger maps took approximately 30 minutes. This reconstruction example involved entirely manual operation for both the FIB cut operations and the interleaved Auger mapping. At present there is no commercial system that integrates image registration between the electron beam required for the Auger and FIB beam required for the cross-sectioning needed to automate this process. In future, volume reconstruction applications could be improved if software and control hardware was developed to automatically integrate the alternating FIB cut and Auger map acquisition.

4. ARTIFACTS IN APPLICATIONS OF FIB AND AUGER

It has been emphasized that one of the key advantages in the combination of Auger and FIB stems from the surface sensitivity of Auger. It should also be recognized that XEDS analysis often provides a useful complement to AES analysis, taking advantage of the different analytical volumes of the techniques to obtain a more complete picture of the structure. For instance, a structure or defect may be coated with a thin layer, which is all the AES will detect while the core and bulk composition of the structure or defect will be revealed by XEDS. In certain situations the surface sensitivity of AES can become a liability in other ways as well. A case in point is described with reference to Figure 10. The SEM images depict a FIB cut through a void in a metal film deposited on top of a silicon substrate. The right panel of the figure shows an Auger map produced on the x-section. The Auger map indicates silicon inside the void. However, the silicon detected in the void is an artifact of "FIB dust" caused by silicon substrate material sputtered during the FIB process being re-deposited within the void. This situation will arise whenever a FIB cross-section is produced on a structure that contains voids or gaps. Therefore it is problematic to examine the unaltered surface by Auger within the void or gap cross-section produced by FIB. It is occasionally possible to sputter etch the void to remove the material within, but this solution is not always viable or practical. Auger is also sensitive to the over spray commonly deposited on the surface that extends several microns surrounding the FIB cut, but this is typically not an issue since the cut face is of primary interest. The FIB cut face will always contain some amount of implanted gallium and this will also be detected by Auger. But, the concentration of implanted gallium can be minimized by using a low current final cleaning step on the cut face and residual amount remaining generally will not interfere with the Auger analysis.

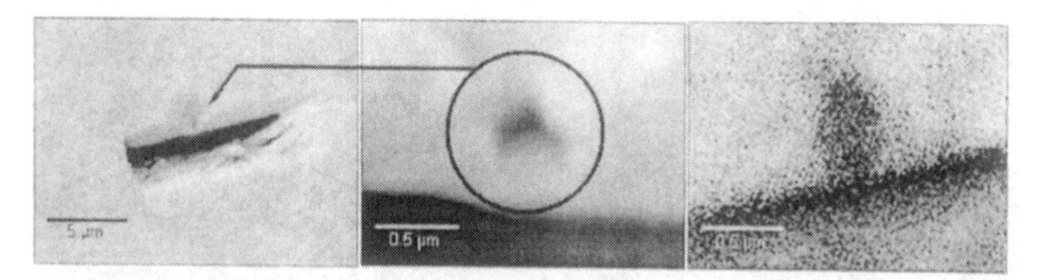

Figure 15-10. A FIB cut through a void in copper is shown in the first two panel and the AES map in the right panel. The green in the Auger map represents copper, while the red represents silicon. The map data suggests the presence of silicon inside the copper void, but this is a "FIB dust" artifact caused by the silicon substrate sputtered during the FIB process re-depositing within the void.

Another type of artifact that may be encountered when combining FIB and AES is illustrated in Figure 11. The cross-section shown was produced in a dual beam FIB and followed by XEDS analysis. The XEDS analysis was not conclusive and the sample was transferred to the AES tool. It is well known that exposure to an electron beam in a SEM can result in hydrocarbon deposition caused by localized breakdown of vapor phase organic materials such as pump oils and other chamber contaminants. The carbon-based material deposited on the sample cut face during the XEDS mapping in the dual beam FIB resulted in a contamination layer that severely attenuated the Auger signal. A brief Ar^+ ion etch using the low energy ion gun in the AES system was used, with the intention intended to sputter clean the surface. Instead, the ion gun caused a small scale "explosion", entirely destroying the x-section, the result shown in the right side of the figure. Previously, the Ar+ ions had been used to clean and lightly etch FIB cut face surfaces on other samples with no ill effect. In this case, the structure was a patterned wafer film stack that included a thick insulating layer. The "explosion" occurred when the 1µA current of 2kV Ar+ ions were grounded through the constricted conducting path created by the FIB. The current overwhelms the dielectric and produces the discharge. Thus, there is a caution in the use of a high flux of ions on insulated surfaces on the FIB cross sections. Note that the much larger FIB etch area above and to the right did not suffer the same fate.

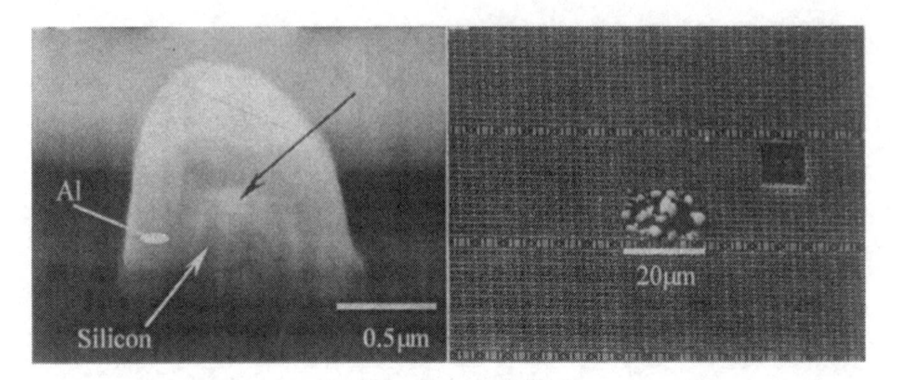

Figure 15-11. The SEM image at left shows a FIB cut through a defect structure. At right the same region is shown at lower magnification following a brief Ar^+ ion etch intended to sputter clean the surface that instead resulted in a "micro explosion."

5. CONCLUSIONS

The analytical power in the combination of FIB and Auger stems from the ability of FIB to produce a controlled surface and the ability of Auger to analyze the composition within the top 10nm of that surface at a spatial resolution of approximately 10-25nm. The surface analyses can be completed in a selective spot mode, as line scans, or as area maps. This tool combination has demonstrated applications in the analysis of defects and structures in the nanometer regime and provides a complement to larger volume analytical techniques based upon primary electron beams, such as XEDS or electron microprobe.

The capabilities of these techniques can be further extended to produce sequential sections through the feature of interest. Using this approach, it is possible to generate three-dimensional compositional volume reconstructions at the nanometer scale. Reconstructions can be processed to produce arbitrary cross-sections, rotated views or animated sequences. Hybrid concepts involving both FIB and low energy Ar+ ion sputtering add another dimension to this capability. Improvements in hardware and software automation in future generation systems should allow the number of practical applications of this sort to grow. The market driver for these applications will arise from a variety of industry and research needs to characterize nanometer scale features.

REFERENCES

Auger P, Compt. Rend., Vol 177, pg. 169 (1923).

Briggs D and Seah MP, Practical Surface Analysis, 2nd Edition. Volume 1-Auger and X-ray Photoelectron Spectroscopy, John Wiley & Sons Ltd. (1990).

Childs KD, et al., Handbook of Auger Electron Spectroscopy, A Book of Reference Data for Identification and Interpretation in Auger Electron Spectroscopy, 3rd edition, Physical Electronics, Mn. (1995).

Feldman LC and Mayer JW, Fundamentals of Surface and Thin Film Analysis. Elsevier Science Publishing Co., Inc. (1986).

Goldstein JI, et al., Scanning Electron Microscopy and X-Ray Microanalysis, 2nd edition. Plenum Press, New York. (1992).

Kisielowski C, Hetherington CJD, Wang YC, Kilass R, O'Deefe MA, Thust A, "Imaging columns of the light elements carbon, nitrogen and oxygen with sub-Ångstrom resolution", Ultramicrsocopy, 89, 4:243-263 (2001).

NIST Standard Reference Database 82. NIST Electron Effective-Attenuation-Length Database. http://www.nist.gov/srd/nist82.htm.

Principe E, Watson DG, and Kisielowski C, "Advancements in the Characterization of "Hyper-Thin Oxynitride Gate Dielectrics Through Exit Wave Reconstruction HRTEM and XPS", 2002 Supplement to the Microelectronic Failure Analysis Desk Reference, ASM International, Materials Park, Ohio (2002).

Williams DB, Carter C,. Transmission Electron Microscopy, A Textbook for Materials Science. Plenum Press, New York. (1996).

Appendix A: Ga Ion Sputter Yields

Z	Element	30 keV 0 degrees	30 keV 88 degrees	5 keV 0 degrees	5 keV 88 degrees
1	H	0.79	90.78	0.22	36.09
2	He	1.20	58.14	0.65	29.00
3	Li	1.73	50.23	0.94	19.55
4	Be	2.22	35.77	1.34	15.37
5	B	1.45	29.36	0.92	11.39
6	C	1.69	22.93	0.81	9.38
7	N	2.61	45.43	1.53	15.35
8	O	4.75	37.00	2.42	17.72
9	F	3.62	39.19	2.05	13.87
10	Ne	5.14	47.28	3.22	18.18
11	Na	4.93	42.25	3.14	16.31
12	Mg	4.91	42.54	3.28	15.61
13	Al	3.47	34.78	2.67	13.29
14	Si	2.78	27.61	1.73	7.98
15	P	2.91	29.10	1.44	9.71
16	S	3.48	27.66	2.37	10.36
17	Cl	4.06	37.49	3.03	13.69
18	Ar	2.69	18.22	1.75	7.41
19	K	2.11	17.39	1.40	10.37
20	Ca	2.35	14.59	1.68	7.47
21	Sc	1.60	11.62	1.52	5.82
22	Ti	2.28	13.28	1.41	5.70
23	V	2.90	13.96	1.32	5.92
24	Cr	5.15	20.61	2.94	8.37

25	Mn	8.17	33.55	4.98	11.66
26	Fe	5.58	24.49	3.74	7.81
27	Co	7.39	24.63	5.37	8.53
28	Ni	9.69	27.13	5.26	9.20
29	Cu	8.37	32.71	8.13	12.04
30	Zn	22.80	51.20	15.29	22.66
31	Ga	5.93	20.57	5.62	8.33
32	Ge	4.54	15.58	2.93	5.82
33	As	13.64	34.74	10.50	15.97
34	Se	7.72	22.57	4.57	8.57
35	Br	4.50	19.50	5.10	6.96
36	Kr	4.42	18.56	4.53	8.58
37	Ru	5.37	18.52	3.73	8.39
38	Sr	3.74	16.08	3.90	7.12
39	Y	4.17	9.85	1.94	4.11
40	Zr	3.87	9.72	2.20	3.97
41	Nb	4.18	10.31	3.05	4.25
42	Mo	6.05	13.63	3.86	4.86
43	Tc	22.27	48.35	16.63	16.13
44	Ru	9.91	13.75	5.96	5.85
45	Rh	9.61	18.90	6.92	7.24
46	Pd	14.23	27.19	9.94	9.39
47	Ag	13.43	26.31	8.51	7.66
48	Cd	18.47	51.00	17.42	15.57
49	In	13.12	19.30	7.31	8.31
50	Sn	8.17	15.88	6.05	6.82
51	Sb	7.88	21.05	5.57	7.79
52	Te	8.37	20.41	7.91	7.61
53	I	5.08	16.92	4.13	7.43
54	Xe	7.66	18.99	5.53	7.40
55	Cs	8.33	19.53	6.57	9.29
56	Ba	7.73	13.89	5.04	6.92
57	La	4.30	10.93	3.24	4.25
58	Ce	5.14	15.40	3.36	4.05
59	Pr	7.28	15.15	4.51	4.30
60	Nd	3.57	14.31	3.64	4.58
61	Pm	9.88	21.37	5.49	8.52
62	Sm	8.40	24.35	7.21	8.91
63	Eu	9.51	17.27	4.94	8.29
64	Gd	4.27	10.79	3.44	5.41
65	Tb	5.34	13.73	3.67	5.44
66	Dy	6.87	15.93	5.23	6.37
67	Ho	7.19	14.04	6.18	6.47

68	Er	8.70	16.35	5.82	6.71
69	Tm	11.12	20.15	7.01	8.18
70	Yb	13.23	20.76	7.99	8.63
71	Lu	4.33	9.21	4.19	4.79
72	Hf	6.55	14.47	4.39	4.29
73	Ta	7.26	11.45	4.49	3.87
74	W	7.59	11.21	4.37	4.37
75	Re	8.74	11.09	5.66	4.47
76	Os	8.30	13.17	6.31	5.18
77	Ir	10.69	15.96	8.22	6.18
78	Pt	9.76	15.88	7.14	6.00
79	Au	15.75	24.13	10.19	8.08
80	Hg	48.16	59.84	26.83	21.04
81	Tl	15.23	26.64	8.76	10.72
82	Pb	15.05	25.95	9.97	8.69
83	Bi	11.92	17.84	7.98	6.34
84	Po	12.13	20.89	6.66	6.60
85	At	11.00	22.66	6.78	7.86
86	Rn	15.18	16.19	7.02	7.15
87	Fr	7.92	19.67	5.66	5.86
88	Ra	4.20	10.18	3.43	4.36
89	Ac	12.48	18.82	6.26	7.46
90	Th	5.94	8.87	3.24	3.52
91	Pa	20.13	21.01	9.94	7.89
92	U	8.28	12.22	6.56	4.56

Appendix B: Backsputtered Ga Ion Fraction

Z	Element	30 keV 0 degrees	30 keV 88 degrees	5 keV 0 degrees	5 keV 88 degrees
1	H	0.00	0.32	0.00	0.36
2	He	0.00	0.48	0.00	0.51
3	Li	0.00	0.55	0.00	0.62
4	Be	0.00	0.60	0.00	0.65
5	B	0.00	0.63	0.00	0.61
6	C	0.00	0.61	0.00	0.64
7	N	0.00	0.57	0.00	0.63
8	O	0.00	0.64	0.00	0.58
9	F	0.00	0.56	0.00	0.64
10	Ne	0.00	0.61	0.00	0.68
11	Na	0.00	0.61	0.00	0.68
12	Mg	0.00	0.56	0.00	0.69
13	Al	0.00	0.56	0.00	0.60
14	Si	0.00	0.58	0.00	0.73
15	P	0.00	0.52	0.00	0.66
16	S	0.00	0.63	0.00	0.73
17	Cl	0.00	0.61	0.00	0.70
18	Ar	0.00	0.67	0.00	0.70
19	K	0.00	0.66	0.00	0.62
20	Ca	0.00	0.74	0.00	0.67
21	Sc	0.00	0.59	0.00	0.72
22	Ti	0.00	0.73	0.00	0.73
23	V	0.00	0.68	0.00	0.66

24	Cr	0.01	0.69	0.00	0.67
25	Mn	0.01	0.62	0.00	0.68
26	Fe	0.00	0.65	0.00	0.69
27	Co	0.01	0.69	0.00	0.73
28	Ni	0.00	0.69	0.00	0.72
29	Cu	0.01	0.68	0.00	0.67
30	Zn	0.00	0.67	0.00	0.71
31	Ga	0.00	0.79	0.00	0.79
32	Ge	0.00	0.74	0.01	0.78
33	As	0.00	0.71	0.02	0.72
34	Se	0.01	0.66	0.01	0.77
35	Br	0.00	0.58	0.03	0.78
36	Kr	0.00	0.73	0.03	0.72
37	Ru	0.01	0.64	0.09	0.70
38	Sr	0.01	0.72	0.07	0.78
39	Y	0.01	0.70	0.01	0.79
40	Zr	0.04	0.68	0.02	0.74
41	Nb	0.05	0.71	0.01	0.75
42	Mo	0.03	0.70	0.03	0.73
43	Tc	0.04	0.72	0.07	0.69
44	Ru	0.03	0.74	0.09	0.69
45	Rh	0.03	0.73	0.05	0.68
46	Pd	0.04	0.67	0.04	0.72
47	Ag	0.04	0.70	0.03	0.78
48	Cd	0.03	0.74	0.05	0.78
49	In	0.09	0.78	0.06	0.70
50	Sn	0.06	0.73	0.07	0.75
51	Sb	0.06	0.66	0.70	0.68
52	Te	0.05	0.71	0.10	0.74
53	I	0.03	0.75	0.09	0.68
54	Xe	0.05	0.73	0.12	0.79
55	Cs	0.02	0.63	0.07	0.72
56	Ba	0.06	0.69	0.15	0.77
57	La	0.07	0.78	0.10	0.75
58	Ce	0.07	0.70	0.11	0.74
59	Pr	0.09	0.71	0.07	0.74
60	Nd	0.05	0.73	0.06	0.78
61	Pm	0.05	0.73	0.09	0.70
62	Sm	0.09	0.66	0.09	0.74
63	Eu	0.10	0.67	0.15	0.69
64	Gd	0.10	0.74	0.16	0.69
65	Tb	0.11	0.75	0.16	0.74
66	Dy	0.08	0.69	0.10	0.76

67	Ho	0.08	0.75	0.16	0.75
68	Er	0.10	0.67	0.15	0.74
69	Tm	0.09	0.72	0.11	0.70
70	Yb	0.09	0.71	0.18	0.75
71	Lu	0.07	0.72	0.15	0.82
72	Hf	0.09	0.69	0.12	0.84
73	Ta	0.12	0.76	0.19	0.73
74	W	0.13	0.77	0.10	0.72
75	Re	0.12	0.77	0.14	0.77
76	Os	0.15	0.74	0.12	0.75
77	Ir	0.10	0.71	0.24	0.75
78	Pt	0.12	0.81	0.14	0.74
79	Au	0.10	0.73	0.22	0.80
80	Hg	0.12	0.80	0.15	0.79
81	Tl	0.09	0.72	0.22	0.76
82	Pb	0.14	0.66	0.23	0.75
83	Bi	0.09	0.80	0.23	0.78
84	Po	0.21	0.76	0.21	0.80
85	At	0.12	0.73	0.14	0.72
86	Rn	0.13	0.81	0.30	0.73
87	Fr	0.11	0.79	0.18	0.84
88	Ra	0.15	0.73	0.25	0.80
89	Ac	0.13	0.68	0.22	0.76
90	Th	0.11	0.80	0.23	0.78
91	Pa	0.20	0.71	0.23	0.82
92	U	0.14	0.74	0.25	0.84

Appendix C: 30 keV Ga Ion Range at 0 degrees

Z	Element	Long-itudinal Range	Long-itudinal Straggle	Lateral Projected Range	Lateral Straggle	Radial Range	Radial Straggle
1	H	339.9	26.3	16.4	20.1	24.6	12.2
2	He	363.3	55.2	30.3	38.2	48.3	24.2
3	Li	101.8	21.4	10.4	12.9	18.4	9.2
4	Be	30.2	6.8	4.0	4.9	5.8	3.3
5	B	25.2	6.2	3.7	4.4	5.3	2.8
6	C	23.6	6.5	3.6	4.4	5.8	2.8
7	N	55.0	14.0	8.7	11.3	14.6	7.5
8	O	39.1	13.0	6.6	8.4	10.9	6.5
9	F	57.2	17.9	9.2	11.9	14.9	7.5
10	Ne	49.5	17.3	9.3	11.7	15.5	9.5
11	Na	63.8	22.1	11.8	15.2	20.8	12.6
12	Mg	34.9	14.5	7.5	9.5	10.9	6.2
13	Al	23.9	9.1	6.3	7.9	9.9	5.4
14	Si	28.6	11.1	6.8	9.0	11.0	6.4
15	P	38.5	16.3	8.7	11.2	13.3	7.1
16	S	33.7	14.6	7.6	10.1	12.9	7.1
17	Cl	39.7	19.3	9.3	12.2	15.5	8.8
18	Ar	48.0	24.5	13.3	16.6	19.5	10.9
19	K	83.8	40.0	22.3	28.0	35.5	19.1
20	Ca	48.4	23.7	13.2	16.8	22.3	13.3
21	Sc	27.0	12.3	8.0	10.9	12.5	7.8
22	Ti	18.5	8.4	6.1	8.2	9.2	5.6
23	V	15.0	7.4	4.4	5.7	7.7	4.5

24	Cr	11.7	5.5	3.8	4.8	6.0	3.3
25	Mn	11.6	5.7	3.0	5.0	6.2	3.4
26	Fe	11.5	5.5	4.0	5.0	6.2	3.3
27	Co	10.2	5.4	3.4	4.3	5.3	3.0
28	Ni	9.2	4.9	2.9	3.8	4.9	3.0
29	Cu	11.0	5.3	3.9	4.9	6.3	3.4
30	Zn	13.0	7.3	5.5	7.3	8.1	5.0
31	Ga	17.3	9.0	6.8	8.3	10.2	6.2
32	Ge	18.2	10.3	6.5	8.6	11.5	7.0
33	As	17.9	9.5	7.2	9.0	11.0	6.6
34	Se	21.1	12.8	8.6	11.3	13.3	7.9
35	Br	31.8	15.8	13.0	17.2	21.6	14.8
36	Kr	38.2	22.0	15.7	20.8	26.8	16.8
37	Ru	41.7	25.4	32.4	42.7	24.7	12.2
38	Sr	20.2	16.7	22.3	26.7	17.8	8.8
39	Y	24.5	14.8	11.9	15.2	16.3	9.9
40	Zr	18.0	10.4	6.4	8.3	10.9	6.4
41	Nb	13.2	6.8	5.6	7.3	8.4	5.1
42	Mo	10.6	6.2	4.1	5.3	6.8	4.2
43	Tc	10.2	5.3	4.3	5.6	6.8	3.9
44	Ru	9.3	5.4	4.0	5.5	6.2	4.0
45	Rh	10.0	5.0	4.5	6.2	6.8	4.4
46	Pd	9.7	5.4	4.2	5.4	6.3	4.0
47	Ag	11.6	7.0	5.0	6.7	8.3	5.2
48	Cd	15.2	8.7	6.7	8.8	10.2	6.7
49	In	16.2	11.6	7.8	9.9	12.3	7.4
50	Sn	17.0	11.3	8.6	11.3	13.0	7.7
51	Sb	18.9	11.7	9.6	12.4	14.9	8.7
52	Te	21.1	12.4	9.9	13.0	15.0	8.5
53	I	25.5	16.3	11.5	15.9	19.5	11.7
54	Xe	47.8	28.4	20.6	27.0	34.7	22.5
55	Cs	72.9	43.2	38.8	49.6	55.8	35.4
56	Ba	38.7	24.1	19.0	24.5	28.8	26.8
57	La	21.2	14.2	11.7	14.9	17.5	11.1
58	Ce	19.9	11.7	10.9	14.5	16.5	10.3
59	Pr	19.7	11.3	10.7	14.0	16.3	9.6
60	Nd	20.1	12.6	10.1	13.1	15.8	9.2
61	Pm	21.0	13.6	11.2	14.6	17.9	11.5
62	Sm	18.7	11.9	9.9	13.4	15.8	10.0
63	Eu	26.6	16.0	13.3	18.3	20.8	14.2
64	Gd	18.3	11.1	11.1	14.3	16.7	9.8
65	Tb	17.2	9.1	8.8	11.9	14.0	9.5
66	Dy	16.7	11.1	9.9	13.6	15.3	9.5

67	Ho	16.7	9.7	9.0	12.4	14.5	9.5
68	Er	17.2	9.7	9.5	13.0	15.0	9.3
69	Tm	15.0	10.3	9.7	12.5	14.2	8.9
70	Yb	22.1	14.0	12.9	17.2	20.1	13.4
71	Lu	18.0	10.5	8.4	11.1	13.8	8.4
72	Hf	12.2	9.2	7.5	9.7	11.5	6.5
73	Ta	10.0	5.9	5.2	7.0	8.5	5.3
74	W	9.5	5.4	4.5	6.0	7.8	4.8
75	Re	8.6	5.4	4.4	6.1	7.4	4.5
76	Os	8.3	4.5	4.8	6.2	7.0	4.4
77	Ir	8.5	4.5	4.2	5.4	6.6	3.8
78	Pt	8.4	5.3	4.3	5.5	6.9	3.9
79	Au	8.7	5.1	5.3	7.1	7.7	5.1
80	Hg	12.7	2.8	7.3	9.6	12.0	6.6
81	Tl	15.0	9.4	8.4	11.3	13.9	9.1
82	Pb	15.7	8.7	9.3	12.4	13.5	8.6
83	Bi	19.4	11.8	11.6	15.5	17.6	11.4
84	Po	20.2	13.9	10.5	13.6	16.7	10.9
85	At	15.9	9.9	8.7	11.2	14.2	9.5
86	Rn	18.8	12.1	11.4	14.0	18.0	10.9
87	Fr	18.6	11.4	10.4	14.0	16.3	9.8
88	Ra	39.7	23.0	22.3	30.0	33.4	19.6
89	Ac	20.8	11.5	12.9	17.2	19.4	11.7
90	Th	17.4	10.3	8.3	10.6	13.7	8.0
91	Pa	12.6	8.6	8.6	11.1	12.4	8.5
92	U	11.2	8.0	6.6	8.5	9.4	6.5

Appendix D: 30 keV Ga Ion Range at 88 degrees

Z	Element	Long-itudinal Range	Long-itudinal Straggle	Lateral Projected Range	Lateral Straggle	Radial Range	Radial Straggle
1	H	19.2	11.5	339.9	340.8	340.3	24.9
2	He	38.4	26.6	341.8	345.6	344.2	50.9
3	Li	15.0	8.0	101.1	102.8	102.2	18.6
4	Be	5.6	3.3	30.7	31.3	31.1	6.1
5	B	4.6	2.7	24.5	25.1	24.8	5.2
6	C	4.5	3.5	24.5	25.1	24.9	5.6
7	N	9.7	5.8	52.9	54.0	54.2	10.6
8	O	8.4	5.6	45.1	47.1	46.1	13.7
9	F	15.9	10.6	64.9	67.9	66.6	20.2
10	Ne	12.5	9.6	47.2	49.5	49.5	15.4
11	Na	14.6	9.2	62.2	66.5	65.3	23.9
12	Mg	10.8	6.4	36.2	39.4	37.6	15.3
13	Al	6.6	5.1	22.8	24.5	24.3	9.4
14	Si	8.1	5.5	27.8	30.3	29.8	12.4
15	P	11.9	8.0	42.7	46.2	44.9	17.6
16	S	11.6	7.3	34.9	37.8	37.1	14.7
17	Cl	13.6	9.8	41.0	45.2	42.8	19.3
18	Ar	15.7	11.9	45.1	50.1	47.4	22.2
19	K	21.3	20.0	92.2	101.1	97.9	43.4
20	Ca	15.7	11.2	47.0	50.2	50.8	17.0
21	Sc	8.6	6.3	22.7	25.6	24.7	11.8
22	Ti	6.3	4.6	17.1	18.8	19.2	8.2
23	V	4.8	4.0	13.6	14.7	14.7	5.5

24	Cr	4.1	3.0	12.1	13.6	12.9	6.2
25	Mn	4.8	3.5	11.2	12.8	12.0	6.0
26	Fe	4.5	3.4	10.1	11.1	10.8	4.6
27	Co	3.1	2.7	8.9	10.3	9.8	5.1
28	Ni	3.6	2.4	8.8	10.0	9.5	4.6
29	Cu	4.1	3.1	9.8	10.9	10.6	4.9
30	Zn	7.2	4.8	14.0	15.9	15.0	7.3
31	Ga	6.7	5.3	16.0	18.9	17.5	9.5
32	Ge	6.2	4.4	17.4	19.7	20.6	9.9
33	As	8.8	5.4	19.7	22.9	21.6	11.0
34	Se	10.6	7.2	26.8	31.6	29.8	16.6
35	Br	14.9	10.8	29.2	35.1	35.3	19.6
36	Kr	19.6	12.8	37.1	43.7	42.5	23.8
37	Ru	27.7	19.3	58.5	65.8	68.3	28.7
38	Sr	17.7	13.2	41.3	47.6	49.2	22.1
39	Y	12.4	7.0	26.2	30.6	29.2	15.6
40	Zr	8.9	6.6	16.5	21.1	19.3	12.4
41	Nb	7.4	5.8	12.0	14.2	13.8	7.3
42	Mo	6.5	4.2	11.5	13.1	13.3	6.4
43	Tc	4.6	4.4	10.6	12.2	12.3	5.7
44	Ru	4.9	3.2	9.9	12.2	11.1	7.0
45	Rh	5.0	3.3	7.5	8.8	8.1	4.5
46	Pd	5.5	3.8	9.8	11.5	10.9	6.0
47	Ag	5.7	3.5	9.2	11.2	10.5	6.6
48	Cd	9.1	5.8	11.3	13.7	14.2	8.2
49	In	7.6	5.6	13.0	17.3	15.0	11.2
50	Sn	7.3	5.1	10.5	13.4	13.5	8.9
51	Sb	8.4	4.8	16.6	19.7	18.8	10.2
52	Te	14.0	10.1	22.1	27.6	24.6	16.5
53	I	15.0	8.9	24.5	31.5	30.2	19.2
54	Xe	25.5	16.8	52.3	65.5	58.0	39.6
55	Cs	39.0	28.6	75.4	87.7	87.0	40.4
56	Ba	22.4	16.7	32.0	40.2	39.6	23.5
57	La	11.4	5.4	21.1	24.3	24.9	11.4
58	Ce	11.5	6.3	22.3	30.1	26.0	18.8
59	Pr	12.5	8.2	24.3	28.5	28.1	13.8
60	Nd	11.5	5.7	21.0	25.6	23.6	15.1
61	Pm	11.9	8.2	20.7	26.0	26.0	15.2
62	Sm	10.3	7.1	16.7	20.0	20.6	10.7
63	Eu	16.6	13.9	24.3	33.4	32.6	23.8
64	Gd	10.1	8.7	17.1	21.5	20.4	12.7
65	Tb	13.3	8.7	15.4	19.2	18.5	10.5
66	Dy	11.7	7.5	17.0	21.3	20.9	12.1

67	Ho	11.3	6.9	19.2	24.9	22.3	14.8
68	Er	10.9	7.6	15.9	21.2	18.0	13.1
69	Tm	9.9	5.9	15.9	20.8	19.2	12.6
70	Yb	16.1	11.5	21.5	29.2	26.4	18.0
71	Lu	10.0	4.5	17.4	20.4	19.6	10.0
72	Hf	7.9	6.1	10.7	14.2	13.4	9.0
73	Ta	5.5	4.1	9.4	11.3	11.0	6.1
74	W	4.7	2.6	8.2	10.0	9.5	5.9
75	Re	4.5	2.7	8.4	10.4	10.0	6.3
76	Os	4.7	3.7	6.9	8.2	8.5	4.6
77	Ir	5.3	3.2	8.6	10.4	10.6	6.8
78	Pt	6.6	4.3	8.4	9.8	11.1	6.3
79	Au	7.0	4.7	8.9	11.2	10.5	7.1
80	Hg	8.8	6.1	11.3	16.3	15.7	12.7
81	Tl	10.3	8.1	16.1	19.7	19.5	10.9
82	Pb	8.6	5.4	12.7	17.1	15.7	11.7
83	Bi	12.2	8.0	14.2	17.5	18.8	10.2
84	Po	14.7	8.3	21.4	26.2	24.4	14.0
85	At	10.5	5.3	13.1	16.3	17.5	10.3
86	Rn	10.4	6.7	16.6	20.0	22.6	10.6
87	Fr	13.3	8.8	18.9	25.2	22.3	16.4
88	Ra	23.0	13.1	27.6	31.7	32.0	15.5
89	Ac	14.9	9.4	16.6	21.8	21.1	13.0
90	Th	11.8	8.0	14.3	18.5	17.2	11.0
91	Pa	8.0	4.3	11.4	13.9	13.5	7.0
92	U	7.9	4.8	10.2	12.5	12.4	7.2

Appendix E: 5 keV Ga Ion Range at 0 degrees

Z	Element	Long-itudinal Range	Long-itudinal Straggle	Lateral Projected Range	Lateral Straggle	Radial Range	Radial Straggle
1	H	117.8	6.0	6.1	7.4	9.7	4.5
2	He	106.1	15.3	9.5	12.2	13.9	7.8
3	Li	30.3	5.2	3.8	4.8	6.0	3.3
4	Be	9.8	1.8	1.5	1.9	2.5	1.2
5	B	7.7	1.5	1.4	1.8	2.1	1.0
6	C	7.8	1.7	1.5	1.9	2.4	1.1
7	N	16.9	4.6	3.4	4.2	4.8	2.3
8	O	12.2	3.3	2.9	3.6	4.1	2.2
9	F	16.8	5.0	3.8	4.6	5.7	3.2
10	Ne	15.1	5.2	3.5	4.4	5.5	3.0
11	Na	19.7	6.6	4.3	5.5	7.2	3.9
12	Mg	1.1	3.8	2.5	3.1	4.0	2.2
13	Al	7.4	2.6	1.6	2.1	3.0	1.7
14	Si	8.7	3.4	2.4	2.9	3.6	2.0
15	P	11.1	4.2	2.9	3.7	4.6	2.4
16	S	10.0	4.1	2.6	3.3	4.4	2.6
17	Cl	11.3	4.7	3.1	3.9	5.1	3.6
18	Ar	14.5	7.4	4.8	6.3	7.6	4.6
19	K	27.6	13.2	8.9	11.3	12.8	7.1
20	Ca	15.0	7.2	5.4	6.6	7.7	4.3
21	Sc	8.2	4.2	3.0	3.7	4.2	2.3
22	Ti	5.2	2.5	1.7	2.3	2.6	1.4
23	V	3.9	1.9	1.5	2.0	2.4	1.4

24	Cr	3.3	1.5	1.4	1.8	2.0	1.2
25	Mn	3.2	1.5	1.3	1.7	2.1	1.2
26	Fe	3.1	1.4	1.3	1.6	2.0	1.1
27	Co	2.8	1.7	1.2	1.4	1.8	1.0
28	Ni	2.9	1.6	1.1	1.4	1.8	0.9
29	Cu	2.8	1.7	1.3	1.7	2.1	1.2
30	Zn	3.6	2.0	1.6	2.0	2.4	1.5
31	Ga	4.6	2.7	1.7	2.2	2.7	1.4
32	Ge	5.1	3.0	2.2	3.0	3.4	2.2
33	As	5.1	2.8	1.9	2.4	2.9	1.6
34	Se	6.4	3.7	2.9	3.7	4.3	2.5
35	Br	9.3	6.3	4.0	5.0	6.0	3.5
36	Kr	1.1	7.7	5.0	6.4	7.2	4.6
37	Ru	21.8	13.6	10.9	14.2	15.4	9.2
38	Sr	11.5	7.7	4.9	6.5	7.9	4.7
39	Y	7.6	4.5	3.5	4.4	5.3	2.9
40	Zr	4.8	2.7	2.3	3.9	3.4	2.3
41	Nb	3.7	2.5	1.6	2.0	2.4	1.3
42	Mo	3.0	1.9	1.5	1.9	2.3	1.5
43	Tc	2.9	1.9	1.2	1.5	2.0	1.3
44	Ru	2.7	1.7	1.3	1.8	2.1	1.3
45	Rh	2.6	1.7	1.2	1.5	2.1	1.3
46	Pd	2.7	1.9	1.4	1.8	2.2	1.3
47	Ag	3.3	2.2	1.4	1.8	2.2	1.1
48	Cd	4.1	2.1	2.1	3.0	3.3	2.3
49	In	5.5	3.6	2.5	3.5	3.9	2.4
50	Sn	5.4	3.5	2.7	3.5	4.1	2.5
51	Sb	5.8	3.8	2.8	3.7	4.6	2.7
52	Te	6.1	4.3	3.1	4.0	4.8	3.0
53	I	8.8	5.3	3.9	5.1	5.9	3.8
54	Xe	13.9	9.5	6.5	8.1	11.4	7.0
55	Cs	20.3	12.5	11.0	14.9	17.1	10.8
56	Ba	15.4	8.8	6.3	8.3	9.6	5.8
57	La	6.4	3.5	3.6	4.3	5.0	2.9
58	Ce	6.6	4.1	3.0	3.8	4.7	3.0
59	Pr	5.8	3.0	2.9	3.7	4.5	2.9
60	Nd	6.2	3.7	3.0	3.9	4.7	3.0
61	Pm	7.4	4.8	3.7	4.8	5.4	3.2
62	Sm	6.2	3.7	3.2	4.1	4.8	2.9
63	Eu	9.5	5.9	4.7	5.7	7.2	4.4
64	Gd	6.2	4.1	2.7	3.7	4.4	3.9
65	Tb	5.6	3.2	3.2	4.2	4.6	3.1
66	Dy	5.7	3.7	2.7	3.5	4.2	2.3

67	Ho	5.2	3.2	2.9	3.9	4.4	2.7
68	Er	5.8	3.5	3.0	4.1	4.5	3.1
69	Tm	5.3	3.5	3.0	3.9	4.3	2.6
70	Yb	7.2	4.5	3.6	4.9	6.1	4.1
71	Lu	5.2	2.8	2.9	3.6	4.3	2.4
72	Hf	3.7	2.4	2.3	3.0	3.5	2.0
73	Ta	3.1	1.6	1.6	2.1	2.6	1.6
74	W	2.6	1.5	1.3	1.7	2.0	1.4
75	Re	2.6	1.6	1.4	1.9	2.3	1.6
76	Os	2.3	1.4	1.1	1.6	1.9	1.1
77	Ir	2.5	1.6	1.2	1.6	2.0	1.2
78	Pt	2.7	1.7	1.3	1.8	2.2	1.3
79	Au	2.9	1.6	1.6	2.1	2.5	1.7
80	Hg	4.3	3.0	2.6	3.5	3.9	2.5
81	Tl	5.3	3.4	2.6	3.4	4.4	2.4
82	Pb	5.2	3.5	2.7	3.4	4.2	2.2
83	Bi	5.3	3.1	3.7	4.7	5.5	3.1
84	Po	6.3	3.6	4.0	5.2	5.6	3.4
85	At	5.9	3.7	3.3	4.3	5.1	3.0
86	Rn	6.7	4.1	3.7	4.7	5.3	2.9
87	Fr	7.0	4.6	3.3	4.2	5.6	3.3
88	Ra	13.7	8.5	7.5	9.6	11.6	6.4
89	Ac	6.8	3.9	3.6	4.7	5.6	3.4
90	Th	6.3	4.2	3.3	4.2	5.1	2.6
91	Pa	4.5	2.6	2.4	3.1	3.8	2.2
92	U	3.2	2.0	2.0	2.7	3.2	1.9

Appendix F: 5 keV Ga Ion Range at 88 degrees

Z	Element	Long-itudinal Range	Long-itudinal Straggle	Lateral Projected Range	Lateral Straggle	Radial Range	Radial Straggle
1	H	8.0	5.3	119.2	119.4	119.4	6.5
2	He	12.9	9.0	104.0	105.0	104.6	14.7
3	Li	4.7	3.4	31.9	32.5	32.4	5.9
4	Be	1.8	1.0	9.6	9.7	9.7	1.4
5	B	1.6	0.9	7.5	7.6	7.6	1.3
6	C	1.8	0.9	7.7	7.9	7.0	1.6
7	N	3.4	2.4	17.2	18.0	17.6	5.1
8	O	2.9	2.2	13.5	14.0	13.9	3.3
9	F	4.3	2.6	16.7	17.6	17.1	5.6
10	Ne	3.8	2.3	15.0	15.9	15.4	5.2
11	Na	5.2	3.2	19.7	20.8	20.3	6.7
12	Mg	3.0	1.6	10.8	11.4	11.3	3.4
13	Al	2.1	1.5	8.3	8.8	8.7	3.1
14	Si	2.6	1.4	8.2	8.7	8.7	2.9
15	P	3.5	2.3	10.7	11.6	11.2	4.4
16	S	3.3	1.8	10.1	10.6	10.8	3.3
17	Cl	4.2	3.0	11.8	12.9	12.4	5.4
18	Ar	6.1	3.3	15.3	16.9	16.2	17.0
19	K	9.9	5.5	23.6	25.7	25.3	10.0
20	Ca	5.5	3.2	14.9	16.8	16.2	7.4
21	Sc	2.5	1.6	8.3	9.2	9.1	4.0
22	Ti	2.1	1.2	4.4	5.1	5.2	2.4
23	V	1.4	1.0	3.9	4.3	4.4	1.8

24	Cr	1.4	0.9	3.3	3.6	3.7	1.5
25	Mn	1.2	0.8	3.2	3.5	3.5	1.4
26	Fe	1.2	0.9	3.0	3.2	3.3	1.3
27	Co	1.2	0.7	3.0	3.2	3.3	1.2
28	Ni	1.2	0.7	2.9	3.0	3.1	1.1
29	Cu	1.1	0.6	2.7	2.9	3.0	1.1
30	Zn	1.4	1.0	4.1	4.6	4.6	2.0
31	Ga	2.8	1.9	4.4	5.1	5.0	2.4
32	Ge	2.2	1.2	5.2	6.1	6.2	3.1
33	As	2.1	1.7	4.6	5.4	5.5	2.9
34	Se	3.4	2.6	6.2	7.2	6.9	4.0
35	Br	4.9	2.4	9.7	11.5	11.2	6.0
36	Kr	6.1	4.5	12.1	14.2	13.1	7.8
37	Ru	9.6	7.5	18.7	22.9	21.0	13.5
38	Sr	6.6	4.5	9.6	10.9	11.8	5.4
39	Y	3.1	1.6	5.5	6.6	6.8	3.8
40	Zr	2.3	1.9	4.2	5.6	5.0	3.7
41	Nb	2.1	1.6	4.2	4.6	4.6	2.1
42	Mo	1.4	0.9	2.5	3.1	3.4	1.8
43	Tc	1.5	0.9	2.3	2.7	3.0	1.7
44	Ru	1.5	1.1	2.6	3.2	3.0	1.8
45	Rh	1.5	1.1	2.6	3.3	3.0	1.9
46	Pd	1.4	1.1	2.3	2.8	3.0	1.7
47	Ag	1.5	1.1	3.1	3.6	3.7	1.9
48	Cd	2.5	1.6	3.4	3.9	4.4	2.4
49	In	2.9	1.9	3.8	4.9	4.8	3.1
50	Sn	2.6	2.0	5.4	6.5	6.0	3.8
51	Sb	3.6	2.7	6.1	7.6	7.5	4.4
52	Te	3.6	2.0	5.2	6.1	6.2	3.4
53	I	3.7	2.5	7.5	9.0	9.2	6.1
54	Xe	9.6	5.5	16.4	18.9	18.8	9.5
55	Cs	13.2	8.4	18.1	23.0	21.9	15.9
56	Ba	6.8	3.6	11.2	14.2	3.9	9.3
57	La	2.9	1.7	5.2	6.2	6.4	3.8
58	Ce	3.7	2.7	5.9	7.3	7.1	4.3
59	Pr	3.8	2.7	5.8	6.9	7.0	4.0
60	Nd	3.4	2.1	5.0	5.9	6.2	3.0
61	Pm	4.1	3.3	7.4	10.0	8.4	6.8
62	Sm	3.2	1.8	5.2	6.1	6.4	3.4
63	Eu	5.0	2.9	8.2	10.1	10.3	5.9
64	Gd	3.3	2.5	5.4	7.0	6.5	4.5
65	Tb	3.1	2.4	5.6	7.3	6.5	4.7
66	Dy	3.3	2.6	4.6	5.5	6.0	3.6

67	Ho	3.0	1.8	4.2	4.9	5.4	2.8
68	Er	2.7	1.5	4.4	5.2	5.4	2.7
69	Tm	2.4	1.5	3.4	4.2	4.4	2.7
70	Yb	3.8	2.5	5.1	6.7	6.7	4.6
71	Lu	2.3	1.4	3.7	4.4	4.7	2.4
72	Hf	2.5	1.4	3.5	4.1	4.4	2.1
73	Ta	2.3	1.4	3.0	3.4	3.3	1.6
74	W	1.9	1.1	3.0	3.6	3.5	2.1
75	Re	1.3	0.8	1.8	2.4	2.5	1.6
76	Os	1.6	1.0	2.0	2.5	2.6	1.6
77	Ir	1.6	0.9	1.9	2.4	2.5	1.6
78	Pt	1.4	1.0	1.7	2.1	2.5	1.4
79	Au	2.1	1.1	2.9	3.3	3.3	1.5
80	Hg	2.5	1.5	3.1	3.8	4.5	1.9
81	Tl	3.1	2.0	3.8	5.1	5.1	3.4
82	Pb	3.2	2.0	3.3	4.1	4.8	2.9
83	Bi	4.4	3.3	5.4	6.8	7.7	3.3
84	Po	4.0	1.7	4.0	5.5	6.2	3.2
85	At	3.4	2.4	5.2	6.2	6.7	3.7
86	Rn	4.7	3.7	6.0	7.9	8.5	5.2
87	Fr	3.5	3.1	4.4	5.2	6.0	2.5
88	Ra	8.5	5.4	8.1	12.0	10.9	7.7
89	Ac	4.0	2.9	6.4	8.1	7.9	4.1
90	Th	4.1	2.9	5.3	7.4	7.3	5.2
91	Pa	2.4	1.6	2.7	3.7	3.6	2.5
92	U	2.9	2.0	3.2	3.5	4.0	1.1

Notes

All data shown in the appendices were run using TRIM calculations from SRIM 2003, using detailed calucation with full damage cascades, and 100 ions. SRIM 2003 was used with permission from James Ziegler.

Index

Lightning Source UK Ltd.
Milton Keynes UK
UKOW05n0622030615

252811UK00012B/255/P